Basic Concepts in Organic Stereochemistry

Sunil Kumar Talapatra · Bani Talapatra

Basic Concepts in Organic Stereochemistry

Springer

Sunil Kumar Talapatra
Department of Chemistry
University of Calcutta
Kolkata, India

Bani Talapatra
Department of Chemistry
University of Calcutta
Kolkata, India

ISBN 978-3-030-95989-0 ISBN 978-3-030-95990-6 (eBook)
https://doi.org/10.1007/978-3-030-95990-6

This Springer imprint is published by the registered company Springer Nature Switzerland AG
The registered company address is: Gewerbestrasse 11, 6330 Cham, Switzerland

Dedicated

To our children
Anupam and Sharmistha
Who have eagerly cooperated
During our professional career

And

To our grandchildren
Ayush and Agneev
As an Inspirational gift

And

To our students
Who have inspired us through years
As a gift with love and best wishes

Foreword

Stereochemistry, dealing with the spatial arrangement of atoms, is at the heart of chemistry and is a molecular science, foundational to understanding the structure, reactivity, and function of molecules. Its origin can be traced back to Louis Pasteur (1848, optical activity in tartaric acid), J. H. van't Hoff, and J.-A. Le Bel (1874, tetrahedral carbon and asymmetry), and the term "stereochemistry" itself is attributed to Viktor Meyer (c. 1878). During the past 170 years, advances in chemistry and insights into stereochemistry have shared complementary and co-evolutionary trajectories. Stereochemistry-based consideration transcends disciplinary boundaries and is pivotal wherever manipulation of carbon atom(s) containing chemical matter in three-dimensional space is implicated. Indeed, many spectacular advances in chemistry—particularly in organic chemistry (stereo- and enantio-selective syntheses), chemical biology (molecular level understanding of biological processes), and material science and nanotechnology (accessing diverse, intelligent, made-to-purpose materials)—have hinged on the smart interplay of stereochemical features and controls. Thus, learning, understanding, and application of stereochemistry harbor interdisciplinary dimension, recognized as molecular sciences. Indeed, stereochemical thinking is essential not only for chemistry/organic chemistry teaching and learning, but it is also an enabler for providing solutions to many current and future challenges involving human health, access to lifestyle materials, and mega issues related to sustainability.

Among the various branches of natural sciences, it is in the arena of organic chemistry, a predominantly carbon atom anchored science, wherein molecular shape, size, and symmetry in spatial domain play a central role in modulating reactivity and properties. Therefore, it is fundamentally important that learners and researchers in the arena are fully conversant and well exposed to its basics at the conceptual and dynamic (reaction outcome and mechanism) level. Indeed, many notable texts and monographs from time to time, starting with Ernest. L. Eliel's seminal book *Stereochemistry of Carbon Compounds* in 1962, have kept pace with the contemporary advances and enriched the learning of stereochemistry. In the present context, a topical undertaking by the duo of Professors Bani Talapatra and Sunil Talapatra, organic chemistry veterans, passionate teachers, and partners in profession and in life, has resulted in a very welcome offering *Basic Concepts in Organic Stereochemistry* to enrich the repertoire of available teaching and learning aids in the area. This book can be regarded as a worthy successor and updated version of an earlier offering *Chemistry of Plant Natural Products: Stereochemistry, Conformation, Synthesis, Biology, and Medicine* by the Talapatra duo in 2015.

Basic Concepts in Organic Stereochemistry is an ambitious undertaking and offers an incisive overview of the basics of stereochemistry for both learners and teachers, and its finer nuances to researchers and interdisciplinary practitioners of chemistry. This comprehensive book is set in a modular framework and comprises 12 chapters, each one devoted to either a classical or an emergent theme like configuration, conformational analysis, concepts of chirality and symmetry, chiroptical properties, asymmetric synthesis, etc., to name a few. Each chapter is further divided into clearly identified sections for smooth, user-friendly, and topic-based navigation. In consonance with the title of the monograph, the focus in each chapter/section is on the "basics" with a learner-centric approach with appropriate illustrations (nicely and imaginatively drawn) and diagrams, easy to conceptualize visual aids. The vocabulary of stereochemistry is full of compound words with very carefully defined meanings, and sometimes it can be a bit tricky to relate to the underlying concepts. However, in the present narrative, this issue is well-handled with smooth contextual build-up in each chapter with definitions, nomenclatures, and relevant conventions/rules preceding the discussion of underlying concepts. There is an appropriate blend of rigor and elements of quantitative analysis while explaining underlying principles and qualitative discussion and interpretations during consideration of the real world examples from the chemical literature. The latter is a special feature of this chronicle, as the illustrative examples (acyclic to cyclic, polycyclic, heterocyclic, and natural products) chosen, cover a wide cross section of organic substrate types. All things considered, this wholesome approach augurs well from the learner's perspective, and to an extent negates the commonly felt student perception that stereochemistry is difficult to grasp and master.

This monograph attempts to highlight the molecular diversity manifestation of chemical matter in spatial environment, thereby providing an opportunity to understand and appreciate the temporal dimension in which molecules exist, act, and react—a prime requirement in learning and advancing organic chemistry. There is a saying that good learning is guided by "where to look rather than what to look"; one hopes that this monograph will find favor with learners, teachers, and researchers, and will figure in their priority list of "where to look" category. I wish to complement Profs. Sunil and Bani Talapatra for this huge effort, executed through thoughtful diligence, leveraging their long experience, in the arena that has won them peer esteem.

Prof. Goverdhan Mehta, FNA, FRS
University Distinguished Professor
Dr. Kallam Anji Reddy Chair, School of Chemistry
University of Hyderabad
Hyderabad, India

Preface

Springer-Verlag published in 2015 ***Chemistry of Plant Natural Products: Stereochemistry, Conformation, Synthesis, Biology, and Medicine*** (Volume 1, pp. 1–624; Volume 2, pp. 625–1180; +LXIII), co-authored by Sunil Kumar Talapatra and Bani Talapatra. This comprehensive book has been well received, serving both as a teaching text and a reference book for established chemical scientists. The full titles of the papers are mentioned in the references. The Chap. 2 of this book (henceforth referred to as the **parent book**) also serves as a reliable source of stereochemical information for a wide spectrum of graduate and postgraduate students and research chemists. In the Chap. 2, titled *Fundamental Stereochemical Concepts and Nomenclatures* of the parent book the core concepts have been presented. In the Foreword of the book Professor K C Nicolaou commented "….. *The latter topic (Stereochemistry) is discussed in considerable detail and its impact on the chemical and biological properties of the molecule and its asymmetric synthesis are articulated and explained. The special emphasis on conformation is particularly informative and educational, and provides useful insights and understanding of the nature of the molecule to students of organic chemistry.*" He also commented that each Chapter of this book is *stunningly illuminating in terms of breadth and depth of coverage.*

The genesis of the term "Stereochemistry" first coined by one of the finest German chemists **Victor Meyer** is fairly detailed below.

It is not much known including ourselves that the genesis of the term 'Stereochemistry' implying space chemistry or 3-dimentional chemistry involving the transition states of chemical reactions was first coined (*vide* Alan Rocke, A Chemist to Remember, *Chemistry in Britain*, Vol. 33, pages 27–29, **1997**, August Issue) by the well-known German chemist Professor Victor Meyer (8 September 1848–8 August 1897), while working in this field in the University of Heidelberg. Victor Meyer made pioneering contributions in both organic chemistry and inorganic chemistry (*vide* the above reference of Alan Rocke). In 1872, at the age of 24 he became the Professor of Chemistry, and subsequently succeeded Friedlich Whöler at Göttingen and Robert Bunsen in Heidelberg. He was the first Jewish chemist hired as a full Professor in a major German University. There were the luminous days in Heidelberg with fabled masters such as Adolf Baeyer and Emil Fischer with outstandingly brilliant brains. Maeyer researched in Heidelberg and realized the steric hindrance in preventing free rotation about carbon-carbon single bond, and one of his main interests was stereochemistry, the term he introduced into the chemists' vocabulary.

Stereochemical thinking is becoming more important day by day not only in the area of chemistry but also in the more recently developed areas of molecular science, such as chemical biology, material science, and nanotechnology. Thus, during one semester each of the advanced UG and the beginning PG and research level, a systematic knowledge through teaching of the often confusing area of stereochemistry has become more essential.

According to many teachers, students, and readers in India Chap. 2 of our parent book is most updated and lucid due to many examples given, compared to the available existing books on advanced level stereochemistry, published about 2 decades or more ago.

With a view to presenting the basic concepts of stereochemistry for students and researchers across a wide spectrum of fields we have prepared a book entitled "**Basic Concepts in Organic Stereochemistry**" for publishing as a monograph by ***Springer Nature***, containing the whole matter of stereochemistry in Chap. 2 (pages 23–201, 179 pages) of our parent book, which have been included in Chaps. 1–5 and 7–12, along with some additional new Sects. 1.5 and 1.6, Sub-Sects. 1.6.1–1.6.3; 2.4.1–2.4.3; Sub-Sub-Sect. 9.1.2.1 and Sub-Sects. 9.2.1–9.2.3. Chapter 6 incorporating new materials, titled *Some Other Aspects of Dynamic Stereochemistry: Conformation and Reactivity* (containing 25 pages & 23 Figures; and 3 Sections, 14 Sub-Sections and 5 Sub-Sub-Sections) of this monograph has significantly strengthened the stereochemical materials of the parent book. This monograph is designed to serve as a textbook for postgraduate and advanced undergraduate Honours **students**. This is expected to be equally useful to the learners in recently developed interdisciplinary areas of chemical biology, material science, and nanotechnology. It might also serve as a guide to subsequent studies and research by those who might wish to have core stereochemical concepts explained with various illustrations in one small updated book in their personal library, because of its much less affordable price.

This monograph offers readers just the right information in a most concise and articulate manner without diluting the importance of the subject matter. The limitation of space persuaded us to exclude many other related important topics. To enable the entry into the literature of stereochemistry, we have included the full titles of all papers and review articles throughout the monograph (excepting a few very old unavailable ones).

Kolkata, India
April 2022

Sunil Kumar Talapatra
talapatrask@gmail.com

Bani Talapatra
Surendranath Co-operative, Unit II
P.O. Kankurgachi
sunilbanit@gmail.com

Acknowledgements

Acknowledgements and thanks are not casual words—they have their origin in our hearts. The job of writing this monograph could not have been accomplished without the help of our supporters and well-wishers from various walks of life.

We express our deepest gratitude and heartfelt thanks to **Professor Goverdhan Mehta** for kindly writing the **Foreword** to this monograph with his candid opinion, in spite of his heavy academic, social, and administrative commitments.

Our affectionate thanks go to our former students Professor Anindya Datta, Professor of Chemistry, IIT Bombay, and to Dr. Sanchita Goswami, Department of Chemistry, University of Calcutta (CU) for continual supply of PDF files of research papers and review articles during the last two years. The library facilities received from CU Chemistry Library (through the courtesy of (Mr. Nandadulal Biswas) and IIT Bombay are acknowledged with gratitude.

We express our affectionate appreciation to our daughter Sharmistha for helping us in the conversion of PDF files to MS word files and *vice versa* whenever needed, and also for collecting needed information through online access. Also we thank our only son Anupam for helping us during our stay with him in San Raman California.

We are solely responsible for errors, inconsistencies, and infelicities that may persist in the text and figures, despite our most sincere efforts to eliminate them. Our endeavors would be meaningful if this monograph benefits the target readers. We welcome any corrective suggestions and recommendations for its future edition/printing.

Structure Diagrams

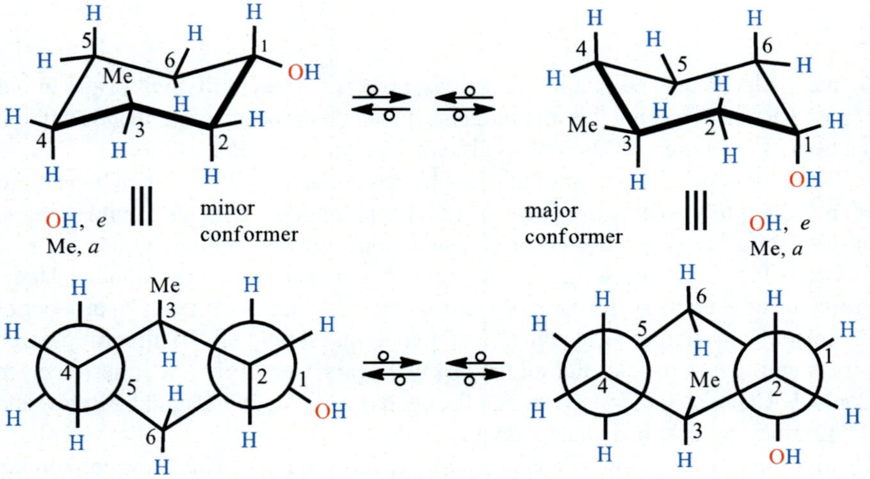

One enantiomer of trans-3-Methylcyclohexanol (cf. p.)

- The above structure diagrams reveal some basic concepts in organic stereochemistry.
- Two conformers of one enantiomer of trans-3-methylcyclohexanol, interconvertible by flipping, are shown in perspective or chair (upper row) and equivalent Newman projection (lower row) formulas.
- The main skeletons, the bonds, and the Me groups are of black color.
- The H atoms are of blue color (usually of nitrogen), to make them appear better.
- The O atoms are of red color.

Structure Diagrams

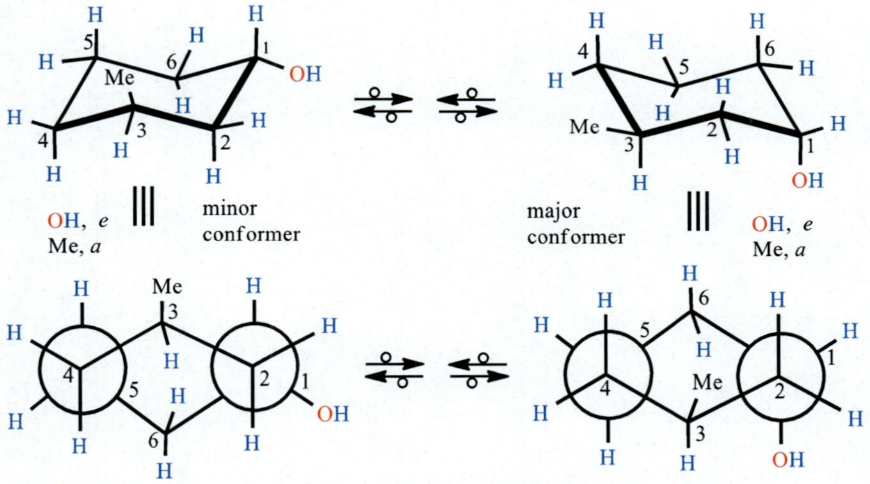

One enantiomer of trans-3-Methylcyclohexanol (cf. p.)

- The above structure diagrams reveal some basic concepts in organic stereochemistry.
- Two conformers of one enantiomer of trans-3-methylcyclohexanol, interconvertible by flipping, are shown in perspective or chair (upper row) and equivalent Newman projection (lower row) formulas.
- The main skeletons, the bonds, and the Me groups are of black color.
- The H atoms are of blue color (usually of nitrogen), to make them appear better.
- The O atoms are of red color.

Contents

1 Symmetry and Molecular Chirality. Conformation, Stability, and Physical Properties 1
1.1 Chirality. Symmetry Elements. Optical Rotation 1
1.1.1 Simple or Proper Axis of Symmetry 2
1.1.2 Plane of Symmetry 3
1.1.3 Center of Symmetry or Inversion Center 4
1.1.4 Alternating or Improper or Rotation-Reflection Axis (S_n) 5
1.1.5 Dissymmetric and Asymmetric Molecules. Chiral and Achiral Point Groups. Central Chirality 6
1.1.6 Symmetry Number, Order of Point Groups, Achiral Point Groups 8
1.1.7 Local Symmetry (or Site Symmetry). Desymmetrization 12
1.1.8 Optical Isomerism. Optical Rotation 13
1.1.9 Specific Rotation. Molecular Rotation. Units (Fig. 1.13) 15
1.2 Conformation of Simple Acyclic Molecules 18
1.2.1 Dihedral Angle. Torsion Angle. Torsional Strength 18
1.2.2 Klyne–Prelog Nomenclature for Torsion Angles. Conformational Chirality 19
1.2.3 Torsional Strain Curve (Potential Energy Diagram) of Ethane 23
1.2.4 Torsional Strain Curve of Propane 23
1.2.5 Torsional Strain Curve of Molecules ACX_2CX_2B, n-Butane 23
1.3 Configuration. Relative Configuration. Absolute Configuration 24

1.4 Relationship Between Two Molecules of Same Molecular Formula. Homomers, Constitutional Isomers, Stereoisomers, Enantiomers, Diastereomers, Configurational/Conformational Enantiomers/Diastereomers 26
1.5 Conformational Effect on Stability of Diastereomers (Fig. 1.25) 26
1.6 Conformation and Physical Properties of Acyclic Molecules and Some Cyclohexane Derivatives 26
1.6.1 Dipole Moment. Three Examples 28
1.6.2 Acid Strength Measurements. Cyclohexane-1,2-Dicarboxylic Acids, 4-T-Butylcyclohexanecarboxylic Acids (Fig. 1.29) 31
1.6.3 Different Affinity for Adsorbents of 4-t-Butyicyclohexanol Isomers 32
1.6.4 Physical Properties of Substituted Cyclohexanes; Von Auwers–Skita Rule or Conformational Rule, Van Arkel Rule 33
References 34

2 Configurational Nomenclature. Physical Properties of Geometrical Isomers 35
2.1 Fischer's D,L Nomenclature 35
2.2 *R,S* Nomenclature for Absolute Configuration 38
2.2.1 *R,S* Nomenclature. Center of Chirality 39
2.2.2 Specification of Center/s of Chirality 39
2.2.3 Priority Sequence of the Application of the CIP Sub-rules 42
2.2.4 Modification of Sub-rule 3 42
2.2.5 R* and S* Nomenclature 43
2.2.6 Specification of Other Tetracovalent Chiral Atoms 46
2.2.7 Specification of Tricovalent Chiral Compounds (*with Pyramidàl Stereocenter*) 46
2.3 Stereochemistry of Alkenes. *E,Z* Nomenclature [9–11] 47
2.4 Physical Properties of Geometrical Isomers [12] 49
2.4.1 Dipole Moments of Geometrical Isomers [13] 49
2.4.2 Physical Properties of Geometrical Isomers: Melting Points, Boiling Points, Densities, and Refractve Index [12] 49
2.4.3 Acid Strengths of Geometrical Isomers [16, 17] 50
References 51

3 Projection (Fischer, Newman, Sawhorse) and Perspective (Flying Wedge and Zigzag) Formulas, Working Out Stereoisomers 53
3.1 Molecules with Two Unlike (Unsymmetrical) Chiral Centers (AB Type) 53
3.1.1 *Erythro* and *Threo* Nomenclature 53
3.1.2 "*Pref*" and "*Parf*" Nomenclature 56
3.1.3 *Syn* and *Anti* System 57
3.1.4 *Like* (*l*) and *Unlike* (*U*) System 58
3.1.5 Brewster's System of Nomenclature 59
3.2 Molecules with Two like (Symmetrical) Chiral Centers (AA Type) 60
3.3 Molecules with Three Unlike Chiral Centers (ABC Type) 61
3.4 Constitutionally Symmetrical Molecules Having Three Chiral Centers (ABA Type) 61
3.5 Stereogenecity and Chirotopicity (Fig. 3.12) 62
3.6 Molecules with Four (ABCD Type) or More Unlike Chiral Centers in a Chain 63
3.7 Constitutionally Symmetrical Molecules with Four or More like Chiral Centers in a Chain (ABBA, ABCBA, Etc., Types) 64
3.8 Chiral Compounds with Asymmetric Carbon Atoms in Branched Chains 65
3.9 Chirality and Dimension. One-, Two-, and Three-Dimensional Chiral Simplexes 67
References 68

4 Prochirality and Prostereoisomerism. Topicity of Ligands and Faces Nomenclature [1–5] 71
4.1 Introduction 71
4.2 Homotopic Ligands 72
4.3 Homotopic Faces 73
4.4 Enantiotopic Ligands 74
4.5 Nomenclature of Geminal Enantiotopic Ligands. Pro-R and Pro-S 74
4.6 Enantiotopic Faces 77
4.7 Nomenclature of Enantiotopic Faces 77
4.8 Diastereotopic Ligands 80
4.9 Nomenclature of Diastereotopic Ligands 80
4.10 Diastereotopic Faces. Nomenclature 81
4.11 Interesting Examples of Topicities of Homomorphic Ligands [8] 81
4.12 Interrelation of Topicity of Ligands with Isomerism 81
4.13 Molecules with Prostereogenic but Proachirotopic Center and Multiprochiral Centers 82

4.14 Topic Relationship of Ligands and Faces 82
4.15 Stereoheterotopic Ligands and NMR Spectroscopy 83
4.15.1 Anisochrony Arising Out of Diastereotopic Faces 83
References 86

5 Asymmetric Synthesis 87
5.1 Introduction. Principles of Stereoselection: Enantioselection. Diastereoselection 87
5.1.1 Lack of Stereoselection 87
5.1.2 Enantioselection 87
5.1.3 Diastereoselection 89
5.2 Asymmetric Synthesis. Definition. Stereoselective and Stereospecific Reactions. Product/Substrate Stereoselectivity. Regioselectivity 90
5.2.1 Enantiomeric Excess. Diastereomeric Excess. Optical Purity 91
5.3 Cram's Rule 92
5.3.1 Cram's Open Chain Model 92
5.3.2 Cram's Chelate or Cyclic Model 93
5.3.3 Cram's Dipolar Model 93
5.4 Felkin–Anh Models [8, 9] 96
5.4.1 Felkin–Anh Open Chain Model 96
5.4.2 Felkin–Anh Dipolar Model 97
5.5 Prelog's Rule 97
5.5.1 Attempted Rationalization of Prelog's Model 99
5.5.2 More Examples of the Application of Prelog's Rule 100
5.5.3 Exception to and Anomalies of Prelog's Rule 101
5.6 Horeau's Rule 101
5.7 Sharpless Enantioselective Epoxidation 102
5.7.1 Kinetic Resolution of Racemate Allyl Alcohols 103
5.7.2 Mechanism of the Sharpless Reaction 104
References 105

6 Some Other Aspects of Dynamic Stereochemistry: Conformation and Reactivity 109
6.1 Introduction 109
6.2 Conformational Analysis of Chemical Reactions 110
6.2.1 Stereoelectronic Effects and Steric Effects 110
6.2.2 Conformationally Rigid Diastereomers 111
6.2.3 Conformational Energy: Anancomeric Systems 112
6.2.4 Conformation and Chemical Reactivity for a Conformationally Mobile Substrate Molecule: The Winstein–Holness Equation and the Curtin–Hammett Principle 113

6.2.5 Curtin–Hammett Principle. Case 1, Predominant Conformer Leading to Predominant Product ($G^oP_2 > G^oP_1$) 116
6.2.6 Elimination Reactions of Two Diastereomers *trans*- and *Cis*-2,3-Dibromobutane Leading to Two Diastereomeric Products 117
6.2.7 Curtin–Hammett Principle. Case 2. Energy Diagram for Predominant Conformer Leading to the Less Predominant Product. Derivation of C-H Principle 117
6.2.8 Curtin–Hammett Principle. Case 3. Different Conformers of *Trans*-2-Phenyl-*Trans*-3-*t*-Butylcyclohexan-*r*-1-ol Xanthate Having Same Free Energy Leading to Different Products. Chugaev Reaction 121
6.2.9 Quantitative Treatment of Mobile Systems. Winstein–Holness and Eliel Equations 121
6.2.10 Application of Winstein–Holness Equation. Examples 1 and 2 122
6.2.11 Application of Eliel Equation. Examples 1 and 2 124
6.2.12 Quaternization of 1-Methyl-2-Phenylpyrrolidine and 1-Methyl-2-Arylpyrrolidine. Product Ratio 126
6.2.13 The Conformational Preference of Substituents by ^{1}H NMR Method. Calculation of Equilibrium Constant K (Cf. Kinetic Method) [18, 19] 127
6.2.14 Absolute Configuration of Allene Molecules (Figs. 6.18–6.22) 127
6.3 Baldwin's Rules for Ring Closure 133
6.4 Conclusions 134
References 135

7 Conformation of Saturated Six-Membered Ring Compounds 137
7.1 Conformational Aspects of Cyclohexane 137
7.1.1 Geometry of Cyclohexane Chair. Bond Lengths. Bond Angles. Torsion Angles 137
7.1.2 Equatorial and Axial Bonds 138
7.1.3 Symmetry of Cyclohexane Conformations 139
7.1.4 Enthalpy (H) or Potential Energy (E) Difference 140
7.1.5 Cyclohexane Ring Inversion 140
7.1.6 Stable Boat or Skew–Boat Conformers 142
7.2 Monosubstituted Cyclohexanes. Conformational Energy 143
7.3 Conformational Energy 143
7.4 1,1-Disubstituted Cyclohexanes 145
7.5 Nongeminal Disubstituted Cyclohexanes 146

7.5.1 Some Typical Disubstituted Cyclohexanes (Fig. 7.10) 146
References 148

8 Cyclohexanone 151
8.1 Torsion Angles, Stability 151
8.2 Ring Inversion 152
8.3 Alkylketone Effects 152
8.3.1 2-Alkylketone Effect (Fig. 8.2) 152
8.3.2 3-Alkylketone Effect 153
8.3.3 4-Alkylketone Effect 154
8.4 Addition of Nucleophiles to Cyclohexanones. Stereochemical Aspects 155
8.4.1 PDC (PSC) and SAC (SSC) 155
8.4.2 Observations Against PSC 156
8.4.3 Torsional Strain. Role of C2 and C6 Axial Hydrogens. Burgi–Dunitz Trajectory 157
8.5 Cieplak Hypothesis 159
8.6 Highly Stereoselective Reduction of Saturated Cyclohexanones by Dissolving Metals. Birch Reduction 160
8.7 Alkylidene Cyclohexanes. Allylic$^{(1,3)}$ Strain 162
8.7.1 Conformational Preference 163
8.7.2 Synthetic Utility [24] of $A^{(1,3)}$ Strain. Stereochemistry of Exocyclic Enolate Anion Protonation 164
8.7.3 Another Example of the Use of $A^{(1,3)}$ Strain Concept 164
8.8 Cyclohexene. Conformation. $A^{1,2}$ Strain 164
8.8.1 Conformation of Cyclohexene. Torsion Angles 164
8.8.2 Allylic 1,2-Strain ($A^{(1,2)}$-Strain) 167
References 171

9 Fused Ring Systems 175
9.1 Decalins 175
9.1.1 Brief History 175
9.1.2 *Trans*-Decalin. Conformation. Torsion Angles. Symmetry 177
9.1.3 *Cis*-Decalin. Conformations. Torsion Angles. Symmetry (Fig. 9.2) 178
9.1.4 Ring Inversion in *Cis*-Decalin 179
9.1.5 Entropy Difference in Decalins 179
9.1.6 Enthalpy and Physical Constants. Auwers-Skita Rule 180
9.1.7 Free Energy Difference in Decalins 180
9.1.8 Effect of Introduction of Angular Methyl Group/s 181
9.1.9 *Cis*-Decalones and *Trans*-Decalones 182

9.1.10 *Trans*-2-Decalols. Conformational Analysis 182
9.1.11 *Cis*-2-Decalols. Conformational Analysis 183
9.2 Octalins (Octahydronaphthalenes)—Their Conformations 183
9.2.1 Six Octalins. $\Delta^{9,10}$-Octalin. $\Delta^{1,9}$-Octalin and Its Conformations and Torsion Angles. 2Δ Acetoxytestosterone (Fig. 9.10) 185
9.2.2 *Trans*-$\Delta^{1,2}$-Octalin and *Trans*-$\Delta^{2,3}$-Octalin. Torsion Angles at the Ring Junction and Their Relative Stability (Fig. 9.11) 185
9.2.3 *Cis*-$\Delta^{1,2}$-Octalin and *Cis*-$\Delta^{2,3}$-Octalin. Their Conformations and Relative Stability (Fig. 9.12) 187
9.3 Perhydrophenanthrenes (PHP's). Stability. Point Groups. Optical Activity 187
9.3.1 Stereochemistry of Some Perhydrophenanthrones and All Perhydrodiphenic Acids (PHDPAs) 190
9.4 Perhydroanthracenes: Relative Stability. Torsion Angles. Point Group. Optical Activity 191
References 193

10 Stereoisomerism: Axial Chirality, Planar Chirality, (*R*, *S*) Notations. Helicity 195
10.1 Introduction 195
10.2 Stereochemistry of Allenes. Configurational Nomenclature 195
10.3 Chiral Spiranes and Analogs. Configurational Nomenclature 198
10.4 Chiral Adamantoids. Configurational Nomenclature 198
10.5 Chiral Catenanes. Configurational Nomenclature 200
10.6 Biphenyl Derivatives and Atropisomerism [8] 200
10.6.1 Introduction 200
10.6.2 Energy Profile Diagram 201
10.6.3 Examples of Atropisomerism 202
10.6.4 Orders of Steric Hindrance and of Buttressing Effect [11] 203
10.6.5 Configurational Nomenclature of Chiral Biphenyls (*R, S* or *aR, aS*) 203
10.6.6 Some Interesting Examples of Axially Chiral Molecules Exhibiting Atropisomerism 205
10.7 Planar Chirality 206
10.7.1 Introduction 206
10.7.2 The (*R, S*) Specification of Planar Chirality 206
10.8 Helicity and *P*,*M*-Designation 207
References 208

11 Chiroptical Properties I: Optical Rotation. ORD, CD 211
11.1 Origin of Optical Rotation. Circular Birefringence, Its Effect 211
11.2 Optical Rotatory Dispersion. Plain Curve 213
11.3 Circular Birefringence and Circular Dichroism. Cotton Effect 214
11.4 The Axial Haloketone Rule and Its Applications 217
11.4.1 Position of the Halogen Substituent 218
11.4.2 Absolute Configuration by Comparison Method 218
11.4.3 Absolute Configuration by Axial Haloketone Rule. Conformational Mobility 219
11.4.4 Boat Form of Ring A of a Steroid Bromoketone 220
11.5 The Octant Rule and Its Applications 221
11.5.1 Determination of the Preferred Conformation 222
11.5.2 Determination of Absolute Configuration of Trans-Decalones 224
11.5.3 Tricyclic Ketones: Perhydrophenanthrenones and Perhydroanthracenones 224
11.5.4 Tetracyclic Ketones: Steroids 226
References 227

12 Chiroptical Properties II: Helicity Rule or Chirality Rule 229
12.1 Introduction 229
12.2 Conjugated Dienes and Enones: Steroids 229
12.3 Biaryl Atropisomers and Helicenes 231
12.4 Correlation of Optical Rotation with Ligand Polarizability: Brewster's Rule 231
12.5 Absolute Configuration of Chiral Allenes: Lowe's Rule 233
12.6 The Exciton Chirality Method or the Dibenzoate Chirality Rule 234
12.7 Absolute Configuration of the 5α-Steroid Diols by Exciton Chirality Method 237
12.8 Absolute Configuration of *Trans*-Cyclohexane-1,2-Diol Enantiomers 237
12.9 Prediction of the First CE Signs of Vicinal and Nonvicinal Dihydroxy-5α-Steroid Diesters 238
12.10 Further Reading References on the Topics of This Monograph 238
References 240

Index 243

About the Authors

Prof. Sunil Kumar Talapatra born in Kalikachha, a village now in Bangladesh, taught stereochemistry, conformational analysis, reaction mechanisms, and alkaloids in the postgraduate classes of Calcutta University (CU) and Presidency College (now University), Kolkata, for more than three decades, and taught stereochemistry for three decades in the Universities of Vidyasagar (till 2012) and Tripura as a *Guest Teacher*, and as a *Resource Person* of *UGC Refresher Courses and Orientation Programmes* held in the *Universities of Calcutta, Jadavpur, Kalyani, Burdwan, Vidyasagar (Midnapur), North Bengal, Tripura, Gauhati, Dibrugarh, Mujjafarpur, Andhra, Osmania, Kurukshetra*, etc.

He did his B.Sc. with Chemistry Honours (1951), M.Sc. (1st Class) (1953), Ph.D. (1958) (Supervisor Professor Asima Chatterjee), PRS and Mouat Gold Medallist (1958), all from CU, post-doctoral work (2 year) with Prof. Asima Chatterjee as a Senior Research Fellow of the National Institute of Sciences (and then post-doctoral work (4 year) with Prof. Michael P. Cava at the Ohio State University, Columbus USA. He stood 1st in 1st Class in German Language Certificate Examination (CU) (1955) and studied in the French Language Certificate Course (CU) (1956–1957), but had to leave before the Examination. In 1960, he was selected for the coveted **1851 Exhibition Scholarship for Ph.D. work in UK, was offered a Post-Doctoral Research Fellowship by Prof. Michael P. Cava at the Ohio State University, and he accepted the latter**. He was also awarded the **Fulbright Travelling Fellowship** by the USEFI, New Delhi, which he availed himself of.

On his return, he joined the Chemistry Department of the CU as a CSIR Scientists Pool Officer on January 1, 1965, and as a Lecturer on June 1, 1965. Eventually he became Reader in 1969, and Professor in January 1975, and served the CU till November 1998.

Professor Talapatra's research contributions include the discovery, structure elucidation, and stereochemistry of more than 100 new bioactive secondary metabolites (from medicinal plants), mostly alkaloids and terpenoids of different classes, coumarins and other O-heterocycles, polyphenolics, cardiac glycosides, one new C26 tetracyclic diamine from a marine sponge, synthesis, stereochemistry, and mechanisms. He published more than 175 research papers. A total of 25 students obtained the Ph.D. degree of the Calcutta University under his guidance, and he co-supervised more than 40 Ph.D. students and postdoctoral fellows, who joined universities, IIT Kharagpur, other research institutes, and colleges, many of whom pursued their own teaching and research careers.

A recipient of several awards, he was a **British Council Guest Visitor** (1973) in UK, under the **INSA and UGC Cultural Exchange Programmes** in Hungary (1990) and Czechoslovakia (1973) and was **elected CChem FRIC (London) in 1974**.

He was a **Senior Visiting Scientist**at the Illinois Univ. (Chicago) (1974), UCSC Santa Cruz (1995), and UCSD San Diego (2000) (for 3 months at each university). As a **UNESCO Senior Visiting Scientist**(1982), he was briefly associated with Profs. E. J. Corey (Nobel Laureate), K. Nakanishi, H. Rapoport, J. Meinwald, N. Takahashi, K. Mori, and Y. Kanaoka (all were nominees for Nobel Prize). He participated in more than 40 International Conferences as invited Speaker and/or Chairman, held worldwide.

He represented India as an elected Member of the Federation of the Asian Chemical Society (FACS) Executive Committee for 1981–1983 and 1987–1988, and acted as a Member of the FACS General Assembly from 1981 to 1987, and contributed significantly to the FACS activities.

He was a nominator for the Nobel Prize Award in Chemistry for 1989 and 1994 (two nominees for each year). Two of his nominees Profs. E. J. Corey and K. B. Sharpless received the Nobel Prize later.

His association with the Indian Chemical Society is stupendous and has no parallel. Since 1974, he has been extending his dedicated service to the society in all the capacities—as a Council Member, Associate Editor, Honorary Treasurer and Honorary Editor (each one for two terms), Honorary Secretary (8 years), Vice-President, and President (2002 and 2003), an Ex-Officio Vice-President in 2004–2019, and currently an Advisor.

In recognition of his outstanding contribution to the Advancement of Chemical Science, **the Indian Chemical Society**presented him the "**Life Time Achievement Award**" during the Annual Convention of Chemists held in Bundelkhand University, Jhansi, UP, in December 2003.

The Indian Chemical Society paid a befitting tribute to Professor Talapatra by publishing his 80th Birthday Commemoration Issue of the *Journal of the Indian Chemical Society*, Vol. 90, #10, Pages 1505–1906, **October, 2013**, comprising 38 invited articles (including 12 review articles on frontline topics of organic chemistry) by eminent organic chemists of India and abroad.

He has been actively associated with another scientific body, the **Indian Science New Association (ISNA)**, Calcutta, as a Council Member (2001–2008), and Vice-President (2008–till date). He edited, along with Professor Biswapati Mukherjee, a book titled *Sir P. C. Ray, the Father of Chemistry Teaching and Research in India, a Philanthropist and Entrepreneur*comprising 15 illuminating articles by erudite Indian and UK scholars, and one article by Sir P. C. Ray, published by ISNA in April, 2016.

He had acted as **the Vice-President of the Prof. Asima Chatterjee Foundation, Kolkata** since its inception and registration on 15 December, 2014, and has been acting competently as **the President of the Foundation** since February 22, 2019.

He co-authored with his wife Prof. Mrs. Bani Talapatra an advanced level classic text-cum-reference book titled *Chemistry of Plant Natural Products: Stereochemistry, Conformation, Synthesis, Biology, and Medicine*", with an illuminating foreword by **Prof. K. C. Nicolaou (a nominee for Nobel Prize)**, published by Springer-Verlag (Heidelberg, Berlin) in March 2015 (two volumes, 34 chapters, three appendices, total pages 1280 + LXIII, along with its e-book downloaded

(Book Matrix) more than 1.4 lakh times till December, 2020 throughout the world).

He has written jointly with Mrs. Talapatra a number of valuable articles in journals and magazines of cultural and social importance.

He is a Life Member of (i) the Indian Science Congress Association, (ii) the National Academy of Sciences of India, Allahabad, (iii) the Chemical Research Society of India, Bangalore, (iv) the Indian Science News Association, (v) the Indian Chemical Society, and (vi) the Indian Association for the Cultivation of Science.

Prof. Bani Talapatra born in Calcutta, taught natural products chemistry at the Calcutta University (CU) for more than three decades, in the Universities of Midnapur, and Tripura as a *Guest Teacher* for more than a decade, and lectured in the Universities of Jadavpur, Burdwan, and Tripura as a *Resource Person* of *UGC Refresher Courses*. She taught general chemistry in Victoria Institution (College) at the B.Sc. and I.Sc. levels for a decade till 1968.

She obtained her B.Sc. with Chemistry Honours (1955) and M.Sc. (1957) degrees of the CU>. She received the *Krishnalal De Memorial Gold Medal* of the Scottish Church College for securing the highest total marks among all the B.Sc. Honours examinees of that college in 1955. She received the D.Sc. degree (2nd lady of CU) working under Professor (Mrs) Asima Chatterjee in 1963. In the same year, she was selected for a Smith–Mundt Scholarship, but she joined a Postdoctoral Fellowship at the Microbiology Department, Vanderbilt University (VU) with the renowned microbiologist, Professor V. Najjar. At VU, she learnt much about organic reaction mechanisms from Sir C K Ingold (Visiting Professor at VU) for few months.

After 3 years of her return, she joined the Chemistry Department of the CU as Lecturer and eventually became Professor and Head of the Department. As a UNESCO Senior Visiting Scientist (in 1976 and 1983), she was briefly associated with about 2 dozens of outstanding scientists, Professors of major universities of USA, UK, Germany, and Japan. She also participated in more than 25 International Conferences in Europe, S-E Asia and India.

Her research interests include various aspects of natural products chemistry including terpenoids, alkaloids, coumarins and other O-heterocycles, polyphenolics, reaction mechanisms, and stereochemistry. She has published more than 140 research papers; 15 students have obtained a Ph.D. degree under her guidance, and about 50 Ph.D. students and postdoctoral fellows were cosupervised by her.

She co-authored with her husband Professor Sunil Kumar Talapatra a unique text-cum-reference book titled *Chemistry of Plant Natural Products: Stereochemistry, Conformation, Synthesis, Biology, and Medicine*, published by Springer-Verlag, Berlin, Heidelberg (March 2015, 2 Vols, total pages 1280+LXIII), which will remain a classic in the field.

She acted as the Organizing Secretary of the *National Symposium on Contributions of Women in Science* (2018), under the aegis of ISNA, first of its kind in India, with Prof. Sunil K. Talapatra as the President.

She has written jointly with S. K. Talapatra a number of valuable articles in journals and magazines of cultural and social importance.

She is a Life Member of (i) the Indian Science Congress Association, (ii) the Chemical Research Society of India, Bangalore, (iii) the Indian Science News Association, (iv) the Indian Chemical Society, and (v) the Indian Association for the Cultivation of Science.

She wrote a book titled *More Anubhabe* containing a collection of Bengali poems (2nd edition, 2001), and another book titled *Aami Hridayer Kathaa Bolitey Byaakul* containing a collection of Bengali and English poems and Bengali articles (2nd edition, 2016).

Chapter 1
Symmetry and Molecular Chirality. Conformation, Stability, and Physical Properties

Abstract This chapter briefly describes chirality, symmetry elements, optical isomerism, optical rotation, conformation of simple acyclic molecules. Conformational chirality, torsional strain curves, relative configuration and absolute configuration; relationship between two molecules of the same molecular formula: Homomers, constitutional isomers, stereoisomers, enantiomers, diastereomers, configurational enantiomers, conformational enantiomers, configurational diastereomers, and conformational diastereomers; conformational effect on stability of diastereomers, stability order of 2,3-butane diol diasteomers; conformation and physical properties of acyclic molecules and some cyclohexane derivatives: dipole moments-1,2-dibromoethane, stilbene-1,2-dichloride, 1,2-dibromocyclohexane, acid strengths—*cis*- and *trans*-cyclohexane-1,2-dicarbocyclic acids, 4-t-butylcyclohexanecarboxylic acid; adsorbing affinity of 4-t-butylcyclohexanol isomers; von Auwers–Skita rule or conformational rule, van Arkel rule, and dipole rule.

1.1 Chirality. Symmetry Elements. Optical Rotation

In a lecture in 1893, Lord Kelvin introduced the term *chirality*. The word *chiral* is from the Greek name *cheir* (for hand or pertaining to hand). Chirality means handedness (topological). It is a purely geometrical property. An object, for that matter, a molecule is chiral if it is not superposable on its mirror image. Organic molecules may be conformationally mobile. *A chiral molecule must be nonsuperposable on its mirror image even after operation of rotation* around single bond/s or translation, i.e., it must be chiral in all conformations. Thus, in order to ascertain whether a flexible molecule is chiral or achiral (not chiral), all possible conformers must be analyzed.

Chirality of a particular molecular conformation of known stereostructure may be decided by

(i) Intuition (not useful for beginners),
(ii) Constructing the molecular models of the stereostructure of the compound and its mirror image and examining their probable superposability (not always possible),
(iii) Looking for *symmetry elements* of that structure by *symmetry operation(s)*.

S. K. Talapatra and B. Talapatra, *Basic Concepts in Organic Stereochemistry*,
https://doi.org/10.1007/978-3-030-95990-6_1

Z

Table 1.1 Symmetry elements (symbols expressed in italics) and symmetry operations

Symbol	Symmetry elements	Symmetry operations
(i) C_n	Simple or proper axis of symmetry	Rotation about an axis through 360°/n
(ii) σ (S_1)	Plane of symmetry	Reflection in a plane
(iii) i (S_2)	Center of symmetry	Inversion through a center
(iv) S_n	Alternating or improper, or rotation-reflection or reflection-rotation axis of symmetry	*Rotation-Reflection*: Rotation about an axis by 360°/*n*, followed by reflection in a plane orthogonal (⊥r) to the axis *Reflection-Rotation*: The order of the above two operations may be reversed to give the same result

Molecules (a particular conformation in case of a flexible molecule) which have a plane of symmetry (σ or S_1), a center of symmetry (i or S_2), or other alternating axis of symmetry (S_n, n even) are said to have reflection symmetry. Such molecules are superposable on their mirror images and are termed *achiral*. Molecules lacking any reflection symmetry are nonsuperposable on their mirror images and are termed *chiral. Thus, nonsuperposability on the mirror image of a molecule is the necessary and sufficient condition for chirality and for displaying optical activity.* The principal symmetry elements and symmetry operations are summarized in Table 1.1 for their convenient application and for subsequent finding out their point groups, which will be illustrated in the sequel.

A preliminary introduction to the different symmetry elements and their operations follow.

1.1.1 Simple or Proper Axis of Symmetry

A simple or proper axis of symmetry of order (multiplicity) n is such that rotation of the model or structure of the molecule around an axis by 360°/n leads to a structure indistinguishable from the original. Such an axis is denoted by *Cn*, expressed in *italics*. Some examples are shown in Fig. 1.1.

Principal Axis.

If a molecule possesses **several** simple axes of symmetry, the axis with maximum multiplicity (order) is designated as the principal axis of symmetry. If the simple axes of symmetry are of the same order, the simple axis passing through the atoms is regarded as the principal or main axis. For an example, see allene (Fig. 1.3).

C_1 Axes. Any object or *molecule* contains an infinite number of C_1 axes since its rotation by 360° around any axis passing through the object or molecule results in the original three-dimensional orientation.

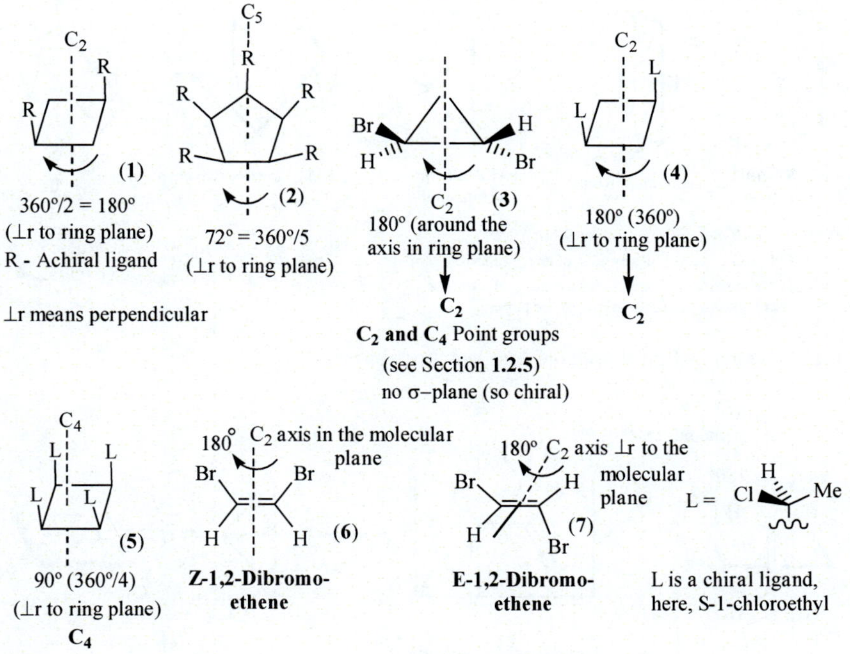

Fig. 1.1 Examples of molecules with C_2, C_4, and point groups $\mathbf{C_2}$ and $\mathbf{C_4}$

1.1.2 Plane of Symmetry

A plane of symmetry or a sigma plane (**σ plane**) is defined as a *mirror plane* which bisects a rigid object or a molecule so that one-half of it coincides with the reflection of the other half in the mirror, i.e., one-half of it reflects its enantiomeric or identical half.

Examples of some common objects with only one plane of symmetry are chairs, cups, file cabinets, spoons, and tooth brushes. Some examples of molecules with one σ-plane are shown in Fig. 1.2.

σ_v and σ_h.

C_n axes and σ planes often occur together in molecules. The principal proper axis of symmetry is conventionally taken as vertical. The symbol σ_v is used to designate a vertical plane of symmetry containing the principal axis; σ_h is used to designate a horizontal plane of symmetry perpendicular to the principal axis of symmetry; and σ_d is used to represent a diagonal plane bisecting the angle between two C_2 axes. The three types of planes are illustrated with examples in Fig. 1.3.

A plane of symmetry is equivalent to onefold alternate axis of symmetry S_1 (vide infra).

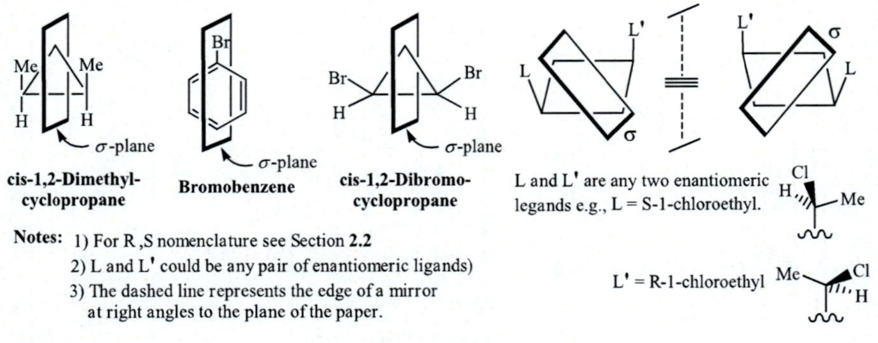

Fig. 1.2 Achiral molecules with only one symmetry plane

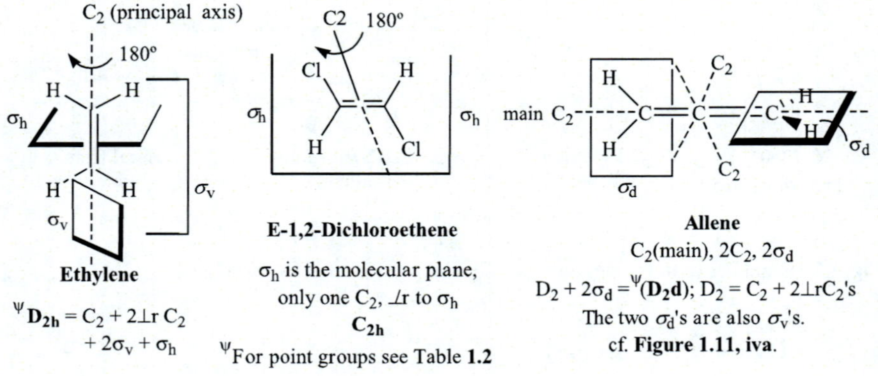

Fig. 1.3 Examples of σ_v, σ_h, and σ_d planes

1.1.3 Center of Symmetry or Inversion Center

The center of symmetry designated *i* is a point such that all imaginary straight lines that can be drawn through it meet identical atoms at equal distance from the point. The objects including molecules with a center of symmetry (point symmetry) are termed *centroasymmetric*.

A center of symmetry is equivalent to twofold alternating axis of symmetry S_2, as illustrated in Fig. 1.4 (see Sect. 1.1.4).

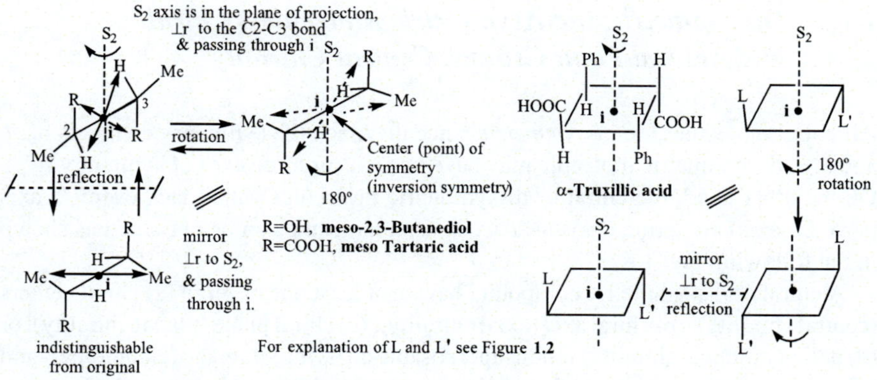

Fig. 1.4 Achiral molecules with center of symmetry (S_2) (inversion symmetry, *i*)

1.1.4 Alternating or Improper or Rotation-Reflection Axis (S_n)

An alternating or an improper or a rotation-reflection axis of order n (S_n) present in a molecule is such that it can be rotated about the axis by an angle of 360°/n and then reflected across a plane perpendicular to the axis to provide a structure indistinguishable from the original. The order of the two operations may be reversed (reflection-rotation) to give the same result. It has been exemplified for *meso*-2,3-butanediol and *meso*-tartaric acid, each having an S_2 axis (Fig. 1.4), and also for substituted cyclobutane with two mirror image (enantiomeric) ligands at 1,3 positions having an S_4 axes (Fig. 1.5). Another example with an S_4 axis is illustrated in Fig. 1.5; see also Reference [1].

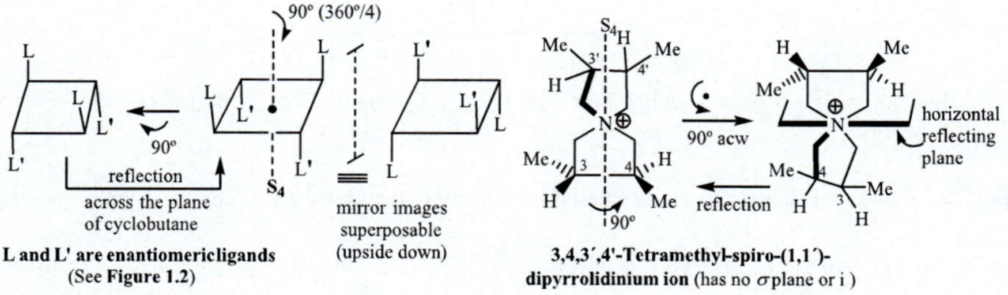

Fig. 1.5 Achiral molecules with S_4 axis explained

1.1.5 Dissymmetric and Asymmetric Molecules. Chiral and Achiral Point Groups. Central Chirality

All chiral molecules are *dissymmetric* since they lack the symmetry elements S_n, *i*, and σ. A dissymmetric molecule may have one axis or more axes of symmetry (C_n). On the other hand, the chiral or dissymmetric molecules which lack even C_n axis (save C_1 axis) are termed *asymmetric*. Thus, a molecule can be classified as shown in the following Fig. 1.6.

A chiral or dissymmetric compound having at least one or more (a) chiral centers (central chirality), (b) chiral axis (axial chirality), (c) chiral plane (planar chirality), or (d) helicity (helical chirality) is nonsuperposable on its minor image (enantiomer) and hence is optically active. Axial chirality, planar chirality, and helicity—lacking any chiral center—will be discussed in Chap. 10, and Sects. 10.6 and 10.7 respectively.

The ***point group*** defines the symmetry class to which a molecule belongs. The symmetry operations (or symmetry elements) are combined to form a point group (notation by Schönfliess), since each operation *leaves a point*, the center of gravity of the molecule, *unchanged*. The C_n and S_n operations organize a point group which is by convention expressed in **bold letters**. The chiral point groups (Fig. 1.6) are illustrated by examples in Figs. 1.7, 1.8, 1.9, and 1.10).

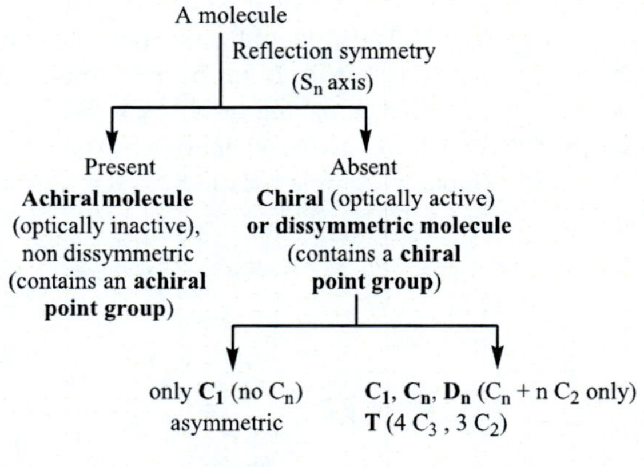

Fig. 1.6 Chiral and achiral molecules based on reflection symmetry. Chiral point groups

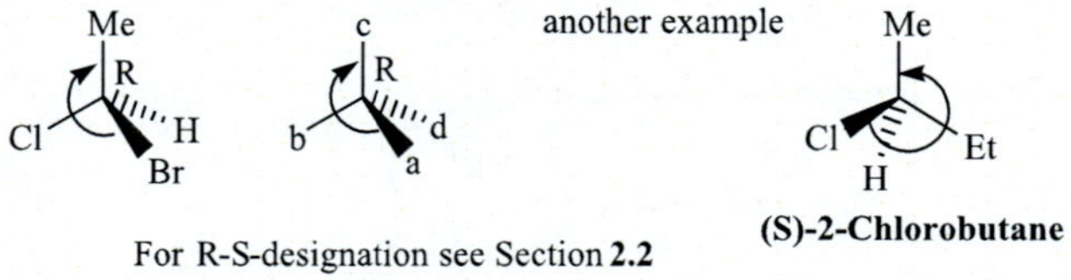

Fig. 1.7 Asymmetric compounds with C_1 point group

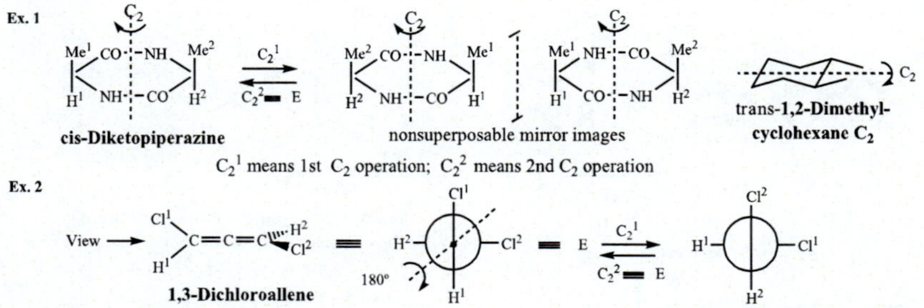

- The C_2 axis passes through the central carbon and bisecting the dihedral angle between the two Cl's and two H's (taking the three allene carbons as a C-C bond in the Newman projection formulas.
- Any operation like C_2^2 in the above examples, producing a shape identical with the original one is called identity or 'E' or 'I' operation.
- The C_2 symmetry of 1,3-dichloroallene is best seen in its Newman projection, as shown.

Fig. 1.8 Asymmetric compounds belonging to $\boldsymbol{C_2}$ point group

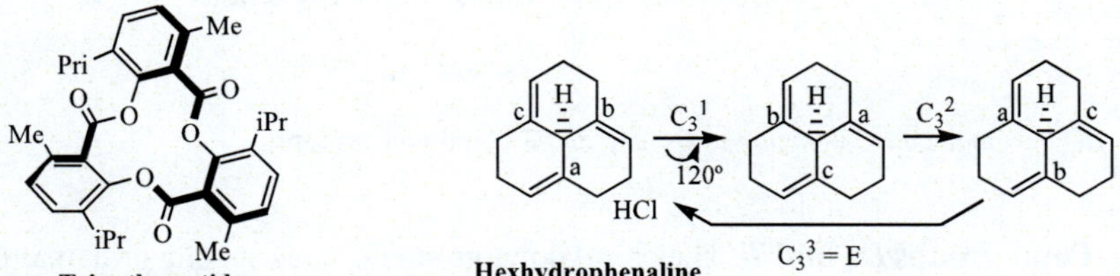

Tri-o-thymotide
(no σ-plane, a known example of $\mathbf{C_3}$ point group having the shape of a 3-bladed propellar)

Hexhydrophenaline
(no σ-plane)

Imaginary hydrocarbon (propellar type) (C_3 axis ⊥r to the paper plane, passing through central H-C bond)

Fig. 1.9 Asymmetric compounds belonging to $\boldsymbol{C_3}$ point group

Some Examples of Chiral Point Groups.

(i) **Point group $\mathbf{C_1}$**

All organic molecules with only one stereogenic center or axis or plane belong to $\mathbf{C_1}$ point group.

C_n axis and C_n operation

The compounds **(3)** and **(4)** of Fig. 1.1 are chiral and belong to the point groups $\mathbf{C_2}$ and $\boldsymbol{C_4}$. Compounds **(1)**, **(2)**, **(6)**, and **(7)** possess σ-planes and hence achiral (Fig. 1.1). A few more compounds belonging to $\boldsymbol{C_2}$ point group are illustrated in Figs. 1.8 and 1.9.

(ii) **Point group $\boldsymbol{C_2}$** (*possesses the only symmetry element*, $\boldsymbol{C_2}$)

(iii) **Point group $\boldsymbol{C_3}$**

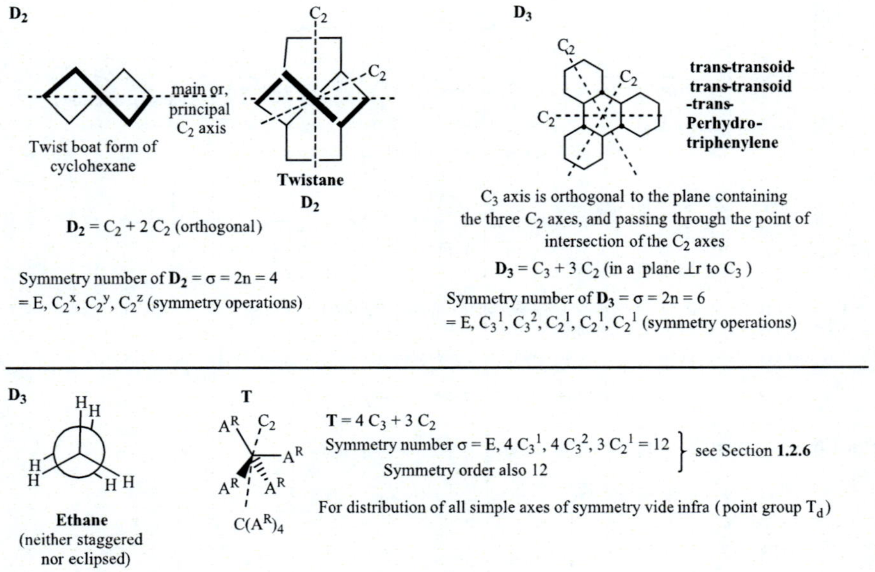

Fig. 1.10 Few molecules belonging to $\boldsymbol{D_2}$, $\boldsymbol{D_3}$, and $\boldsymbol{T}$ chiral point groups

(iv) **Point groups $\boldsymbol{D_n}$ and $\boldsymbol{T}$**. The chiral point group $\boldsymbol{D_n}$ consists of a principal C_n axis of maximum multiplicity (or passing through some atoms of the molecule) and n C_2 axes orthogonal to the principal axis (Fig. 1.10).

T* point group.** Tetrahedral (T***) point group is also chiral, possessing four C_3 and three C_2 axes, but no σ-plane. The four chiral ligands are necessarily having same structure and absolute configuration (Fig. 1.10).

1.1.6 Symmetry Number, Order of Point Groups, Achiral Point Groups

Point group notations of *achiral* molecules are enumerated in Table 1.2.

Symmetry number σ of a molecule is the number of indistinguishable but nonidentical positions into which the molecule can be turned by rigid rotation (around all simple axes of symmetry), taking the parent position as 1.

Table 1.2 Point group notations of achiral molecules

C_s	S_n	C_{nv}	C_{nh}	D_{nd}	D_{nh}	T_d
One σ only	(no σ) n even	$Cn + n\,\sigma_v$ only	$C_n + \sigma_h$ no σ_v	$C_n + n\,C_2$ (⊥r) $+ n\,\sigma_v$ but no σ_h	$C_n + n\,C_2$ (⊥r) $+ n\,\sigma_v + \sigma_h$	$4\,C_3 + 3\,C_2$ $+ 6\,\sigma$

Table 1.3 Symmetry number and order of point groups

Pt. group	C_1	C_n	D_n	C_S	S_n^{ψ}	C_{nv}; C_{nh}	$C_{\propto v}$	D_{nd}; D_{nh}	T_d	O_h
Order	1	N	$2n$	2	n	$2n$	$\propto$	$4n$	24	48
Symmetry number, σ	1	N	$2n$	1	$n/2$	n	1	$2n$	12	24

$\mathbf{S_n}^{\psi}$ includes $\mathbf{S_2} = \mathbf{C_i}$, $\sigma = 1$, order 2

Order of the point group is defined as the total number of symmetry operations that can be performed in that point group. Table 1.3 lists the order and symmetry number (σ) of various chiral and achiral point groups.

Achiral Point Groups: Several examples of achiral molecules belonging to different point groups (Table 1.2) are displayed in Fig. 1.11. The point groups could be allocated to each molecule by finding out the C_n, S_1 (σ-plane), and S_2 (center of symmetry) possessed by the molecule.

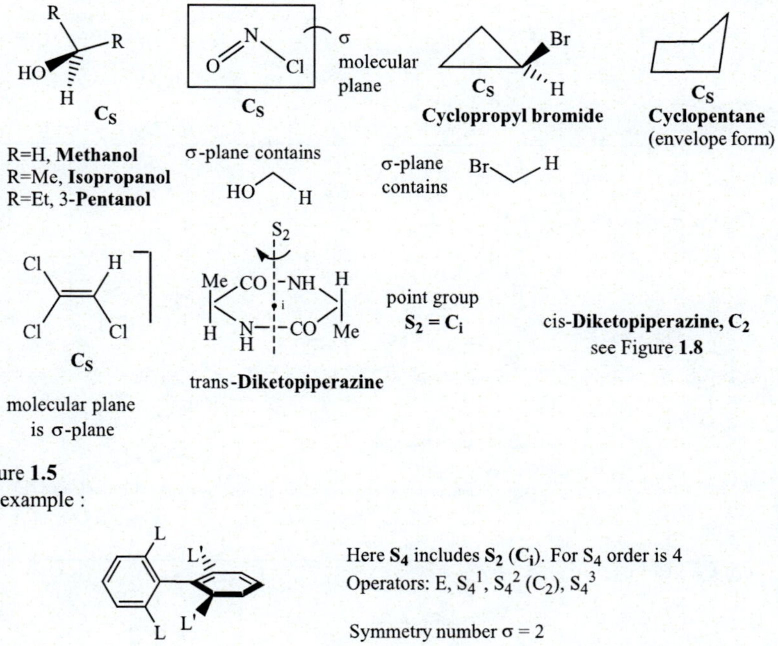

Fig. 1.11 Examples of some achiral compounds belonging to different achiral point groups

(ii) $\mathbf{C_{nv}} = C_n + n\,\sigma_v$; no σ_h

Each example contains a C_2 (shown) and two σ_v planes, as shown in the 1st example.

(a) $\mathbf{C_{2v}} = C_2 + 2\,\sigma_v$

Other examples:

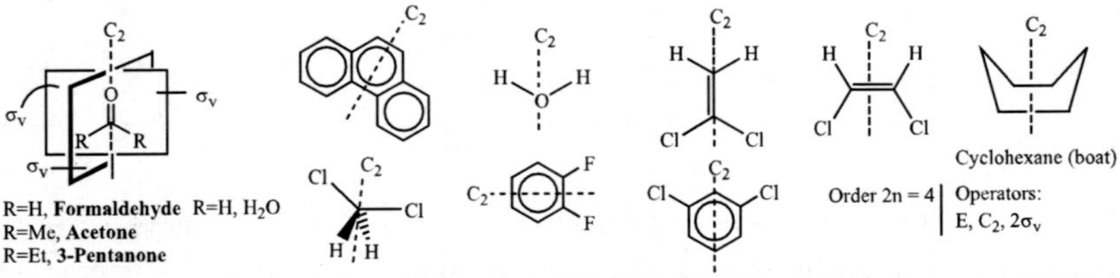

(b) $\mathbf{C_{3v}} = C_3 + 3\,\sigma_v$ Examples:

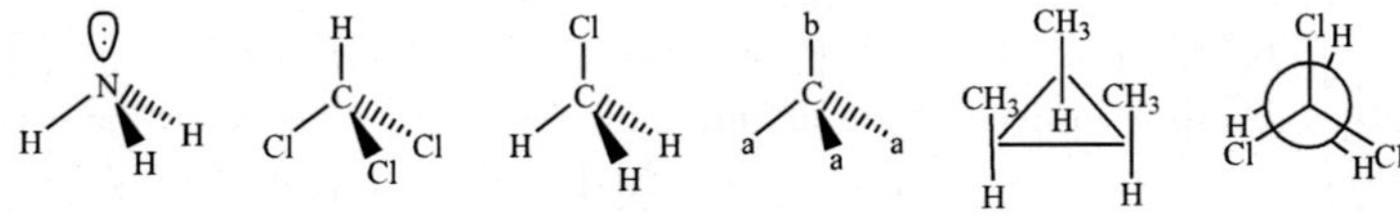

Order is 6: E, C_3, 3 σ_v's, C_3^2

(c) $\mathbf{C_{4v}} = C_4 + 4\,\sigma_v$

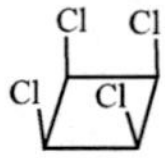

all cis

(d) $C_{5v} = C_5 + 5\,\sigma_v$

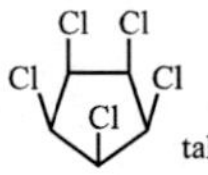

the ring is taken as planar

(e) $C_{\mu v} - C_\mu + \mu\,\sigma_v$

Conical symmetry. H-Cl; C=O; H-C≡N; H-C≡C-Cl

(iii) $\mathbf{C_{nh}} = C_n + \sigma_h$; no σ_v.

(a) $\mathbf{C_{2h}} \rightarrow$ order is 4: E, C_2^1 σ_h, i

$\sigma = 2$

Examples of point group $\mathbf{C_{2h}} = C_2 + \sigma_h$

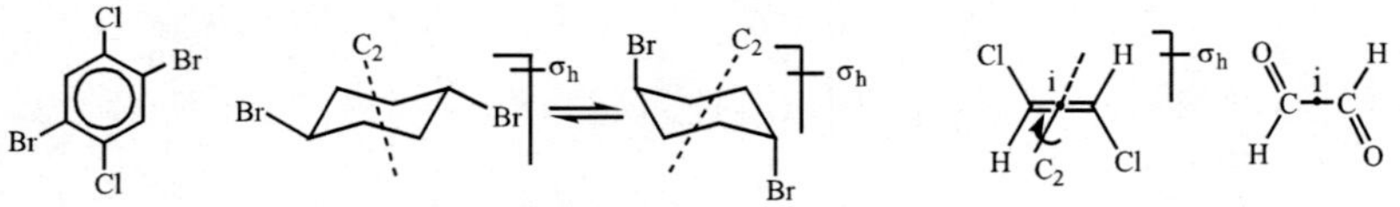

trans-**1,4-Dibromocyclohexane**

σ_h passing through 2C-Br bonds

C_2 passing through the mid points of opposite bonds shown

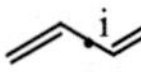

s-trans-**1,3-Butadiene**

(b) $\mathbf{C_{3h}} = C_3 + \sigma_h$

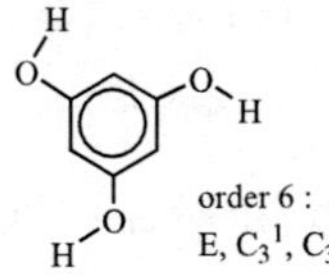

order 6 :

E, C_3^1, C_3^2, σ, S_3^1, S_3^2

$\sigma = 3$

This particular conformation has point group $\mathbf{C_{3h}}$

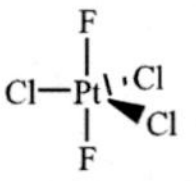

$\mathbf{PtCl_3F_2}$

Fig. 1.11 (continued)

(iv) D_{nd} = $C_n + n\,C_2\,(\perp r) + n\,\sigma_v\,(\sigma_d)$; no σ_h

(a) D_{2d} = $C_2 + 2 \perp r\, C_2 + 2\,\sigma_d$
also S_4 (prin. C_2)

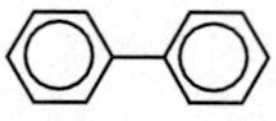

Biphenyl

Spirane

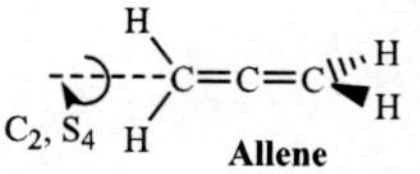

Allene

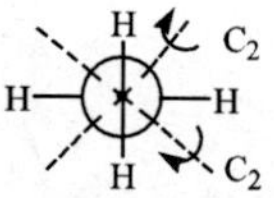

(also see **Figure 1.3**)

Order 8 :
E, $3C_2$ (mutually $\perp$r), S_4^1, S_4^3, $2\sigma_d$
$\sigma = 4$

(b) D_{3d} = $C_3 + 3\,C_2\,(\perp r) + 3\,\sigma_d$

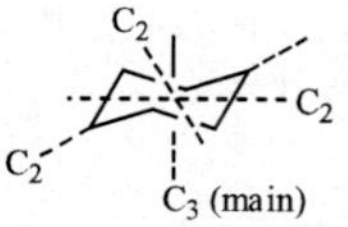

Cyclohexane (chair)

See **1.3.2**

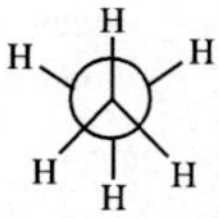

Staggered Ethane

Order 12 ; $\sigma = 6$
E, C_3^1, C_3^2, $3\,C_2^1$,
S_6^1 (coincident with C_3)
S_6^5, i, $3\sigma_d$

(v) D_{nh} = $C_n + n \perp r\, C_2 + n\,\sigma_v + \sigma_h$

(a) D_{2h} = $C_2 + 2\,C_2\,(\perp r) + 2\,\sigma_v + \sigma_h$

Examples:

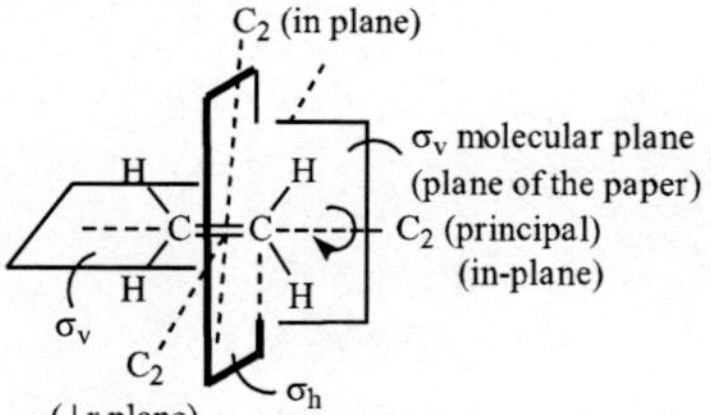

Ethylene

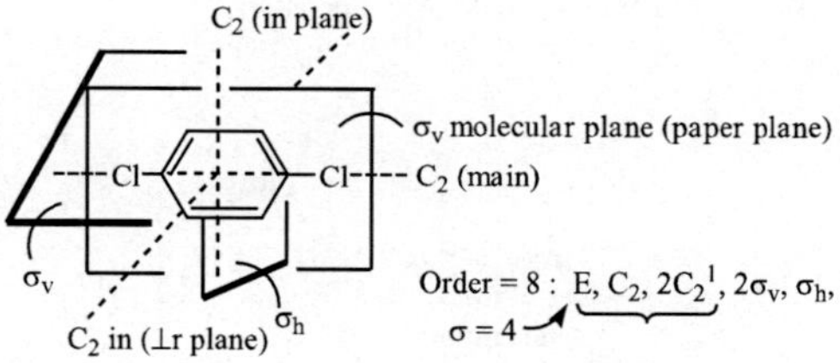

Order = 8 : E, C_2, $2C_2^1$, $2\sigma_v$, σ_h, i
$\sigma = 4$

D_{2h} = C_2 + 2 orthogonal C_2 + $2\,\sigma_v + \sigma_h$
D_2
(also see **Figure 1.3**)

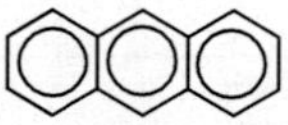

Anthracene

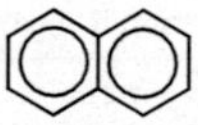

Naphthalene

(b) D_{3h} = $C_3 + 3\,C_2\,(\perp r) + 3\,\sigma_v + \sigma_h$

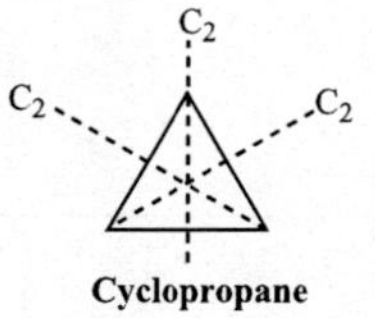

Cyclopropane

C_3 axis $\perp$r to the ring plane
3 C_2 axes in the ring plane
3 σ_v planes each containing a C_2 axis, the ring plane is σ_h

BF_3,

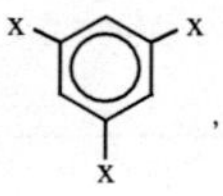

,

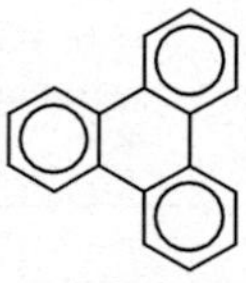

Triphenylene

(c) D_{4h} = $C_4 + 4\,C_2\,(\perp r) + 4\,\sigma_v + \sigma_h$

cyclobutane

Pt Cl_4 PtX_4
square planar

Order = 16 : E, C_4^1, C_4^2, C_4^3, $4\,C_2^1$, $4\,\sigma_v$, σ_h S_4^1, I, S_4^3
$\sigma = 8$

Fig. 1.11 (continued)

(d) D_{6h}

Benzene

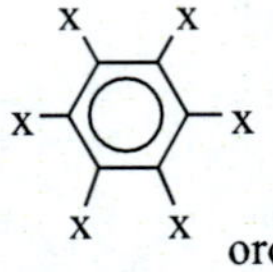

order = 24, σ = 12
sym. no.

$\mathbf{D_{6h}} = C_6 + 6\,C_2 + 6\,\sigma_v + \sigma_h$

$\sigma = E, C_6^1, C_6^2, C_6^3, C_6^4, C_6^5, 6\,C_2^1 = 12$

order $= \sigma + 6\,\sigma_v, \sigma_h, S_6^5, S_6^4, S_6^3, S_6^2, S_6^1 = 24$
sym. no. $(i = S_6^3)$

(vi) $D_{\infty h}$ E, C_∞, $\infty\,C_2$ ($\perp$r C_∞ axis), $\infty\,\sigma_v$, σ_h cylindrical H–H, O=C=O, HC≡CH

(vii) T_d

Pt gr. of regular of tetrahedron
A = achiral group like H, Me, Cl etc.

E, 3 C_2 (mutually $\perp$r) passing through pairs of opposite edges
4 C_3 → passing though each apex and the center of opp. face (containing A-C bond)

(i)

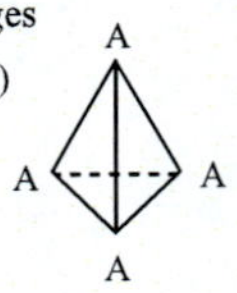

6 σ_d → Each σ_d contains each A–C–A plane; ∴ 4 C_2 = 6 combinations
3S_4 → coincident with C_2 axes

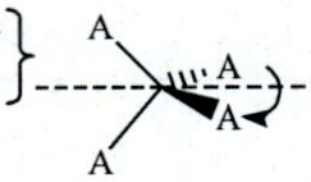

σ = 12; E, $4\,C_3^1, 4\,C_3^2, 3\,C_2^1$ = 12. order = $12 + \underline{6\,\sigma_d + 3\,S_4^1 + 3\,S_4^3}$ = 24

Underlined ones are additional operations

(ii)

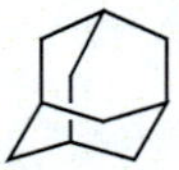

Adamantane
Chiral center in the center, not at any carbon

(iii)

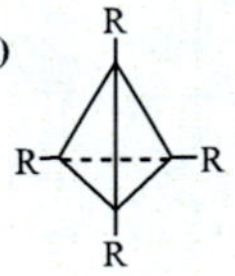

R = t-Bu, synthesised by Maier in 1978
R = H, could not be synthesised as yet

(viii) O_h octahedral symmetry cubic pt. gr.

Sym. operations : E, 3 C_4 (mutually perp.) + 4 C_3, 3 S_4, 3 C_2 (coincident with 3 C_4 axes) 6 C_2, 9 σ, 4 S_6 (coincident with 4C_3 axes)

σ = 1 + 9 + 8 + 6 = 24
order = 24 x 2 = 48
(double by the existence of symmetry planes

Other examples:

$[PtCl_6]^-$; $[Co(NH_3)_6]^{3+}$

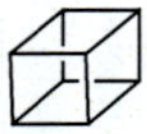

Cubane

first synthesised by Eaton in 1964

Fig. 1.11 (continued)

1.1.7 Local Symmetry (or Site Symmetry). Desymmetrization

Local symmetry shows the symmetry properties of atoms or groups within a molecule (Fig. 1.12).

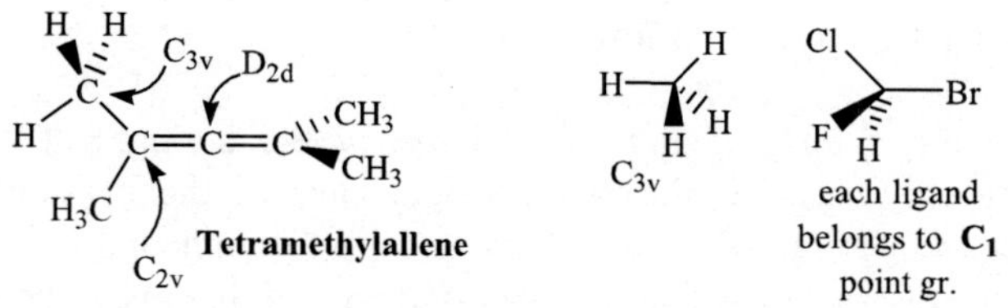

Desymmetrization results from successive substitution on a symmetrical lattice framework

Ex.1 CH_4 → CH_3Cl → CH_2Cl_2 → CH_2ClBr → CHFClBr

T_d C_{3v} C_{2v} C_s C_1

Desymmetrization results in distortion of the framework, small but real, as demanded by the symmetry of the system.

Ex.2 $Cl_2C{=}C{=}CH_2$ ← $H_2C{=}C{=}CH_2$ → $ClCH{=}C{=}CHBr$ ← $ClCH{=}C{=}CH_2$ → $ClCH{=}C{=}CHCl$

C_{2v} D_{2d} (Elongated tetrahedron) C_1 C_s C_2

Ex.3

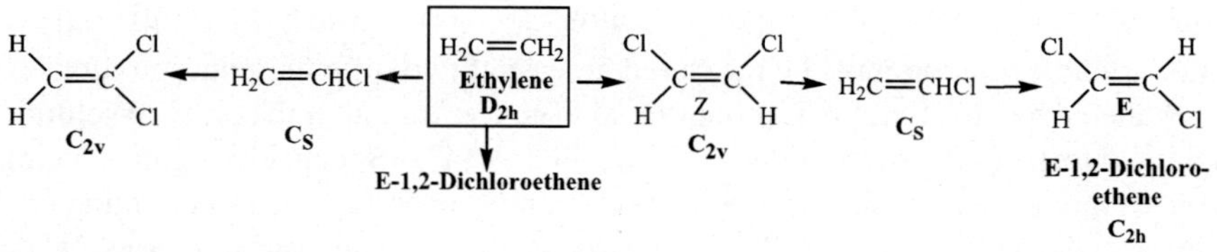

Fig. 1.12 Local symmetry and desymmetrization illustrated

1.1.8 Optical Isomerism. Optical Rotation

Optical Activity Due to Chiral Molecular Structure

A chiral compound in the solid, fused, gaseous (rarely) liquid, or dissolved state, e.g., lactic acid, tartaric acid, glucose, etc., exhibits optical activity, i.e., rotates the plane of polarization of an incident beam of polarized monochromatic light. During the early nineteenth century, the French physicist Jean Baptiste Biot (1774–1862) discovered that many natural organic compounds exhibit optical activity. Since this optical rotation occurs in the liquid, dissolved, or gaseous phase, it must be a molecular phenomenon. Here, the optical activity is entirely due to the *dissymmetry of the molecular structure.* The original molecule and its nonsuperposable mirror images are known as *enantiomers, enantiomorphs* (this name is taken from crystallography), or optical antipodes, one of which is dextrorotatory (rotates the plane of polarization in the clockwise direction by a positive (+)-angle) and the other levorotary (rotates in the anticlockwise direction by a negative (−)-angle). For the measurement of optical rotation, earlier polarimeters have now been replaced by digital polarimeters.

Optical Activity Due to Crystalline Structure

Some crystals, e.g., *quartz, sodium chloride, benzil*, etc. are optically active in the crystalline state only, i.e., rotate the polarized light. Quartz is the first substance shown by Biot in 1812 to be optically active. It exists in two hemihedral nonsuperposable mirror image forms, one of which is *dextrorotatory* and the other *levorotary*. Here, the optical activity is a property of the chiral crystal and not of the molecule, as the rotation disappears when the crystal is melted or dissolved to give the achiral molecules. Such pairs of nonsuperposable crystals are said to be *enantiomorphous*.

Dependence of Rotation (α) on Concentration and Cell Length. Value of α

The observed angle of rotation of the plane of polarization is denoted by α, which can be recorded in a range of $-90°$ and $+90°$. For example, distinction cannot be made between α and $\alpha \pm n$ 180°; the plane when rotated by $\pm 180°$ or its integer *n*, the new plane will coincide with the old one. Thus theoretically, no difference appears between $+40°$, $+220°$ ($40° + 180°$), $+400°$ ($40° + 2 \times 180°$), or $-140°$ ($40° - 180°$). Hence, rotation must be measured at least at two different concentrations to get α unequivocally. Since α is proportional to concentration, if the original solution is diluted to 1/10, α would be $+4°$, $+22°$, $+40°$, $-14°$, respectively. Again α being proportional to length (discovered by Biot) can be unambiguously determined by using smaller cells, e.g., of 0.25 *dm* length for the original concentrations, when observed α will be $+10°$, $+55°$, $+100°$ (equivalent to $-80°$), or $-35°$, respectively, again all clearly distinguishable.

Dependence of Sign of $[\alpha]$ of Polar Compounds on Solvent, Concentration, and pH

The rotation of compounds, especially the polar ones, is affected by the solvent because of its participation in solvation and association phenomena. It has been reported by Winther [2] in 1907 that the specific rotation $[\alpha]_D$ (Sect. 1.1.9) of (+)-nicotine is positive in polar solvents like formamide (maximum positive specific rotation) and water, $[\alpha]_D$ decreasing with increase of concentration. It remains almost same in MeOH and EtOH and increases in benzene, *o*-, *m*-, and *p*-xylenes, and mesitylene with increase of concentration. Whereas, the rotation of natural nicotine in ethylene bromide and chloroform at lower concentrations (>0.6 g/g) was observed to be negative. Specific rotation of nicotine at infinite dilution of the solvent (meaning no solvent) obtained by extrapolation of the specific rotations in solvents of different polarities has been found to be ~8 units (Sect. 1.1.9); this is known as *intrinsic rotation* $\{\alpha\}$. As the concentration of the solute increases, solute–solute interactions are maximized, which can differ greatly from one solvent to another. In polar solvents, solute–solute association effect may be suppressed by competition with concentration-independent solvent association, which appears to increase the rotation.

Reversal of the sign of the rotation of (+)-nicotine takes place in less polar solvents like $CHCl_3$ and ethylene bromide at much less concentrations. Other examples: 2-methyl-2-ethylsuccinic acid displays positive rotation in $CHCl_3$ containing 0.7% EtOH at >6.3%, no rotation (null) at ~6.3%, and negative rotation at <6.3% concentration; but no reversal is observed in alcohol solvents, pyridine, diglyme, and acetonitrile.

Rotation of acids and bases is dependent upon the pH, e.g., (*S*)-(+)-lactic acid is dextrorotatory in water, but its sodium salt is levorotatory. Another example: L-Leucine is levorotatory in water but dextrorotatory in aqueous hydrochloric acid.

1.1.9 Specific Rotation. Molecular Rotation. Units (Fig. 1.13)

The rotation per unit length in dm and unit concentration in g/ml is called *specific rotation* expressed as follows:

Specific rotation $[\alpha]_D^t = \frac{\alpha}{l.c} = \frac{\alpha.100}{l.c'}$ where α is the observed rotation, l = length of the cell in dm, c = conc. in g/ml, t = temperature, c′ = conc. in g/100 ml, d = density of the neat liquid

(Rotation is measured at the wavelength of Na-D line (589 nm)

For liquids $[\alpha]_D^t = \frac{\alpha}{l.d}$

A new term "molecular rotation" [M] or [ϕ] also called molar rotation is defined to compensate for the effect of differing mol. wts. as follows:

$$[M] = [\varphi] = \frac{[\alpha]_D^t.M}{100^a} = \frac{\alpha.M}{l.c.100^a} = \frac{\alpha}{l.\frac{c}{M}(100\,ml)^{-1}} = \frac{\alpha}{l.c''}$$

where M = mol. wt., c″ = mole per 100 ml

[a]The division by 100 is arbitrary in order to keep its numerical value on the same approximate scale as that of specific rotationl.

Note: Thus, optical rotation is proportional to the number of molecules encountered. Hence, for a comp ound with mol. wt. 100, [ϕ] = [α]

Units or dimensions of [α] and [ϕ]

$[\alpha] = \frac{\alpha}{l.c}$ Thus units of [α] are $\frac{degree}{dm.g\,cm^{-3}} = \frac{degree}{100\,cm.g.cm^{-3}} = \frac{degree}{100\,cm^{-2}\,g} = 10^{-1}$ deg. $cm^2\,g^{-1}$

So, while α is given in degrees, [α] should not be expressed in degree alone and should always be given without the units (understood to be 10^{-1} deg $cm^2\,g^{-1}$), as has been done in this book [cf. references in further reading (iii) and (v)].

$$Again, [\varphi] = \frac{[\alpha]M}{100} = \frac{\alpha}{l.mole\ per\ 100\,ml}$$

∴ Units of molar rotation are $\frac{degree}{10\,cm.mole.10^{-2}cm^{-3}} = \frac{degree}{10^{-1}cm^{-2}\,mole} = 10$ deg $cm^2\,mole^{-1}$

[ϕ] is also expressed **without any unit**, understood to be 10 deg $cm^2\,mole^{-1}$

Fig. 1.13 Specific and molecular rotations and their units

Fischer Projection. Flying Wedge Formulas. Tetrahedral Representations of C_{abcd}

A common representation of a tetracoordinated compound with one chiral center was first proposed by Emil Fischer [3] in 1891, which is much easier than tetrahedral representation. While writing a Fischer projection formula of an organic compound having one or more chiral center/s, each sp^3 carbon atom in the chain is written as a cross + . Thus, the *agreed modes of projection* of the three-dimensional molecule on two-dimensional plane (projection plane, PP) to give the Fisher projection (FP) formula are as follows:

1) The asymmetric (or dissymmetric) carbon should be in the projection plane.

The main chain must be in the vertical plane and pointing away from the observer, so that the vertical or up and down bonds are below the projection plane extending toward rear. The other two sideways substituents (ligands) of each carbon must be in the horizontal plane (above the PP) and pointing forward toward the observer.

Corollaries

1. One pair of ligands when exchanged gives the enantiomer.[1]
2. Two exchanges give another FP formula of the same enantiomer.
3. Ligands may be rotated in groups of three.

Limitations.

1. FP formula being two-dimensional cannot be lifted out of the P.P. and turned over.
2. FP formula may be rotated in the projection plane by even multiple of 90° (i.e., 180°), but not by odd multiple of 90° (i.e., 90° or 270°).

Three-dimensional (flying wedge) and tetrahedral representations of enantiomers with one chiral center, C_{abcd}, and their two-dimensional representations as Fischer projection formulae are exemplified in Fig. 1.14.

There are 12 ways of writing the Fischer projection formula of an enantiomer of a compound with one chiral center, as depicted in Fig. 1.16. For detailed discussion on (D,L) and (*R*,*S*) nomenclatures, see Sects. 2.1 and 2.2.

The best way to confirm whether the two-dimensional (Fischer projection) or three-dimensional (flying wedge or tetrahedral) representations are identical or mirror images is to specify their (*R*,*S*) nomenclature (Sect. 2.2). In Fig. 1.14, all representations on the left side or right side of the mirror are of (*R*)-glyceraldehyde or (*S*)-glyceraldehyde. The names are given under the Fischer projection formulas specifying the (D,L) nomenclature. Each of the 12 FP formulae of (*R*)-C_{abcd} (Fig. 1.16) can give many more flying wedge formulas by tilting and rotating to different extents and

[1] It may be noted that a tetrahedron is the only skeleton in which every transposition of ligands is equivalent to reversal of the ligated assembly, in which the tetrahedral atom is at the center of the tetrahedron and the four bonds of that atom are directed toward the four vertices of the tetrahedron carrying the four ligand atoms.

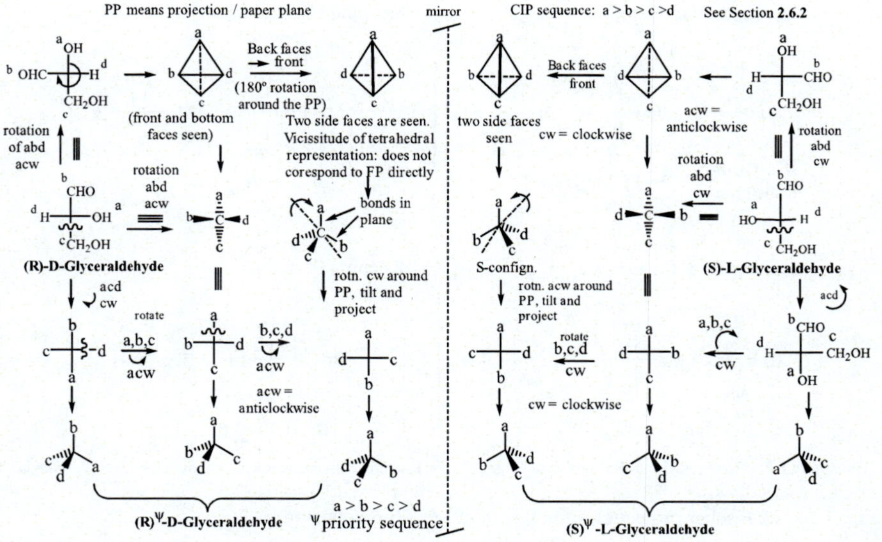

Fig. 1.14 Fisher projection, flying wedge, and tetrahedral representations of (*R*)-D-glyceraldehyde and (*S*)-L-glyceraldehyde

can be confirmed by specifying the absolute configuration according to CIP sequence rules (Sect. 2.2.2).

The tetrahedral representation of (*R*)-C_{abcd} can also be written as shown in Fig. 1.15, tilting the bottom side of the FP formula downwards so that the bottom ligand goes to the rear side of the triangle (front face) made by the other three ligands; the other three faces of the tetrahedron go to the rear side. A few other common compounds with one chiral center are also shown in this figure.

For a chiral compound with one chiral center, there are factorial four (4!) = 4 × 3 × 2 = 24 ways of writing the Fischer projection formula, this being the permutations of four ligands among four sites. Of these, 12 FP formulas will correspond to the *R*-enantiomer and the remaining 12 to the (*S*)-enantiomer.

The 12 Fischer projection formulas for (*S*)-**D**-2-bromobutane are depicted in Fig. 1.16.

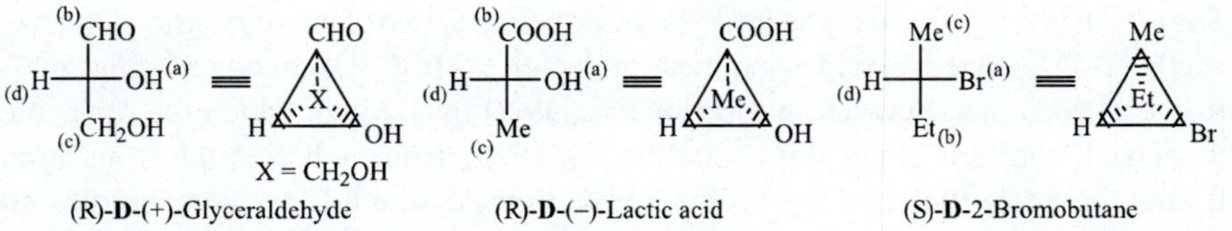

Fig. 1.15 Different tetrahedral representation of a few compounds, C_{abcd} (one enantiomer)

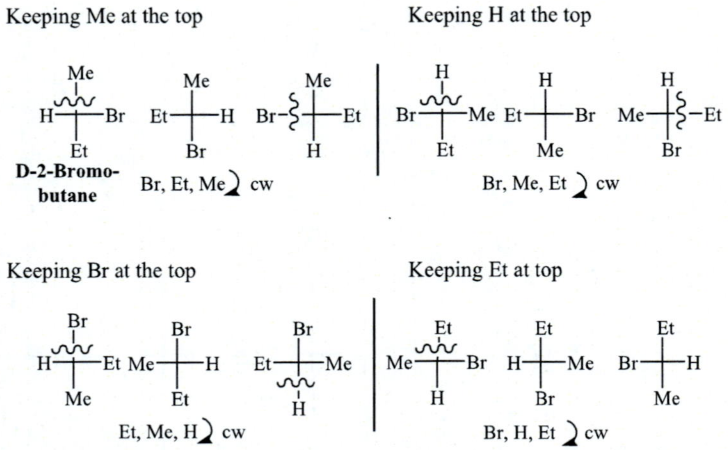

Note: Each F.P. is having (S-) and **D**-absolute configuration. The **D**-configuration is revealed only from the first F.P. formula.

Fig. 1.16 12 Fischer projection formulae of (*S*)-**d**-2-bromobutane

1.2 Conformation of Simple Acyclic Molecules

The term *conformation* of a molecule (excepting the diatomic ones) signifies any one of the infinite number of momentary arrangements of its atoms in space that result from rotation around single bonds and also from twisting around bonds. The conformations of a molecule that correspond to the minimum energy in its potential energy diagram are known as conformers or *conformational isomers*. Thus, any point on the curves (Figs. 1.20, 1.21 and 1.22) corresponds to some conformation of the concerned molecule. Conformations are not superposable upon each other.

Conformational analysis involves the interpretation or prediction of the physical (including spectral) properties, thermodynamic stabilities, and reactivities of substances in terms of the conformation or conformations of their molecules [4].

1.2.1 Dihedral Angle. Torsion Angle. Torsional Strength

The *dihedral angle* θ (*theta*) denotes the angle between two planes containing A–B–C and B–C–D, respectively, in a nonlinear molecule, A–B–C–D, as shown in Fig. 1.17. It is best seen in a Newman projection formula (Fig. 1.17) in which the molecule is viewed from left along B–C axis; the dot in the front indicates the front atom B, and the circle indicates the back atom C. Here, if B and C are tetracoordinated atoms, the remaining two bonds of each one are not shown. Thus, dihedral angle is a three-dimensional parameter of a molecule involving four atoms. The dihedral angle if given *a directional sense* becomes torsion angle denoted by Greek letter ω (omega) or ϕ (phi). If looking along the B–C axis *in either direction,* the turn from A

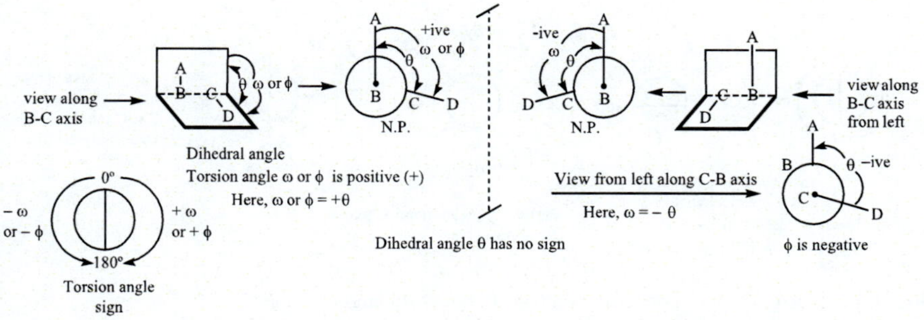

Fig. 1.17 Torsion angles and dihedral angles for two enantiomers

to D or D to A is clockwise, ω is positive; if the turn is anticlockwise, ω is negative. The torsion angle ω is best represented by Newman projection (NP) formula for each enantiomer (Fig. 1.17).

Torsional strain or Pitzer strain (V_ϕ) or potential energy (E) is caused by the rotational motion around the bond axis. Torsional strain is represented by the following equation:

$$V_\phi = \frac{1}{2}\, V_O(1 + \cos n\Delta\phi) \text{ kcal/mol}$$

where $\Delta\phi$ is the displacement of the dihedral angle (torsional displacement), V_O is the torsional energy barrier or Pitzer strain, and n is the periodicity, i.e., the number of times that a given conformation recurs during a complete revolution ($\Delta\phi = 360°$).

1.2.2 *Klyne–Prelog Nomenclature for Torsion Angles. Conformational Chirality*

A general method of nomenclature has been worked out by Klyne and Prelog [5] in 1960 to describe the steric relationship across a single bond in a molecule or a part of a molecule. The following rules are followed.

1. It has been mentioned earlier that in contrast to dihedral angle, torsion angle has a *directional property*, being mentioned as (+) when measured in a clockwise direction, and as (−) when measured in an anticlockwise direction. In a molecule of the type A–B–C–D, the measurement is to be started from the *front substituent* A at 0° ending at the *rear substituent D* at 180° after rotation of the back substituent *D* around *B–C* bond.
2. *Conformation Selection Rule* (CSR): The two fiducial (reference) groups A and D are specified according to the *Conformation Selection Rule* by Cahn, Ingold, and Prelog [6] in 1966 as follows:

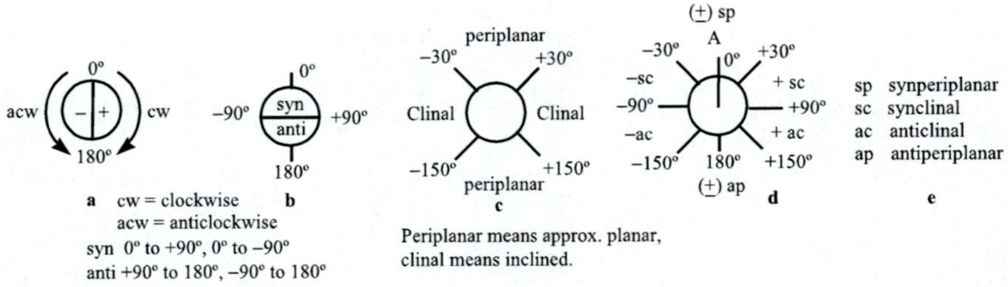

Fig. 1.18 Designation of conformation based on torsion angle ω or ϕ

(i) If all the atoms or groups (ligands) of the set on both carbons (front and back) are different, the ligand most preferred by the standard *sequence sub-rules* (Sect. 1.6.2) is fiducial.

(ii) If two ligands of a set on one or both carbon/s are identical, the nonidentical (unique) group is fiducial, irrespective of the sequence rule.

(iii) If all or two ligands of the set on carbons are identical, that which provides the smallest torsion angle is fiducial.

3. The fiducial group A at the front atom is preferably (but not necessarily) placed at the top of the Newman projection formula, and the torsion angle is named in terms of the three designations **a**, **b**, and **c** (Fig. 1.18). The circle is divided into six segments by combination of **a**, **b**, and **c**, as shown in **d** in the figure. The conformation of any molecule bears the designation of the torsion angle. *Synperiplanar* and *antiperiplanar* conformations are expressed with (±) sign because of probable variation or libration of ω around 0° and 180°, respectively, by upto +30° or −30°.

The sign of a torsion angle in any conformation remains unchanged whether the molecule is viewed from the front or from the rear. Since the exact values of torsion angles are often not known, the (+) and (−) signs immediately show the direction of a torsion angle, and the symbols *sp*, *sc*, *ac*, *ap* in the Klyne–Prelog system show the range of a torsion angle. The system is applicable to any molecule A–B–C–D, whether B and C are tetrahedral or trigonal atom; this system may also be used to describe partial conformation of ring compounds and polymer chains.

Figure 1.19 depicts a few examples of different types. Example 1 illustrates the torsion angle nomenclatures of the conformers and also the higher dipole moment of active stilbene dichloride than that of its meso-isomer based on the analysis of the three conformers in each case.

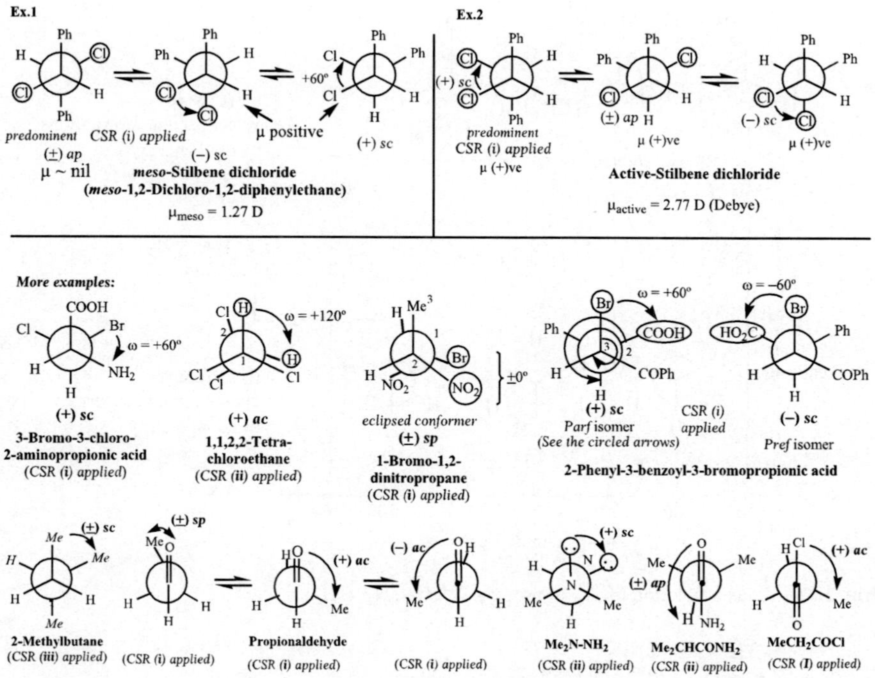

Fig. 1.19 Few examples of Klyne–Prelog nomenclature of torsion angles (*conformational chirality*)

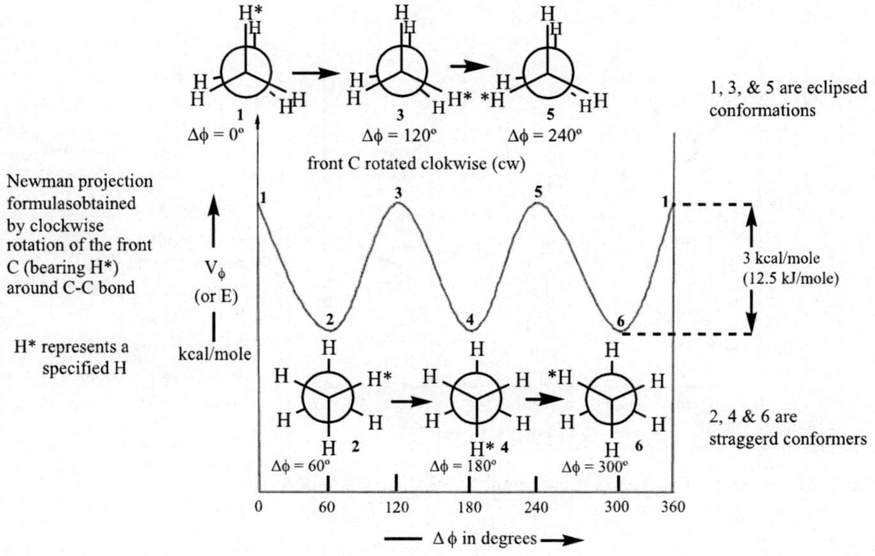

Fig. 1.20 V_{ϕ} as a function of Δ_{ϕ} in ethane ($H_3C{-}CH_3$)

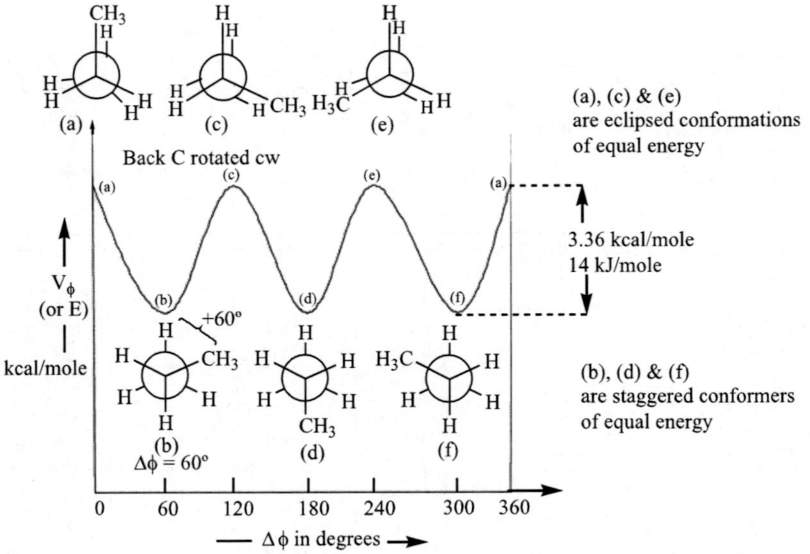

Fig. 1.21 V_ϕ as a function of Δ_ϕ in propane ($H_3C–CH_2–CH_3$)

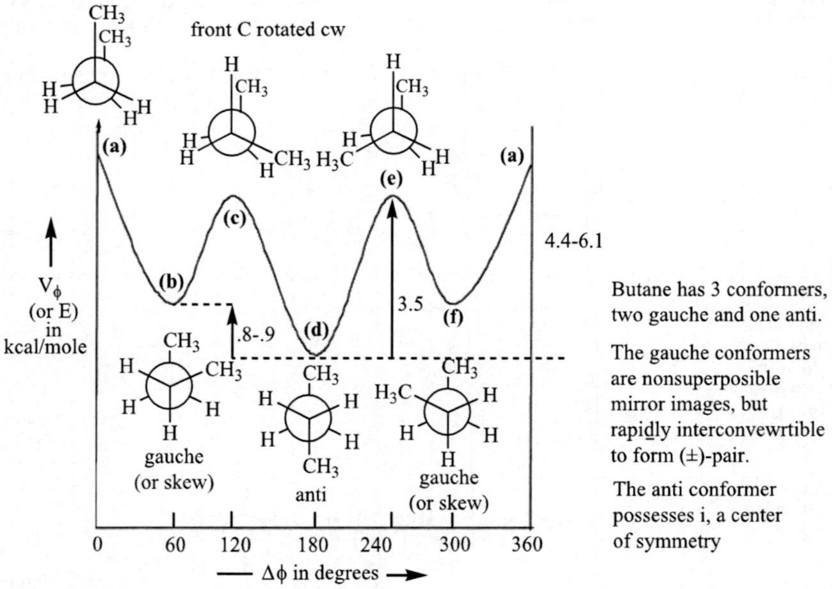

Fig. 1.22 V_ϕ or E of butane as function of Δ_ϕ (displacement of torsion angle)

1.2.3 Torsional Strain Curve (Potential Energy Diagram) of Ethane

The change of V_{ϕ} with ϕ or better with $\Delta\phi$ in case of ethane with the conformations having maximum and minimum energies is represented graphically in Fig. 1.20. The energy difference between the eclipsed conformer of maximum torsional strain or potential energy and the staggered conformer of minimum energy, the so-called *torsional or rotational energy barrier*, is only about 3 kcal/mol (12.5 kJ/mol). Hence, rotation around C–C single bond, though not free, is quite facile. Here, the greater the periodicity, the smaller the torsional barrier. For example, in case of nitromethane ($MeNO_2$) periodicity $n = 6$; the torsional energy barrier is only 0.006 kcal/mol.

Recent studies have shown that the high torsional energy of the eclipsed form of ethane and hence *the torsional energy barrier* is not due to van der Waals repulsive steric interaction between the two eclipsed H atoms (the distance between them is 2.3 Å = 0.23 nm, whereas the van der Waals radii of two H atoms is 1.83 Å), or not due to electrostatic interaction between weakly polarized C–H bonds, but it *is due to torsional strain caused by unfavorable overlap interaction between the C–H bond orbitals*. In case of the staggered form (the distance between two nearest H atoms on C1 and C2 is 2.5 Å), favorable interaction between the bonding–antibonding orbitals makes its torsional strain 3 kcal/mol less than the eclipsed conformation and makes it more stable. This energy difference leads at 25 °C to the fact that for each 160 staggered ethane molecules, there is only one molecule of eclipsed ethane, i.e., in negligible proportion.

1.2.4 Torsional Strain Curve of Propane

The torsional strain curve of propane is similar to that of ethane with a slightly higher torsional strain barrier (3.36 kcal/mol = 14 kJ/mol). The small difference between the energy barriers of ethane (Fig. 1.20) and propane (Fig. 1.21) indicates clearly that the torsional energy does not originate from steric effects, as already mentioned. The eclipsing of H with CH_3 in propane is hardly more unfavorable than the eclipsing of H and H in ethane, despite the bulkiness of the methyl group. In case of propane also, the periodicity n is 3.

1.2.5 Torsional Strain Curve of Molecules ACX_2CX_2B, n-Butane

Torsional strain curve (V_{ϕ} as a function of $\Delta\phi$) of a molecule AC(*XY*)C(*XY*)B becomes more complex when a given conformation does not recur within a complete torsional displacement ($\Delta\phi = 360°$), i.e.., n (periodicity) = 1. The torsional strain curve of the molecules of butane, $CH_3CH_2CH_2CH_3$, (where $A = B = CH_3$ and $X = Y = H$) giving the Newman projection formula of the maximum and minimum energy conformations, is displayed schematically in Fig. 1.20. The diagram shows two types of conformations having maxima: (**a**) and (**c**) = (**e**). The conformation (**d**), called *anti*, represents the energy minimum corresponding to the lowest valley (minimum); the energies of the other conformers (**b**) and (**f**), called *gauche*, corresponding to the other valleys (minima) are measured relative to that of this conformer (**d**), as shown in Fig. 1.22. The conformers, (**c**) and (**e**), representing the steric interaction energy of two CH_3 groups at $\Delta\phi = 60°$ or 300° are destabilized by 0.8–0.9 kcal/mol (~3.3 kJ/mol) at room temperature relative to the *anti*-conformer (**a**).

This energy difference corresponds to about 1 mol of butane as the *gauche* conformer for every 2 moles of butane as the *anti* conformer. This may be calculated for the equilibrium [*gauche*] ⇆ [*anti*], as follows:

$\Delta G° = \Delta H° - T\Delta S° = -0.8$ kcal/mol$-(-$TRln2) = (0.8$-$0.41 = 0.39 kcal/mol at 27 °C, T = 300 °K, and

Rln2 = 1.38 cal/deg/mol.

Putting the value of $\Delta G°$ in the equation $\Delta G = -$RTlnK

$K =$ [*anti*]/[*gauche*] ≈ 2

Thus, it is possible to predict the most stable conformer of a long chain $(-CH_2)_n$, which possesses repeated butane chain, as a *zigzag* planar arrangement of the chain.

The torsional strain curve of other molecules having the same molecular formula ACX_2CX_2B will be of same pattern as that of butane. Only the energy barriers will be different. In *n*-propyl chloride, $CH_3CH_2CH_2Cl$ (A = CH_3, B = Cl, and X = H), and 1,2-dichloroethane (A = B = Cl, X = H) the *anti* and *gauche* conformers have approximately the same energy, whereas in 1,1,2,2-tetrachloroethane and 1,1,2,2-tetrabromoethane (A = B = H, X = Cl or Br), the *gauche* conformer is known to be more stable than the *anti*-conformer by ~1.0 kcal/mol, perhaps due to the stabilizing *gauche effect of the two halogen atoms.*

1.3 Configuration. Relative Configuration. Absolute Configuration

Molecules with the same molecular formula and same constitution or bonding connectivities may still be different. The arrangement of the ligands (atoms or groups) in space around the dissymmetric part of a molecule—in the simplest case around an asymmetric center or around the rigid part of a molecule like a double bond—is termed as *configuration*. If there is any dissymmetry in the molecule (lacking a reflection symmetry)—possessing a chiral point group (Sect. 1.1.5)—the molecule will be nonsuperposable on its mirror image and the two mirror image forms (enantiomers) will exhibit opposite optical rotation. When molecules differ only in relative orientation of ligands in space, stereoisomerism results. Stereoisomers having same bond connectivity but different absolute configuration are often called configurational isomers. Most of the natural products and biologically active molecules occur in specific stereoisomeric forms; their chemical and biological behaviors are controlled by their absolute configuration and also conformation.

In 1780, Scheele isolated (−)-lactic acid [= (−)-α-hydroxypropionic acid] from sour milk. Berzelius isolated (+)-lactic acid from muscle tissues in 1807 (Fig. 1.23). Actually, in 1848, Berzelius established that although they have same structure and same properties, they were not same, but enantiomers (nonsuperposable mirror images) exhibit opposite optical rotation. *Configuration is a theoretical concept* related to the molecular architecture expressed by a three-dimensional or projection formula, *while rotation is an experimental property*, measured in a polarimeter. Configuration is more fundamental than rotation and remain unchanged as long as any bond to the chiral center is not broken; but specific rotation changes with concentration and with solvent; even the sign of rotation may sometimes change with some solvent at different concentrations (*vide* Sects. 1.1.8.3 and 1.1.8.4). Again (+)-lactic acid when dissolved in aqueous sodium hydroxide, the resulting solution is levorotatory and would have to be called a solution of (−)-sodium lactate, although the configuration remains unchanged.

The simplest way to explain relative and absolute configuration is to state that two enantiomers (e.g., of lactic acid) have got the same relative configuration but have different absolute configurations.

Absolute configuration is specified by (D,L) notation (Sect. 2.1) in some cases and by (*R*,*S*) notation universally (Sect. 2.2), to be discussed in the respective sections. In a compound with several asymmetric centers, configuration at each center is to be specified in order to specify its configuration completely.

The arrangement of groups (or ligands) about the rigid double bond or ring (i.e., the configuration) is expressed as *cis* or *trans* (*Z* or *E*) for double bond compounds (Sect. 2.3) and *cis* or *trans* for cyclic compounds (Sect. 7.4).

The Fischer projection and tetrahedral representations of two common chiral molecules are shown in Fig. 1.23).

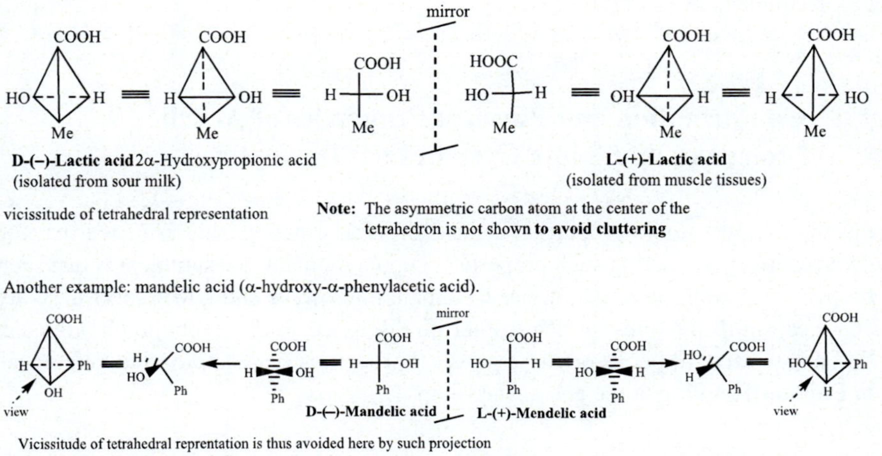

Fig. 1.23 Enantiomers of lactic acid and mandelic acid; each pair has the same relative configuration but different absolute configuration

1.4 Relationship Between Two Molecules of Same Molecular Formula. Homomers, Constitutional Isomers, Stereoisomers, Enantiomers, Diastereomers, Configurational/Conformational Enantiomers/Diastereomers

Relationship between two molecules of same molecular formula giving rise to different stereochemical or structural terms is delineated schematically with examples in Fig. 1.24. The definition of each term, expressed in bold face, is understood from this figure.

1.5 Conformational Effect on Stability of Diastereomers (Fig. 1.25)

For determining the stability of any diastereomer only the most stable conformer having large (L) groups anti is to be considered.

In Fig. 1.25 it has been shown in terms of gauche interactions between large (L), medium (M), and small (S) groups that (**A**) generally the meso diastereomer is more stable than the active (+)- or (–)-diastereomer; (**B**) the racemic (+)-variety is more stable than the active variety due to entropy of mixing; and (**C**) on the contrary, the active (+)- or (–)-diastereomer of vicinal diols may be more stable than the meso variety due to the presence of the H-bonding (attractive interaction) occurring in the gauche conformer of the OH groups, as shown in Fig. 1.25. This reversal of stability will be more prominent in such compounds having vicinal OH groups and bulkier groups in place of methyl/s.

1.6 Conformation and Physical Properties of Acyclic Molecules and Some Cyclohexane Derivatives

Physically measurable properties depend on conformation, which hence may be determined by measuring such properties. Dipole moment measurement is one such property—in addition to the infrared, Raman, microwave and NMR spectra, X-ray and electron diffraction—which cannot be discussed in this monograph. One can find many standard books on these physical properties (*vide* appropriate references in **Further Reading** at the end of the Chap. 12).

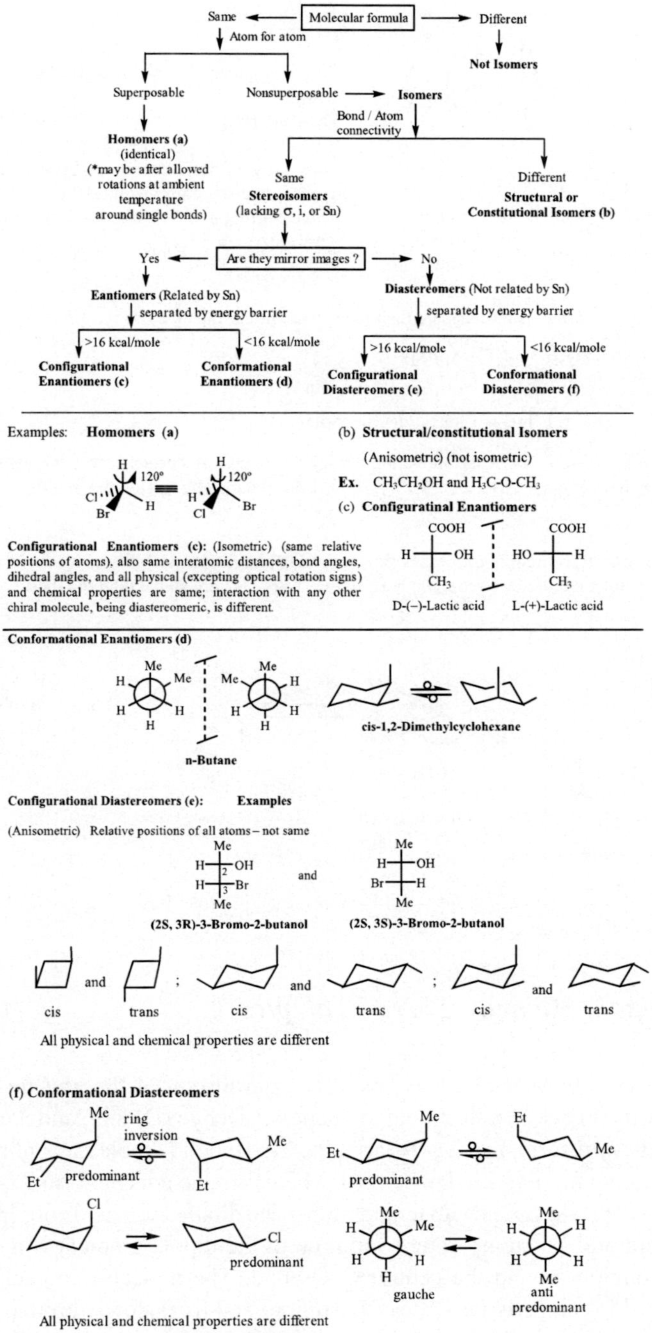

Fig. 1.24 Relationship between two molecules of same molecular formula: different terms with examples

(A)

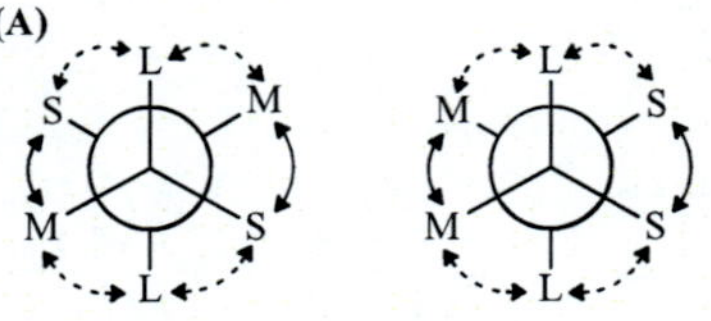

meso S_2 most stable conformer (L, L anti)

active C_1 most stable conformer (L, L anti)

Common gauche interactions are shown by dotted arrows.
Thus, interactions are: two M-S versus M-M+S-S.
We know, two M-S < (M–M+S-S).
Thus, crossed steric interations between two groups of different sizes are less than the sum of interactions between groups of like size - if interactions are purely steric in origin.

(B) Racemic (±)-variety is more stable than the active variety due to entropy of mixing (ΔS)

We known that $\Delta S = -R(\sum x_1 \ln x_1)$

$$= -R\,(\tfrac{1}{2}\ln\tfrac{1}{2} + \tfrac{1}{2}\ln\tfrac{1}{2}) = -R\ln\tfrac{1}{2}$$

$$= R\ln 2 = 1.4\ \text{kcal/degree/mole}$$

Thus, the difference between activation energy difference and ground state energy difference of such diasteromers, $\Delta H - \Delta G = T\Delta S = 300 \times 1.4 = 0.42$ kcal/degree/mole at 27°C

(C) In case, where H-bonding (attractive interaction) is present, the stability order may be reversed, e.g., 2,3-butane diol diastereomers; here active variety is nore stable than the meso.

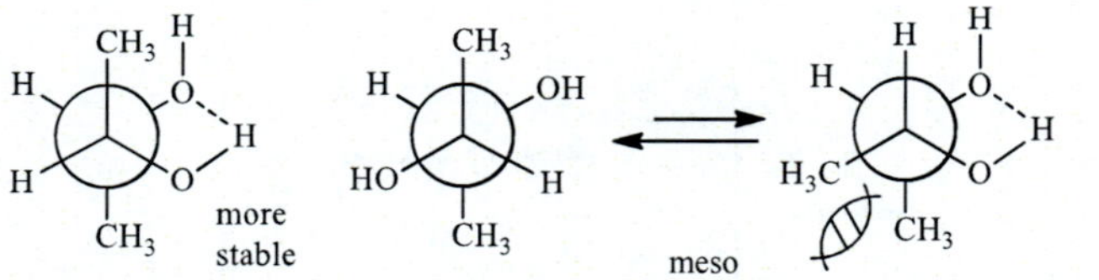

active C_1 (H-bonding stabilizes)

anti conformer (no H-bonding)

S_2

gauche conformer

Here, H-bonding involves gauche butane interaction, so less stable

Fig. 1.25 Conformational effect on stability of *meso* and *active* diastereomers. Stability order of 2,3-butane diol diastereomers

1.6.1 Dipole Moment. Three Examples

Example 1 1,2-dibromoethane exists in an equilibrium of the *gauche* (I) and the *anti* (II) forms (Fig. 1.26). Its dipole moment is 1 Debye (D) unit, and it cannot exist entirely in the *anti*-form. In the *anti*-form, the C—Br bond dipoles are antiparallel and the dipole moment is expected to be zero, whereas in the *gauche* form it is finite. The equilibrium constant varies with temperature. The dipole moment is observed to vary with temperature. The temperature variation of the dipole moment is incompatible with free rotation around the central C–C bond. The molecule in each form also librates [7, 8] [torsionally oscillates (twisting to and fro)] slowly around the central C–C bond without passing into the other form. The dipole moment is likely to vary with temperature and hence the dipole moment of each form may vary, though to a small extent. The overall dipole moment of each form will be equal to the square root

Example 1

1,2-Dibromoethane ($BrCH_2-CH2Br$)

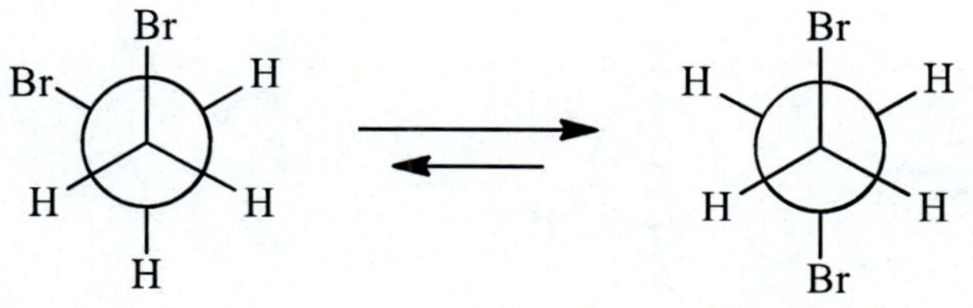

most stable conformer

Gauche (1) Anti (II)

Overall μ = 1D

Finite μ is expected

Expected μ = O

The dipole moment μ (experimental) of the molecule (1,2=dibromoethane) is related to the dipole moments of the conformers by the following equation:

$\mu^2 = n_a \mu_g{}^2 + n_g \mu_g{}^2$ (for two confomers)
where n_a and n_g are the mole fractions of the two conformers *anti* and *gauche* (or, any such pair) and μ_a and μ_g are the diple moments of the conformers.

Fig. 1.26 Dipole moment of 1,2-dibromoethane

of the sum of the squares of the instantaneous dipole moments (Fig. 1.26). Thus, the *anti*-form might have a small dipole moment. The frequency of libration might vary with temperature which will have, however, very little effect on the equilibrium.

Example 2 The dipole moments of diastereomeric stilbene-1,2-dichloride (meso and dl-pair) allow one to conclude reliably their configurations through careful consideration of their conformations (Fig. 1.27). The dipole moment of the meso-form is 1.27 D, whereas that of the dl-pair (one enantiomer shown) is 2.75 D. Their dipole moments are incompatible with free rotation around central C—C bond, in which case the individual bond moments would have averaged out to the same value for the two isomers.

The most favorable conformational isomer of the meso-form is (III), having the bulky phenyl groups *anti*, and the repulsion of the C–Cl dipoles minimum. Since this conformer has zero dipole moment, the overall dipole moment of the meso-form is relatively low.

On the contrary, in the d- or l- isomer, the preferred conformer (IV) with the bulky phenyl groups *anti* to each other, has, of course, appreciable dipole moment, and hence the overall dipole moment is relatively large. The conformer (V) having

Example 2

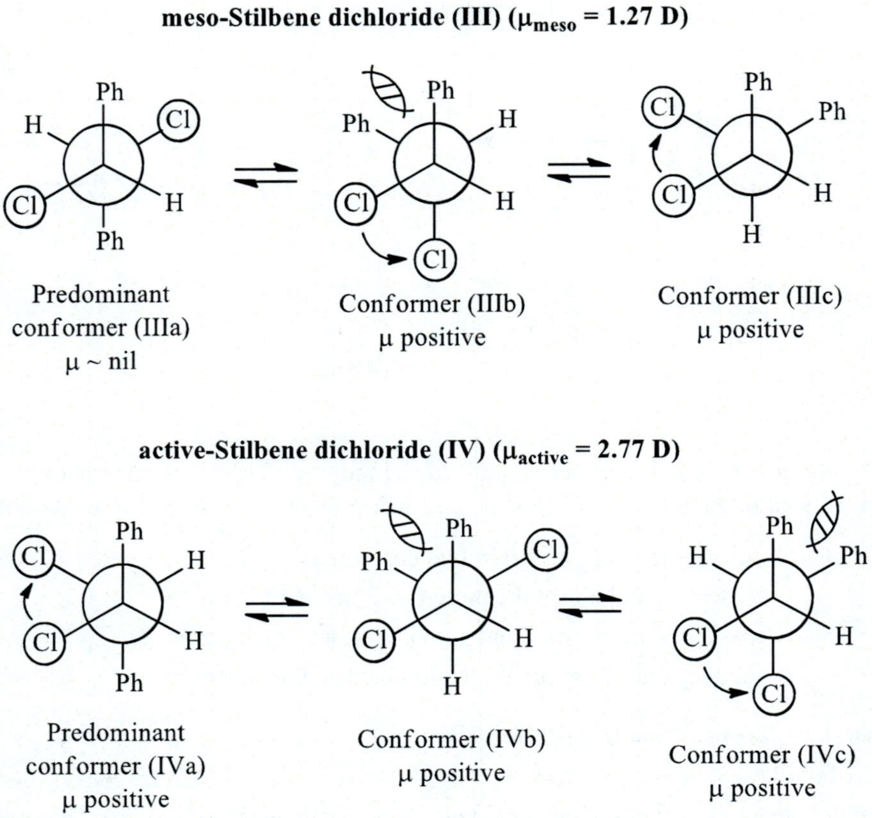

Fig. 1.27 Dipole moments of diastereomeric stilbene-1,2-dichlorides

the C–Cl dipoles antiparallel (having no dipole moment) requires the bulky phenyl groups to be gauche, and hence contributes relatively little.

Example 3. 1,2-dibromocyclohexane exists as two diastereomers, namely, the *cis*-isomer (VI) and the *trans*-isomer (VII) (Sect. 7.4; Fig. 7.9). The *cis*-isomer exists as the two flipped enantiomeric conformers (VIa) and (VIb) (Fig. 1.28), whereas the *trans*-isomer exists as the flipped *ee* conformer (VIIa) and the *aa* conformer (VIIb) (Fig. 1.28). The observed [9] dipole moments of the *cis*-isomer and the *trans*-isomer are 3.09 D and 2.11 D, respectively. The dipole moment of the *aa* conformer (VIIb) is assumed to be zero (the two C–Br dipoles being antiparallel), whereas the calculated dipole moment for either the *ae* conformer (VIa), or the *ea* conformer, or the *ee* conformer is ~3.09 D, the two C–Br dipoles being gauche in each case. It is obvious that the *ee* conformer (VIIb) although favored sterically is destabilized by dipole repulsion. Thus, the configuration of the *cis*- and *trans*-isomers

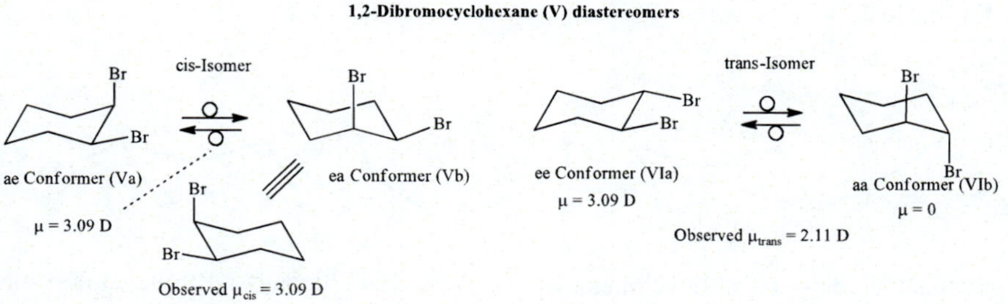

Fig. 1.28 Dipole moments of diastereomeric 1,2-dibromocyclohexanes

of 1,2-dibromocyclohexane can be diagnized reliably from their dipole moment measurements, the lower one being that of the *trans*-isomer.

For the conformers and dipole moments of some typical disubstituted cyclohexanes, Sect. 7.4.1 may be visited.

1.6.2 Acid Strength Measurements. Cyclohexane-1,2-Dicarboxylic Acids, 4-T-Butylcyclohexanecarboxylic Acids (Fig. 1.29)

Example 1. In case of the cyclohexane-*cis*-1,2-dicarboxylic acid, which exists in equilibrium of the two enantiomeric *ae* and *ea* conformers, the difference between the first and second ionization constants corresponds to 2.42 pK units, whereas the corresponding difference for the *trans*-isomer, which exists in equilibrium of the two diastereomeric *ee* and *aa* forms, is 1.75 pK units (Fig. 1.29). In general, this difference is greater, closer together the carboxyl groups are; since the C=O dipole

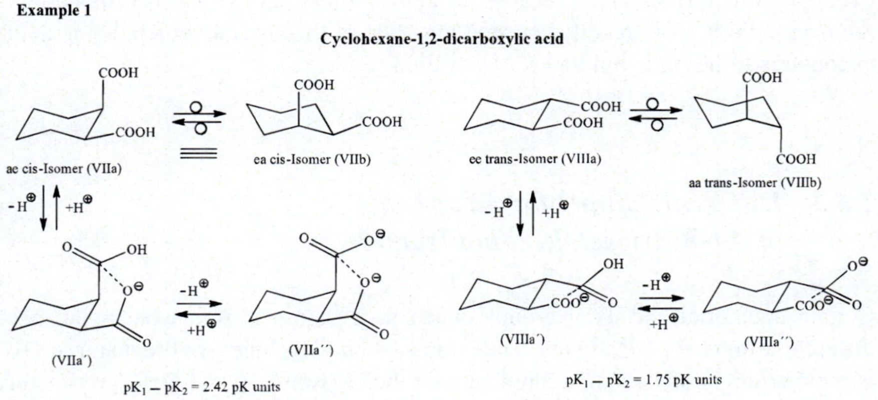

Fig. 1.29 Dipole moments of enantiomeric cyclohexane 1,2-dicarboxylic acids

Example 2

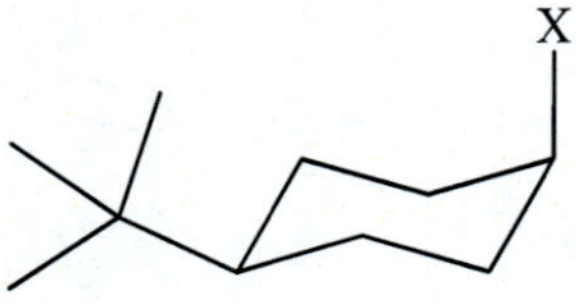

X=COOH, cis-4-t-Butylcyclohexane carboxylic acid

pK_a : 8.23 pK units

X=COOH, trans -4-t-Butylcyclohexane carboxylic acid

pK_a : 7.79 units, stronger

X = OH, cis-4-t-Butylcyclohexanol

$pK_2 - pK_1 = 2.45$ pK units

X = OH, trans -4-t-Butylcyclohexanol

pK_2 – t-pK_1 = 1.76 units

Fig. 1.30 Acid strengths of *cis*- and *trans*-4-*t*-butylcyclohexanols and *cis*- and *trans*-4-*t*-butylcyclohexanol carboxylic acids

of one COOH group facilitates the loss of the proton from a closely proximate OH group by a field effect (induction through space), but the negative charge of the COO^- ion in the already ionized acid prevents the loss of the second proton, if the other COOH group is close by (Fig. 1.29). Hence, it follows that the COOH groups in the cis-isomer are closer together than in the *trans*-isomer, a reasonable conclusion from the fact that the *trans*-isomer exists partly in the *aa* form due to dipole repulsion in the *ee* form. Thus, the configuration of the diastereomers could also be derived from the pK data.

Example 2 A subtle difference is found between the acid strengths [10, 11] of *cis*- and *trans*-4-*t*-butylcyclohexanecarbocyclic acids (Fig. 1.30). The bulkiness of the *t*-butyl group is such that it almost always exists in the equatorial conformation and not in the alternative axial conformation. The *trans*-acid (pKa 7.79) is stronger than the *cis-acid* (pKa 8.23), because the anion (equatorial COO^-) corresponding to the *trans*-acid is more readily solvated than that of the *cis*-acid which, being axial, encounters some steric hindrance to solvation.

1.6.3 Different Affinity for Adsorbents of 4-t-Butyicyclohexanol Isomers

Certain equatorial and axial groups, when their nature is somewhat polar, have different affinity for adsorbents. Thus trans-4-*t*-butylcyclohexanol (equatorial OH) is more strongly adsorbed on alumina than the *cis*-isomer (axial OH), presumably because the OH group with the specific affinity for the adsorbent is more exposed in the equatorial isomer (*ee*). This physical difference may be useful in chromatographic separation for analytical or preparative purposes.

It may be mentioned here that the physical and chemical properties of axially and equatorially substituted cyclohexane derivatives (especially of steroids) were first pointed out by D. H. R. Barton in his classical epoch-making paper [13] in 1950, which should be read by any interested reader.

1.6.4 Physical Properties of Substituted Cyclohexanes; Von Auwers–Skita Rule or Conformational Rule, Van Arkel Rule

The relationship between certain physical properties and conformation was first enuntiated by von Auwers and Skita [14, 15] and is generally known as "von Auwers–Skita rule" or "conformational rule." Allinger made its up-to-date statement as follows: "Among alicyclic epimers not differing in dipole moment the isomer of highest heat content (enthalpy) has the higher density, refractive index and melt boiling point."

van Bekkam et al. [16] suggested that since the rule frequently fails for the boiling point, it should be deleted from the rule (for detailed discussion with examples of the rule see [17, 18]). Allinger [17] stated the *conformational rule* as follows: *"For stereoisomers in cyclic systems, which do not differ in dipole moment, the isomer which has the smaller molecular volume has the higher heat content."*

All these physical properties are inversely related to molecular volume. In general, *the conformational rule* may be more precisely stated as follows: *"For stereoisomers in alicyclic and cyclic systems, which do not differ in dipole moment, the isomer which has the smaller molecular volume has the higher heat content and the higher physical constants.*" This statement entails several modifications of the earlier versions [14–16].

A few exceptions to the von Auwers–Skita rule are known. For example, it does not apply to the boiling points of the alkylcyclohexdanols. Here, the *ee* isomers having lower enthalpy, refractive index, and densityare more strongly hydrogen bonded than the *ea* isomers (the axdial OH group being less well disposed for hydrogen bonding, and therefore, boil at higher temperature [19]. Such isomers follow the "van Arkel rule" or "dipole rule [20], which states that "the isomer of higher dipole moment has the higher physical constants, regardless of the heat content [20, 21]." 1,2-dibromocyclohexane has been cited as an example [9] (Fig. 1.28).

The aforesaid rules, whichever is relevant, may be used to assign configurations to epimers when other physical methods like IR, Raman, microwave, and NMR spectra, or X-ray and electron diffraction cannot be readily used.

References

1. G.E. McCasland, R. Horvat, M.R. Roth, Substances whose optical inactivity can be attributed only to fourfold alternating axial molecular symmetry. III. A second example (*meso*-Tetramenthyloxyacetate. J. Am. Chem. Soc. **81**, 2399–2402 (1959)
2. C. Winther, Zur Theorie der optischen Drehnng. III.(Beiträge zur allgemeinen Lösungstheorie. I.). Z. Phys. Chem. 621
3. E. Fisoher, Ueber die Configuration des Traubenzuckers und seiner Isomerea. 11. Ber **24**, 2683–2687 (1891)
4. D.H.R. Barton, The principles of conformational analysis. An Inaugural Lecture delivered at Berbeck College, London, 11 th February, 10 pages (1954)
5. W. Klyne, V. Prelog, Conformational terminology. Experientia **16**, 521 (1960)
6. R.S. Cahn, S. Christopher, K. Ingold, V. Prelog, Specification of molecular chirality. Angew. Chem. Int. Edn. Engl. **5**, 385–415. For Conformational chirality see p.406 (1966)
7. C.P. Smyth, S.E. Kamerling, Elecric moment and molecular structure. V. The polymethylene bromide. J. Am. Chem. Soc. **53**, 2988–2998 (1931)
8. S. Mizushima, *The structure of molecules and internal rotation* (Academic Press Inc., New York, 1954)
9. P. Bender, D.L. Flowers, H.L. Goering, Dipole moment studies of 1,1- and 1,2-Dihalocyclohexanes. J. Am. Chem. Soc. **77**, 3463–3465 (1955)
10. R. Kuhn, A. Wassermann, Zur Konfiguration der Polymethylen-dicarbonsauren I. Die Dissoziationkonstanten der Cyclohexan-dicarbonsauren. Helv. Chim. Acta, **11**, 50–70 (1928)
11. R.D. Stolow, Quantitative relationship between dissociation constants and conformational equilibria. Cyclohexanecarboxylic acids. J. Am. Chem. Soc. **81**, 5806–5811 (1959)
12. S. Winstein, N.J. Holness, Neighboring carbon and hydrogen. XIX. f-Butylcyclohexyl derivatives. Quantitative conformational analysis. J. Am. Chem. Soc. **77**, 5562–5578 (1955)
13. D.H.R. Barton, Conformation of steroid nucleus. Experientia 316–320 (1950); reprinted in *Top. Stereochem.* **6**, 1–5 (1971)
14. K. von Auwers, Ann **420**, 84 (1920)
15. A. Skita, *Ber*, tfber die Stereoohemie der Hexahydro-toluidine **56**, 1014–1023 (1923)
16. H. van Bekkum, A. van Veen, P.E. Verkade, B.M. Wepster, Rec. Trav. Chim. **80**, 1310 (1961)
17. N.L. Allinger, The relative stabilities of *cis* and *trans isomers*. III. The Cyclodecenes. J. Am. Chem. Soc. **79**, 3443 (1957)
18. E.L. Eliel, N.L. Allinger, S.J. Angyal, G.A. Morrison, *Conformational Analysis* (Interscience Wiley, New York, 1965); reprinted by American Chemical Society, Washington, DC, 1981
19. E.L. Eliel, R.G. Haber, The boiling points of the methylcyclohexanols—an exception to the conformational rule. J. Org. Chem. **53**, 2041–2042 (1958)
20. A.E. van Arkel, Zusmmenhang Zwischen DipolMoment und Kohaesionskraeften. Rec. Trav. Chim. **51**, 1081–1107 (1932)
21. A.E. van Arkel, Zusmmenhang Zwischen DipolMoment und Kohaesionskraeften. VI. Rec. Trav. Chim. **53**, 246–256 (1934)

Chapter 2
Configurational Nomenclature. Physical Properties of Geometrical Isomers

Abstract This chapter depicts the different aspects of configurational nomenclature: **D,L** nomenclature and ambiguities/shortcomings, R,S nomenclature, center/s of chirality, CIP sequence rule, CIP chirality rule—its corollaries, priority of common ligands, modification of sub-rule 3, R^*,S^* nomenclature; specifications of other tetracovalent and tricovalent chiral atoms; E,Z nomenclature; physical properties of geometrical isomers: dipole moments of cis- and trans-1,2-dihaloethylenes, cis- and trans-1-chloro-2-iodoethylenes, diethyl maleate, diethyl fumarate, cis- and trans-1-chloro-1-propenes, cyclohexene, cis- and trans-cyclodecenes, cis- and trans-cyclooctenes; melting points of maleic and fumaric acids and their methyl esters, cis- and trans-cinnamic acids, cis- and trans-1,2-dihaloethylenes; solubilities (in water) of maleic and fumric acids, cis- and trans-crotonic acids, cis- and trans-cinnamic acids; boiling points, densities, refractive indices of geometrical isomers; and acid strengths of monobasic acids: cis- and trans-crotonic acids, dibasic acids: maleic acid and fumaric acid.

Two different methods are in use for completely specifying the absolute configuration of a chiral center in a molecule.

2.1 Fischer's D,L Nomenclature

This oldest system of nomenclature of chiral compounds was introduced by Emil Fischer [1] in 1891, while working with carbohydrates. Rosanoff [2] modified the system in 1906 and suggested the *following conventions* for a *projection* nomenclature of D,L system:

(i) As in Fischer's system, the molecule is written with the longest carbon chain placed vertically.
(ii) The C1 carbon or the most highly oxidized end of the chain is placed at the top, following Fischer's convention, *e.g.*, COOH > CHO > CH_2OH (according to IUPAC also).

S. K. Talapatra and B. Talapatra, *Basic Concepts in Organic Stereochemistry*,
https://doi.org/10.1007/978-3-030-95990-6_2

Some examples of the use of **D,L** nomenclature and its extension in some cases are delineated in Fig. 2.1).

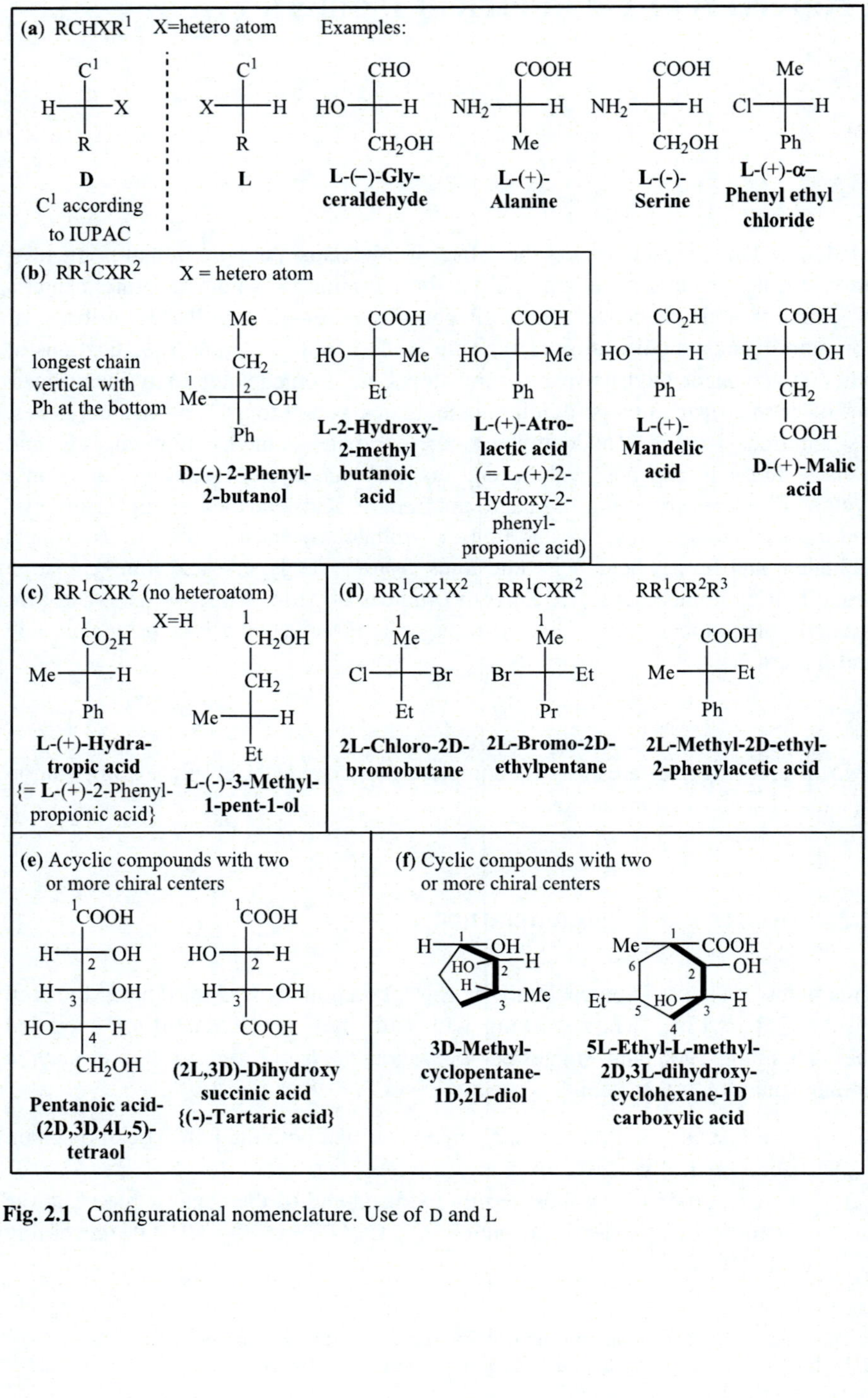

Fig. 2.1 Configurational nomenclature. Use of D and L

(a) The system works well for compounds $RCHXR^1$, when X is a hetero-atom (Fig. 2.1). The molecule is **D** if X is on the right, and **l** if X is on the left. While writing the Fischer projection phenyl or aryl group should be at the bottom.
(b) In compounds of the type RR^1CXR^2, this system is applicable; **d** means that *X* is on the right, and the small alkyl group R^1 is to the left. The enantiomer of **D** is **L**.
(c) If there is no hetero-atom, then the smallest alkyl group (usually Me) becomes the fiducial group; if it is on the right it is **D**, if on the left it is **L**.
(d) When the chiral center carries two different substituents of comparable electronegativity, **D**- and **L**- may be used separately for each substituent to specify the configuration. Thus, the system may be applicable to compounds of the types $RR^1CR^2R^3$, having asymmetric quaternary carbon, *e.g.*, 2**L**-methyl-2**D**-ethyl-2-phenylacetic acid.
(e) For compounds containing more than one asymmetric carbon and
(f) for compounds possessing asymmetric carbons in rings the ***D*,L** nomenclature has been extended by Klyne [3], *e.g.*, 3**D**-methylcyclopentane-1**D**,2**L**-diol. Rings are oriented with the edge of the lowest numbers toward the viewer, the main chain being numbered from the top bearing the more oxidized carbon [2] or C1 (Fig. 2.1).
(g) In carbohydrate chemistry, **D,L** nomenclature is assigned on the basis of configuration of the last chiral center of the chain written vertically counting from the top in the Fischer projection, more oxidized carbon being at the top (Fig. 2.1) [2] (Fig. 2.2).

It is to be noted that **D,L** *(like R, S) nomenclature has no genetic relationship.* Thus, the COOH group of **D**-hydratropic acid when converted to NH_2 gives **L**-1-phenylethylamine, although no bond to the chiral center is broken (Fig. 2.3).

Ambiguities/Shortcomings of* D,L *Nomenclature (Fig. 2.4)

(i) As already mentioned, too many arbitrary conventions are to be remembered for **D,L** nomenclature of different types of compounds.
(ii) This system cannot be conveniently used for polycyclic compounds.
(iii) Ambiguity arises for **D** and **L** designation of tartaric acid diastereomers (Fig. 2.4).

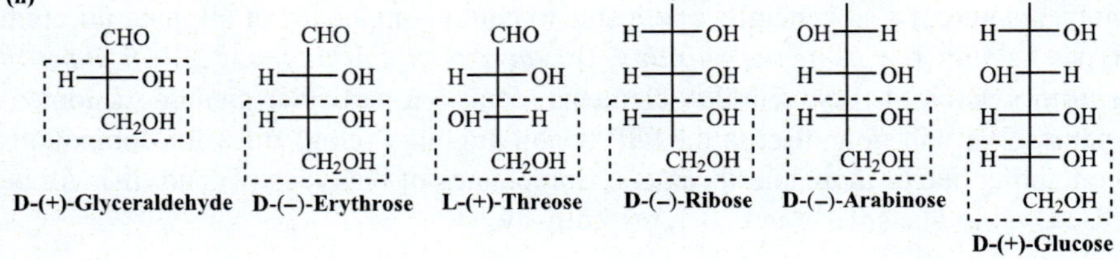

Fig. 2.2 Configurational (D,L) nomenclature, glyceraldehyde, tetroses, pentoses, and glucose [*vide* (g)]

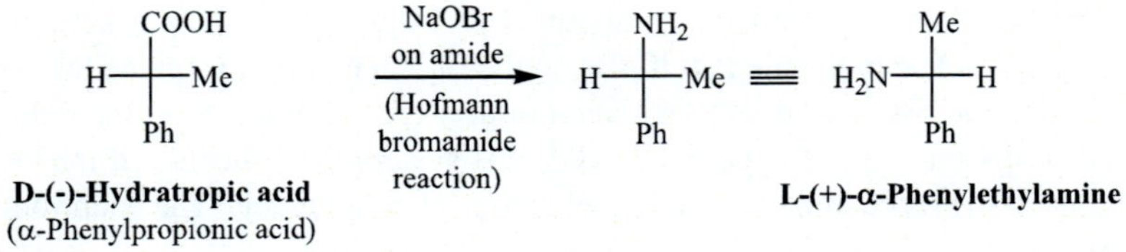

Fig. 2.3 D,L Nomenclature has no genetic relationship. Illustration

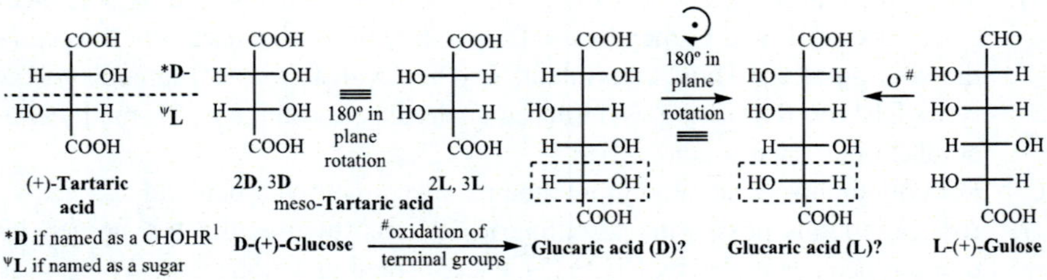

Fig. 2.4 Few ambiguities of D,L nomenclature

(iv) The oxidation product of D-(+)-glucose may be termed both D-glucaric acid and L-glucaric acid which is also obtained by oxidation of L-(+)-gulose (Fig. 2.4).

2.2 *R,S* Nomenclature for Absolute Configuration

The D,L nomenclature of Fischer (Sect. 2.1) applies to the Fischer projection. In spite of the shortcomings/ambiguities already discussed, the Fischer nomenclature has proved particularly useful for sugars or amino acids. However, one should avoid confusion between *d*, *l* (dextrorotatory or levorotatory) and D,L (absolute configuration). A *d* compound may have the absolute configuration either D or L, *e.g.*, (*d*)-L-alanine and (*d*)-D-glyceraldehyde, which are now expressed as L-(+)-alanine and D-(+)-glyceraldehyde.

As D,L nomenclature is not generally useful to some types of compounds with chiral centers mentioned earlier. Cahn, Ingold, and Prelog (CIP) developed *R*-*S* nomenclature, [4–6] generally applicable to chiral compounds of all three different types having (i) *a center of chirality*, (ii) *an axis of chirality*, and (iii) *a plane of chirality*. Each of these chirality elements requires a particular method of nomenclature. We will now discuss the CIP (Cahn–Ingold–Prelog) rules for compounds containing one or more chiral centers. Compounds of the types (ii) and (iii) will be discussed in Sects. 10.1 and 10.2, respectively.

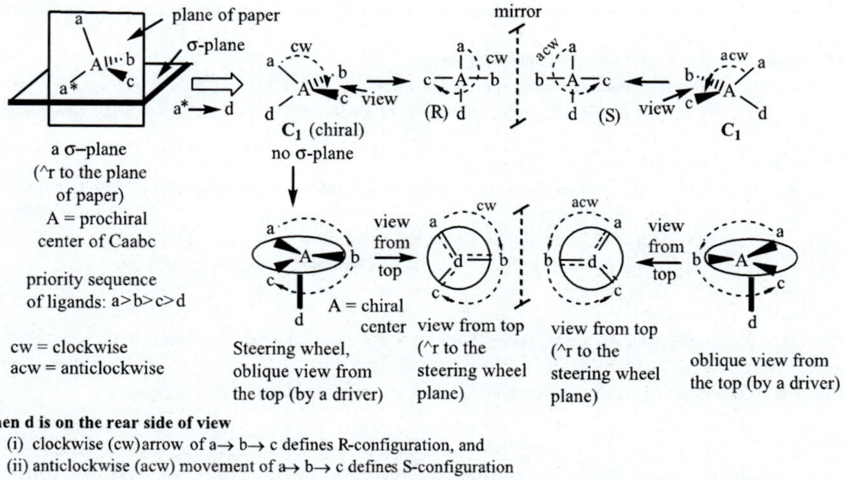

Fig. 2.5 Prochiral center. *R,S*-designation of a chiral center (*cf*. steering wheel)

2.2.1 R,S *Nomenclature. Center of Chirality*

A center of chirality is usually associated with a nonplanar tetracoordinated atom "A" bonded (or ligated) to four different ligands (atoms or groups) or an asymmetric carbon atom, A*abcd*. The latter could be derived from an *achiral* or a *prochiral precursor* A*aabc* by changing a ligand *a* to a new ligand *d* (different from the existing ligands). This precursor possesses a plane of symmetry containing A–b and A–c bonds (Fig. 2.5); the chirality of A*abcd* (point group C_1) is the consequence of the destruction of the plane of symmetry of A*aabc*.

2.2.2 Specification of Center/s of Chirality

Two CIP rules are involved: (i) CIP *sequence rule* and (ii) CIP *chirality rule.*

(i) ***CIP Sequence rule***: Ligands are sequenced by comparing them at each step in bond-to-bond exploration in branched ligands *along the branch path of highest precedence*. This rule will be illustrated by examples.

The following "standard sub-rules" are used for finding out the priority sequence of all possible types of compounds—*each to exhaustion in turn* (Fig. 2.6).

(ii) ***CIP chirality rule*** (Fig. 2.5): The path of the sequence of precedence/priority $a \rightarrow b \rightarrow c$ is followed from the preferred side of the model (containing the three preferred ligands a, b, and c), *i.e.*, remote from the ligand of the lowest precedence *d*. If the path turns *right* (traces clockwise, which is the entire sense

Sub-rule 0: Nearer end of axis or side of plane precedes the further end (applicable to chiral axis or chiral plane).

Sub-rule 1: Higher atomic number precedes lower.

Sub-rule 2: Higher atomic mass number precedes lower, e.g., T > D > H $^{14}C > ^{13}C$

Two ligands varying only by isotopes produce marginal chirality and hence small optical rotation

Sub-rule 3: Seq cis (Z) precedes seq trans (E)

Sub-rule 4: Like pair precedes unlike pair, > stands for 'precedes' or 'has priority over'; M = minus helicity

RR or SS > SR or RS; P = plus helicity

MM or PP > MP or PM; RM or SP > RP or SM

MR or PS > MS or PR; r > s (in case of achirotopic but stereogenic center)

(Like and unlike pairs are gleaned from the sub-rule 5)

Sub-rule 5: R > S; M > P

The priority sequence in the decreasing order, a > b > c > d is assigned to the four ligands of the chiral center.

Fig. 2.6 CIP standard sub-rules for ascertaining the priority sequence of ligands

of direction of abc), the element is assigned chiral label "*R*" (*rectus*, Latin for right). If the said path turns *left* (traces anticlockwise, which is the entire sense of direction of *abc*), then the stereogenic center is assigned the chiral label "*S*" (*sinister*, Latin for left).

One can imagine that *a, b*, and *c* are placed on a steering wheel of a car, and *d* is placed on the shaft (the axis of the wheel is A-d); if from a position above the wheel one rotates the wheel on the right side (in the clockwise sense a→b→c), the car moves to the *right*, and the absolute configuration is called *R*, whereas in the opposite case, the car moves to the *left* and the absolute configuration is called *S*.

It is evident from Fig. 2.5 that in the Fischer projection formula, if *d* (the least priority group) is at the *bottom* or *top*: *a*→*b*→*c* makes a clockwise movement in the *R*-configuration; *a*→*b*→*c* makes an anticlockwise movement in the *S*-configuration.

R,S-configurational specification of some common compounds is shown in Fig. 2.7.

Corollaries of the Chirality Rule

1. If *d* is on the *right* or *left* side of the Fischer projection formula, or *above* the plane containing the two in-plane bonds in *the flying wedge* formula, the *anticlockwise* movement from *a* → *b* → *c* (or 1 → 2 → 3) will define *R*-configuration, and *clockwise* movement will define *S*-configuration.
2. In the above case, clockwise movement from *c* → *b* → *a* (or 3 → 2 → 1) will define *R*-configuration, and anticlockwise movement from *c* → *b* → *a* (or 3 → 2 → 1) will define *S*-configuration.

The sequence sub-rules are applied considering different aspects, as needed:

1. Sub-rule 0 gets the topmost priority but is applicable to axially dissymmetric compounds to be discussed later (Chap. 10).
2. If the atoms of two ligands attached to the asymmetric tetracovalent atom (like C, N, P, Si, etc.) are the same, their respective states of substitution are considered.

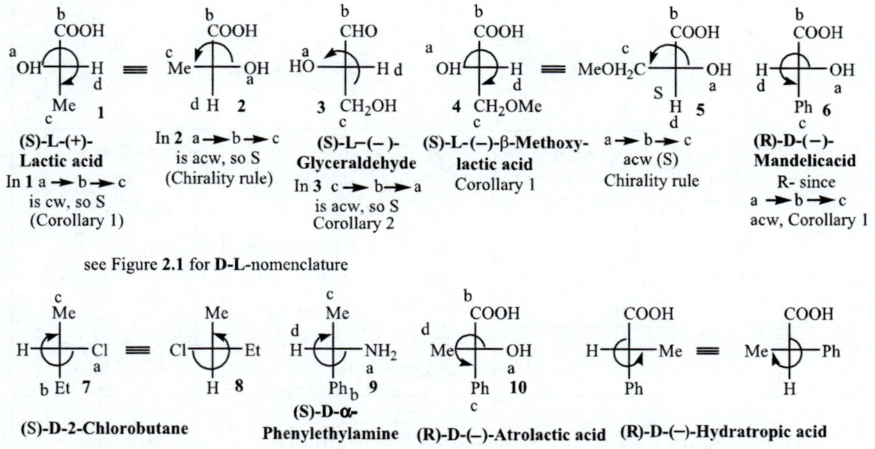

Fig. 2.7 *R-S*-configurational designation of some common chiral compounds. Use of corollaries (**1**) and (**2**) also

If the second atom gives no choice, third, etc., is to be considered, following *along the branches with closer atoms of higher atomic number*.

3. If the two atoms directly linked to the asymmetric atom are different, the atoms of higher atomic number gets precedence, *e.g.*, I > Br > Cl > S > P > Si > F > O > N > C.
4. If the atoms of two ligands attached to the asymmetric atom are same, that with more substituents of higher atomic number gets precedence
5. If the ligands attached to the asymmetric center are composed of only C and H atoms, then *the ligand with less number of H atom gets priority.*

   For example: $-CR^{1R2R3} > -CHR^{1R2} > -CH_{2R}{}^{1} > CH_3$

   (here R^1, R^2, R^3 may be same or different alkyl or aryl substituent)
6. Multiple bonded atoms are treated as four-coordinated ones by adding replica atoms of the same type (duplications or triplications) which are bracketed to signify that they are surrounded by phantom atoms of atomic number zero. This is illustrated by some examples given in Fig. 2.8.

Table 2.1 contains some of the most common groups in the decreasing order of precedence obtained by applications (2–6) of the sequence sub-rules 1 and 2.

7. A few more examples of *R-S* designation of compounds with several asymmetric atoms (written as Fischer projection or flying wedge formulas) have been included in Fig. 2.9.

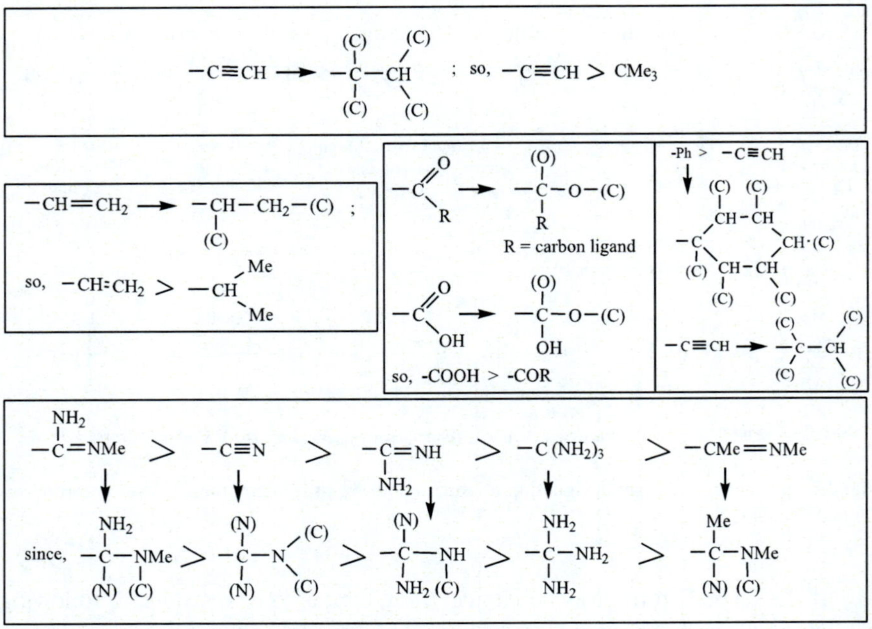

Fig. 2.8 Multiple bonded atoms treated as four-coordinated ones for determining priority sequence

2.2.3 Priority Sequence of the Application of the CIP Sub-rules

Sub-rule O (proximity rule) enjoys the topmost priority in cases of axial chirality.

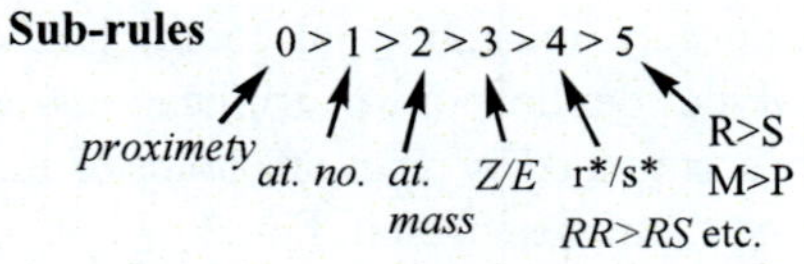

Few *examples of the priorities of the sub-rules are provided in Fig.* 2.10.

2.2.4 Modification of Sub-rule 3

The sub-rule 3 has been modified by Prelog and Helmchen [7] in 1982 by an alternative proposal. The modified sub-rule and some examples illustrating it are given in Fig. 2.11.

Table 2.1 Descending order of priority of some common ligands according to the sequence rules

1. –I	25. $–NH_2$	49. –Ph
2. –Br	26. $–CX_3$	50. –C≡CH
3. –Cl	27. COX	51. $–CMe_3$
4. $–PR_2$	28. $–CO_2CMe_3$	52. –CH = CH–Me
5. $–SO_3H$	29. $–CO_2Ph$	53. $–C_6H_{11}$
6. $–SO_2R$	30. $–CO_2CH_2Me$	54. –CH(Me)Et
7. –S(O)R	31. $–CO_2Me$	55. $–CH = CH_2$
8. –SR	32. $–CO_2H$	56. $–CHMe_2$
9. –SH	33. $CONH_2$	57. $–CH_2Ph$
10. –F	34. –COPh	58. $–CH_2C≡CH$
11. $–OSO_2R$	35. –COMe	59. $–CH_2–CMe_3$
12. –OCOR	36. –CHO	60. $–CH_2CH = CH_2$
13. –OR	37. $–CR_2OH$	61. $–CH_2CHMe_2$
14. –OH	38. –CH(OH)R	62. $–CH_2CH_2Et$
15. $–NO_2$	39. $–CH_2OR$	63. $–CH_2CH_2Me$
16. –NO	40. $–CH_2OH$	64. $–CD_2Me$
17. $–N^+R_3$	41. –CN	65. –CHDMe
18. $–NR_2$	42. $–CH_2NH_2$	66. $–CH_2Me$
19. –NHCOPh	43. $–C_6H_4Me(o)$	67. $–CD_3$
20. –NHCOR	44. –C≡CMe	68. –Me
21. $–NHCH_2Me$	45. $–C_6H_4NO_2(m)$	69. –T
22. $–NMe_2$	46. $–C_6H_4Me(m)$	70. –D
23. –NHR	47. $–C_6H_4NO_2(p)$	71. –H
24. $–N^+H_3$	48. $–C_6H_4Me(p)$	72. Electron pair

2.2.5 *R** and *S** Nomenclature

When *R-S* nomenclature is applied, in a molecule with multiple chiral centers both the relative and the absolute configurations are fixed, as it has been illustrated in many chiral molecules so far. For examples, 2(*R*)-bromo-3(*S*)-hydroxybutane refers to the enantiomer (**1**) (Fig. 2.12).

It often happens that *a pure enantiomer is reported with known relative configuration but unknown absolute configuration*. In such cases, the *R-S* system of nomenclature is modified as follows (IUPAC Commission 1976) [8]. The atoms are numbered such that the chiral center carrying the highest priority ligand, *e.g.*, C–Br in (**1**) or (**2**) or C-I in (**3**) is given the lowest number (lowest locant). The molecule is written in such a way that the lowest chiral locant gets the *R* configuration. The other chiral centers are then assigned in the usual way, and each descriptor is asterisked (pronounced as *R*-star or *S*-star, indicating thereby that they represent relative configuration). Thus, 2(*R**)-bromo-3(S*)-hydroxybutane or simply *erythro*-3-bromo-2-butanol represents either compound (**1**) or its mirror image (**1′**); 2(*R**)-bromo-3(*R**)-hydroxybutane (or threo-3-bromo-2-butanol) represents either (**2**) or

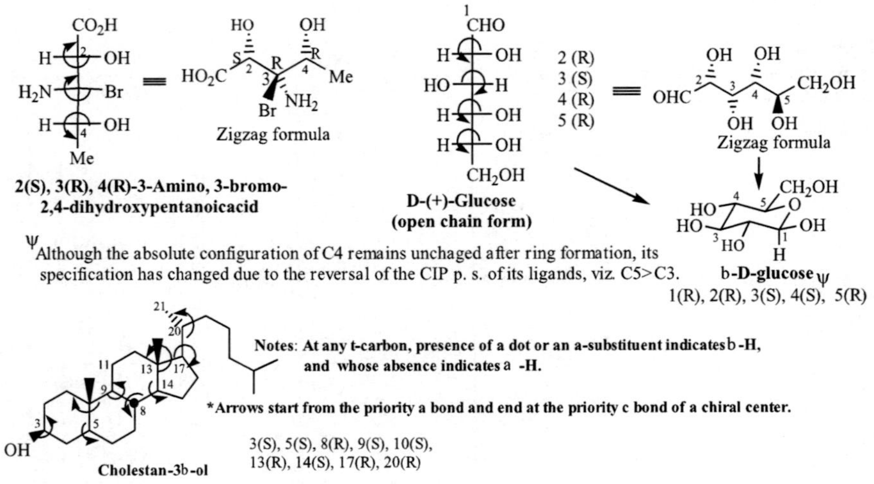

Fig. 2.9 *R-S* designation of compounds with multi-chiral centers and monocyclic and polycyclic compounds

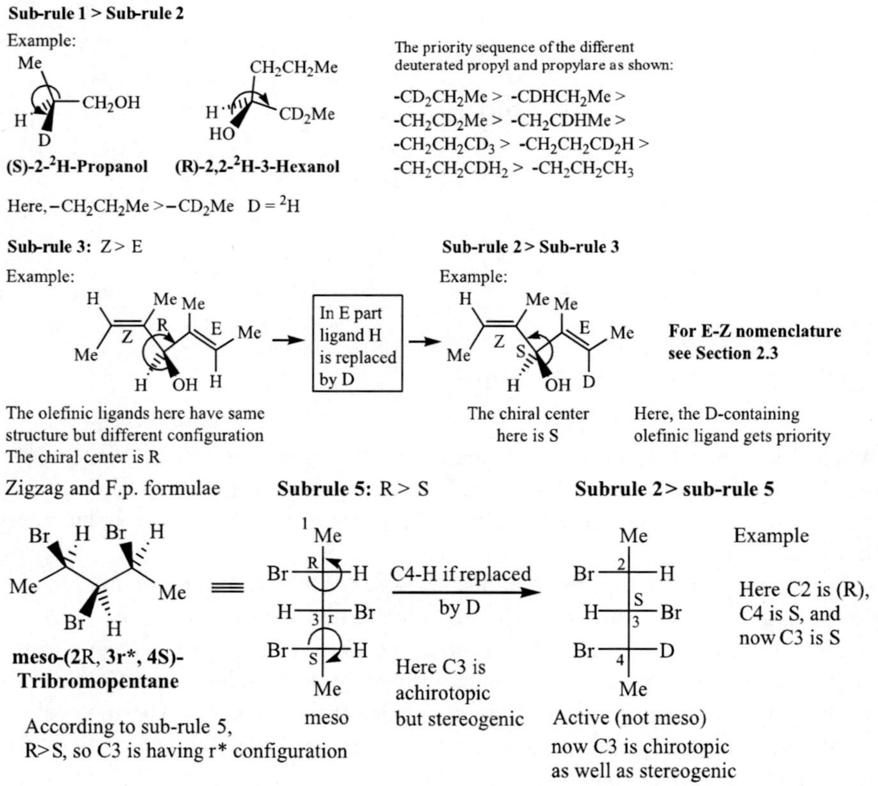

Fig. 2.10 Priority of some sub-rules over others. *R-S*-nomenclature of compounds with multichiral (here 3) centers from zigzag formula

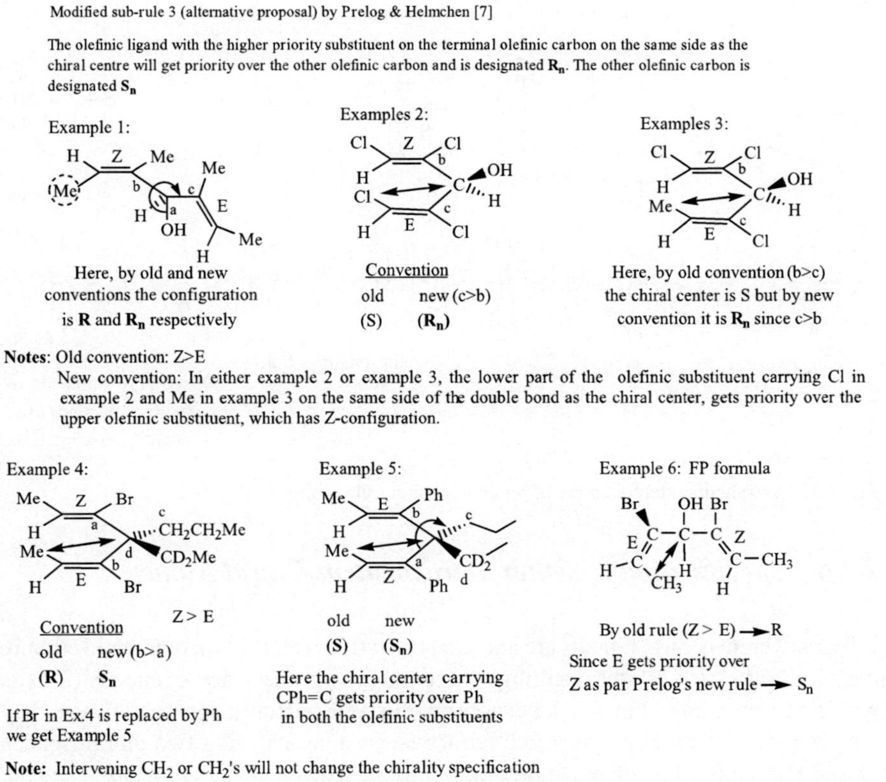

Fig. 2.11 Modification of sub-rule 3 by Prelog and Helmchen. Several examples to show chirality specifications

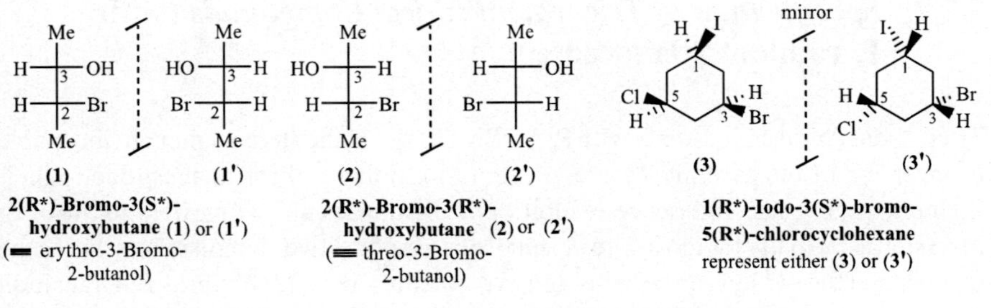

Fig. 2.12 *R*S** nomenclature of some compounds of known relative configuration

(**2′**); and 1(*R**)-Iodo-3-(*S**)-bromo-5(*R**)-chlorocyclohexane represents either (**3**) or (**3′**).

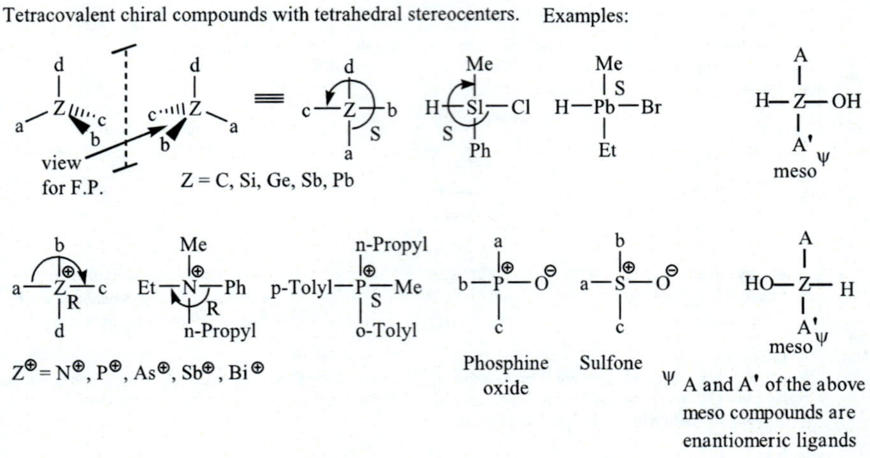

Fig. 2.13 Tetrahedral stereocenters of tetracovalent chiral compounds

2.2.6 *Specification of Other Tetracovalent Chiral Atoms*

If four different achiral ligands are attached to a tetracovalent atom other than carbon, *e.g.*, Si, Ge, As^+, P^+, N^+, the resulting compounds also possess stereogenecity (*R,S*) as well as chirotopicity; they will have nonsuperposable mirror images, and hence, they are optically active. However, such tetracovalent atom carrying two chiral ligands of same structure but of opposite configuration will be still *stereogenic* (*r,s*) but *achirotopic*, making the compound *meso* (Fig. 2.13).

2.2.7 *Specification of Tricovalent Chiral Compounds* (**with Pyramidal Stereocenter**)

Tricovalent chiral compounds with P, As, Sb, Bi, etc. at the stereocenter are incapable of inversion of the pyramidal stereocenter to form their mirror images (due to their high energy barrier) and hence exhibit optical isomerism. The barriers to nitrogen inversion are usually far too low to permit isolation of the two stereoisomers. Since the inversion of the sp^3 hybridized nitrogen pyramid involves an sp^2 hybridized transition state, increase of *p* character of the bonds to N slows down the rate of inversion as in substituted aziridines (internuclear angle 60°–90°); the rate of inversion is also slowed down by electronegative substituent like Cl. These two factors cooperate in substituted aziridines to increase the energy barrier between the two isomers, which are also called *invertomers* (Fig. 2.14), permitting their isolation. In Troger's base (Fig. 2.14), the N atoms are present at the bridgeheads, and hence, pyramidal inversion is not possible without bond cleavage.

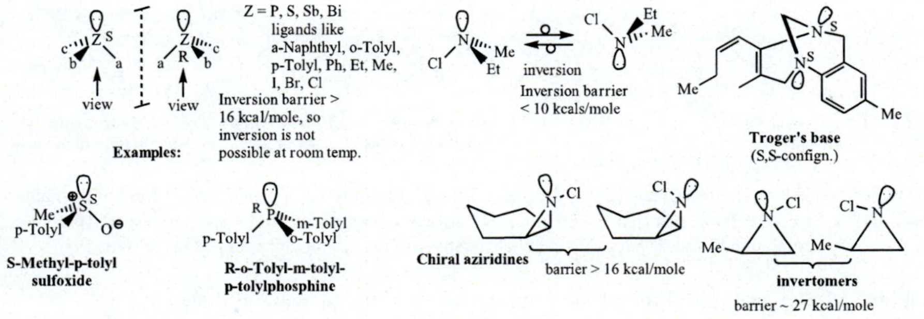

Fig. 2.14 Pyramidal stereocenters of tricovalent chiral compounds

2.3 Stereochemistry of Alkenes. *E,Z* Nomenclature [9–11]

(a) **Stereochemistry**. The C–C σ bond strength is about 83 kcal mol^{-1}, and the strength of the π bond is only 62 $^{kcal\ mol-1}$ due to its less favorable lateral overlap; addition of these two numbers gives the generally accepted total energy of a C = C double bond as 145 kcal mol^{-1}, much more than the rotational barriers in alkanes (*e.g.*, 3.6 kcal mol^{-1} for C2–C3 bond in butane). During the process of rotation of the *Z* to the *E* isomer or *vice versa*, the *p* orbitals of the two olefinic carbon atoms become orthogonal, with no overlap. Thus, the π-bond is completely broken in the transition state.

The bond length of C = C in unstrained unconjugated ethenes ranges from 1.335 to 1.35 Å (133.5–135 pm), but is extended in conjugated alkenes in which C = C bond is weakened. The bond angle of substituted olefins is not 120° and varies. Electron diffraction showed that H–C–H angle of ethane is 117.8°. In propene the C = C–C angle is 124.3° and the C = C–H angle is 119°. In *cis*-2-butene, the C = C–C angle is 125.8°. The Me–C–Me angle in 2-methylpropene, Me–C(Me) = CH_2, is 115.3°. Apparently, R–C–R′ angle in RR′C = C is generally smaller than 120°, while R–C = C angle is larger.

(b) **Nomenclature.** *Cis–trans*-isomerism in substituted olefins, earlier called geometrical isomerism, is now regarded as a type of diastereomerism, since *cis*- and *trans*-isomers are optically inactive (if all olefinic ligands are achiral) stereoisomers and are not enantiomers. Since this isomerism owes its existence to the presence of a π-bond, it is called *π-diastereomerism*, to distinguish it from *σ-diastereomerism*, exhibited by cyclic compounds. This terminology, however, is not commonly used.

For molecules of the type $C_{ab} = C_{ab}$ or $C_{ab} = C_{ac}$, the terms *cis* and *trans* are unambiguous. But if all four substituents are different, this nomenclature leads to ambiguity. This problem is solved by arranging the pair of ligands at each olefinic carbon in CIP sequence. If the groups of higher priority are on the same side the configuration is *seq-cis*, later replaced by the symbol Z (from the German *zusammen*

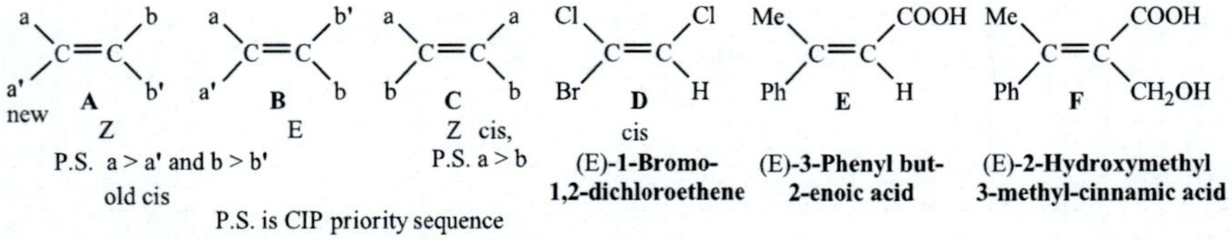

Notes: • The descriptors (E) and (Z) are always italicized and is placed in parenthesis in front of the name
• Structure **D** reveals that E and Z do not always correspond to trans and cis respectively. Compound **E** and **F** have (E) stereochemistry, but they cannot be assigned cis or trans .

Fig. 2.15 (*Z*)- and (*E*)-alkenes. Unambiguous nomenclature for any substituted olefin

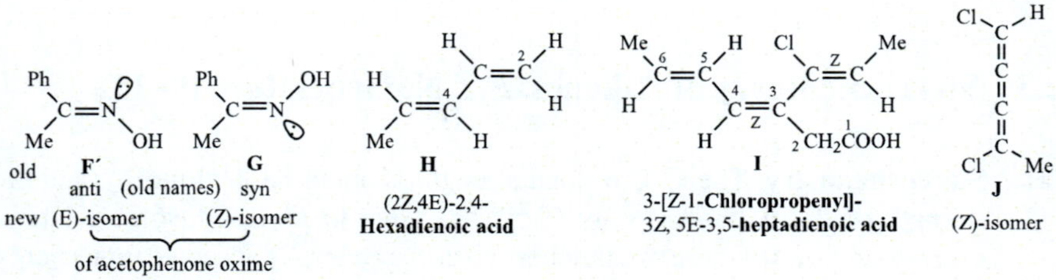

Fig. 2.16 Additional examples of *E-Z* nomenclature

meaning "together"); if they are on the opposite sides, the configuration is *seq-trans*, replaced by the symbol *E* (from the German *entgegen* meaning "opposite" [9]. Examples illustrating application of the rules for assignment of *E* and *Z* are shown in Fig. 2.15.

The *E-Z* nomenclature is always applicable and unequivocal, including cases where *cis*- and *trans*-nomenclature becomes ambiguous (Fig. 2.16). The *E-Z* system is useful for unambiguous nomenclature [10, 11] of oximes (such as **F′** and **G**) and of compounds containing noncumulated (conjugated) double bonds (such as **H** and **I**), and of cumulenes with odd number of cumulated double bonds with two $= C_{ab}$ as terminal groups (such as **J**), as illustrated in Fig. 2.16. The compounds **F′** and **G** are *E* and *Z* isomers rather than *anti* and *syn* isomers (old nomenclature), respectively. In case of compounds **H** and **I**, the *locants must be used in conjunction with the E,Z-descriptors.* In the cumulenes with odd number of double bonds (planar, achiral) (three double bonds in case of **J**), the two ligands in each terminal carbon are in the same plane (cf. olefin), and hence, *E-Z* nomenclature is applicable to them also. Thus, the cumulene **J** is a *Z*-isomer.

2.4 Physical Properties of Geometrical Isomers [12]

2.4.1 Dipole Moments of Geometrical Isomers [13]

A compound of the type abC = Cab (I) has two geometrical isomers: the *cis*-isomer having *a,a* as well as *b,b* on the same side of the double bond, and the *trans*-isomer having the same paurs on the opposite sides of the double bond. If in the *cis* isomer of (I), C—*a* has a strong bond moment, but C—*b* has not, the cis-isomer should have a considerable overall dipole moment (μ). Examples; *cis*-1,2-dichloro-, *cis*-1,2-dibromo-, and *cis*-1,2-diiodoethylene have dipole moments 1.89, 1.35, and 0.75 D, respectively (Fig. 2.17).

On the other hand, in the centrosymmetric *trans*-isomer of (I), the bond moments are opposed, and hence, the overall dipole moment of the *trans*-isomers of 1,2-dichloro-, 1,2-dibromo-, and 1,2-dibromoethylene is zero (Fig. 2.17).

The second example is the dipole moments of abC = Cbd (II), having the *cis*- and *trans*-isomers, where C—a and C—d have appreciable bond moments, and C—b does not. Here, the dipole moment of the *trans*-isomer is expected to be zero, if C—a and C—d have same bond moments. If not, the trans-isomer will have small dipole moment corresponding to the difference of these bond moments. Example: The *cis*-isomer of CHCl = CHBr will have μ 1.55 D, and the *trans*-isomer will have μ ~ 0 (the bond moments of C—Cl and C—Br are nearly identical) (Fig. 2.17).

The third example is a special case of (II). The dipole moments of the isomers of CHCl = CHI are $\mu_{cis=}$0.57 D, μ_{trans} = 1.27 D (a probable explanation is that iodine may be electron donating rather than electron withdrawing. For further explanations and discussion, reference [12] may be seen.

The dipole moments for the following pairs are known [12].

Diethyl maleate, 2.54 D and diethyl fumarate, 2.38 D; *cis*-2-butene-1,4-diol, 2.48 D and the *trans*-isomer, 2.45 D; 1-chloro-1-propene, 1.71 D and the *trans*-isomer, 1.97 D; *trans*-2-butene, zero and cyclohexene, 0.75 D; *trans*-cyclodecene, 0–0.15 D and *cis*-cyclodecene, 0.44 D; *cis*-cyclooctene, 0.43 D and the strained *trans*-isomer, 0.82 D.

2.4.2 Physical Properties of Geometrical Isomers: Melting Points, Boiling Points, Densities, and Refractve Index [12]

The *trans*-isomer having, in general, greater symmetry than the *cis*-isomer, fits into the crystal lattice more easily and hence has a higher melting point. Examples are maleic acid, m.p. 130°, and fumaric acid, m.p. 300°; their methyl esters, m.p. – 8.4° and 101°, respectively; cinnamic cid, cis, m.p. 68°, trans, m.p. 133°; crotonic acid: cis, m.p.15° and trans, m.p.72°; oleic acid, m.p. 16°, and its trans-isomer elaidic acid, m.p. 51°; 1,2-dihaloetylenes: *cis*-dichloro, m.p. – 80.5° and *trans*-dichloro, m.p. – 80.5°; *cis*-dibromo, m.p. – 53° and trans-dibromo, m.p. – 6.5°; *cis*-diiodo, m.p. – 14° and *trans*-diiodo m.p. 72°; stilbene: cis, m.p. 1° and trans, m.p. 125°; 2-butene: cis, m.p. – 139° and trans, m.p. – 106°.

Again, the trans-isomer having better fit in the crystal lattice has lower solubility. For example (in water at 25°), maleic acid, 78.8 g/100 ml and fumaric acid, 0.7 g/100 ml; *cis*-crotonic acid, 40.0 g/100 ml. and *trans*-crotonic acid, 8.3 g/100 ml; *cis*-cinnamic acid, 14.4 g/100 ml and *trans*-cinnamic acid, 0.1 g/100 ml.

The boiling point, density, and refractive index, however, do not have direct relation to configuration as melting point and solubility. Since the boiling point, density, and refractive index are inverse functions of molecular volume (cf. Sect. 1.5.4), the isomer that has the higher value for one of the three propertiesusually has the higher value for the other two also. As stated in Sect. 1.5.4, these properties can often be predicted on the basis of either the *conformational rule* or the *dipole rule* [14, 15]. Since cis- and trans-olefins almost always differ in dipole moment. The dipole rule should apply, i.e., yhe isomer of higher dipole moment should have the higher physical constants. Afew exceptions have been known [12].

2.4.3 Acid Strengths of Geometrical Isomers [16, 17]

The acid strengths of dibasic ethylenic acid isomers are strongly dependent on configuration. In fact, the K_1/K_2 always exceeds 4, since the two carboxyl groups are close enough together so that the electrostatic of one C = O dipole eases the departure of the first proton from the other carboxyl group, increasing K_1,whereas the electrostatic effect of the CO_2^- group in the monoanion retards the departure of the second proton, thus decreasing K_2. Since these electrostatic effects are greater, the closer together the two carboxyl groups are, a cis-dicarboxylic acid generally has a larger K_1 and a smaller K_2, and a much larger K_1/K_2 ratio than its trans-isomer. A most suitable example: HOOC—CH = CH—COOH: cis-acid, maleic acid, K_1 is 1.3 × 10^{-2} and K_2 is 3.2 × 10^{-7}, K_1/K_2 being 40,000; trans-acid, fumaric acid, K_1 is 1.0 × 10^{-3} and K_2 is 3.2 × 10^{-5}, K_1/K_2 being 32.

More subtle differences are found in the isomers of monobasic acids. For example, crotonic acid ($CH_3CH = CH—COOH$), cis-isomer, pKa 4.44, whereas trans-isomer, pKa 4.70; the ratio of the two acidity constants is less than 2. Example 2, cinnamic acid (PhCH = CH–COOH, cis-isomer, pKa 3.96, and trans-ismer, pKa 4.44; here, a slightly greater ratio (~3) of the two acidity constants is found. Steric inhibition of resonance of the type RCH = CH—C(OH) = O ↔ RCH—CH = C(OH)–O^- is more important in the free acid than is the anion, where it would cause the accumulation of two negative charges on the carbonyl oxygen. Hence, the acid is more stabilized by this type of resonance than in the anion, i.e., the effect of the resonance is acid weakening.

When the resonance is inhibited, as by an ortho substituent in benzoic acid, or a cis-substituent in acrylic acid (CH_2 = CH—COOH), as in *cis*-crotonic acid, the acid becomes stronger. The effect is greater in cinnamic acid than in *cis*-crotonic acid, since in cinnamic acid, the phenyl ring is also involved in the resonance, and the total effect of the resonance (and therefore of its inhibition) becomes larger.

References

1. D.H.R. Barton, *The principles of conformational analysis.* (An Inaugural Lecture delivered at Berbeck College, London, 1954), p. 10
2. M.F. Rosanoff, On Fischer's classification of Stereo-Isomers. J. Am. Chem. Soc. **28**, 114–121 (1906)
3. W. Klyne, Nomenclature of some stereoisomeric compounds. Specification of configurations. Chem. Ind. (London), 1022–1025 (1951)
4. R.S. Cahn, C.K. Ingold, V. Prelog, The specification of asymmetric configuration in organic chemistry. Experientia **12**, 81–94 (1956)
5. R.S. Cahn, S. Christopher, K. Ingold, V. Prelog, Specification of molecular chirality. Angew. Chem. Int. Edn. Eng. **5**, 385–415 (1966). For Conformational chirality see p. 406
6. R.S. Cahn, C.K. Ingold, Specification of configuration about quadrivalent asymmetric atoms. J. Chem. Soc. (London) 612–622 (1951)
7. V. Prelog, G. Helmchen, Basic principles of the CIP-system and proposals for a revision. Angew. Chem. Int. Edn. Eng. **21**, 567–583 (1982)
8. *IUPAC Commission on Nomenclature of Organic Chemistry, "Rules for the Nomenclature of Organic Chemistry: Section E, Stereochemistry"*, vol. 45. (Pergamon Press, Oxford, 1976), pp. 11–30. Pure Appl. Chem.; pertinent page 22
9. J.E. Blackwood, C.L. Gladys, K.L. Loening, A.E. Petrarca, J.E. Rush, Unambiguous specification of stereoisomerism about a double bond. J. Am. Chem. Soc. **90**, 509–510 (1968)
10. IUPAC Tentative Rules for the Nomenclature of Chemistry Section E. Fundamental Stereochemistry. J. Org. Chem. **35**, 2849–2867 (1970)
11. L.C. Cross, W. Klyne, Rules for the nomenclature of organic chemistry. Pure Appl. Chem. **45**, 11 (1976)
12. E.L. Eliel, *Stereochemistry of Carbon Compounds.* (Mcgraw-Hill, 1962); (Tata Mcgraw-Hill Edition, 1975), pp. 325–329
13. For sources of data see L.G. Wesson, *Tables of Electric Dipole Moments.* (Technology Press, MIT, Cambridge, Mass, 1948)
14. N.L. Allinger, The relative stabilities of *cis* and *trans Isomers.* III. The cyclodecenes. J. Am. Chem. Soc. **79**, 3443 (1957)
15. A.E. van Arkel, Zusammenhang Zwischen Dipolmoment und Kohaesionskraeften. VI. Rec. trav. Chim. **51**, 1081–1107 (1932). Zusammenhang Zwischen Dipolmoment und Kohaesionskraeften VI. **53**, 246–256 (1934)
16. H.C. Brown, D.H. McDaniel, O. Hafliger, *Dissociation Constants,* Chapter 14
17. V. Gold, Steric effects and acid-base equilibria, in *Progress in Stereochemistry,* ed. by P.B.D. de la Mare, W. Klyne, vol. 3. (Academc Press, New York, 1962). Chapter 5

Chapter 3
Projection (Fischer, Newman, Sawhorse) and Perspective (Flying Wedge and Zigzag) Formulas, Working Out Stereoisomers

Abstract This chapter briefly describes AB-type Molecules, Erythro and Threo Isomers—their ambiguity, Interconversion of Fischer, Newman, Sawhorse, and Flying Wedge formulas, Pref-Parf, Syn-Anti, Like-Unlike. and Brewester's Systems; Stereogenecity and Chirotopicity; Working out Stereoisomeric Molecules with AA, ABC, ABA, ABCD, ABCDE, ABBA, and ABCBA types; and 1-, 2-, and 3-Dimensional Chiral Simplexes.

3.1 Molecules with Two Unlike (Unsymmetrical) Chiral Centers (AB Type)

Many natural products, e.g., steroids, terpenoids, alkaloids, and carbohydrates contain two or more chiral centers, the stereochemistry of which should be thoroughly understood. An acyclic molecule containing two or more chiral carbons is constitutionally unsymmetrical if the two end groups are nonequivalent, $R(Cab)_n R^1$, where $n \geq 2$. Interconversion of Fischer projection, Newman projection, sawhorses, and flying-wedge formulas of $RCabCabR^1$ is depicted in a tabular form in Fig. 3.1. The conventions followed in writing the sawhorse and flying-wedge formulas will be revealed from their careful inspection.

Designation of Diastereomers. Several systems of designation of AB-type diastereomers (Fig. 3.2) are now known of which a few are discussed below.

3.1.1 Erythro *and* Threo *Nomenclature*

A compound containing two adjacent nonequivalent chiral centers, $RCabCabR^1$, i.e., of AB type, gives rise to two diastereomeric (±)-pairs, in total 4 stereoisomers (Fig. 3.2), e.g., the aldotetroses, erythrose and threose (Fig. 3.2), and 3-bromo-2-butanol (Fig. 3.3), each one contains two nonequivalent chiral centers.

S. K. Talapatra and B. Talapatra, *Basic Concepts in Organic Stereochemistry*,
https://doi.org/10.1007/978-3-030-95990-6_3

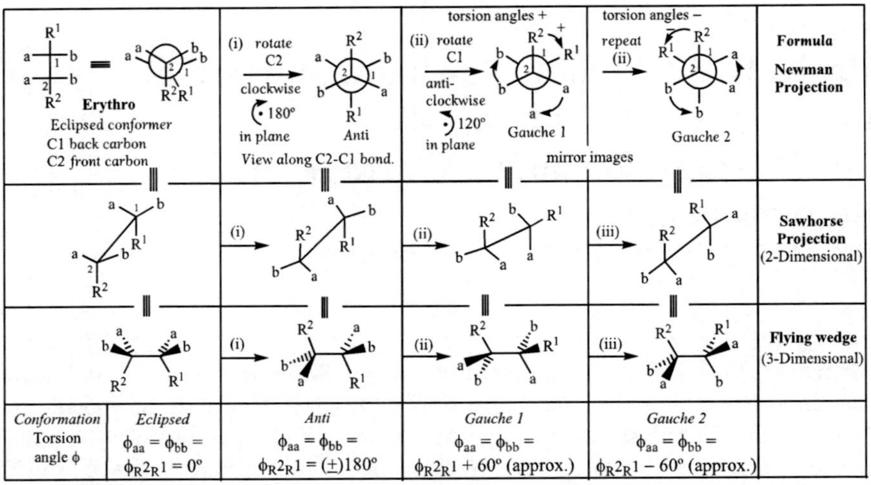

Fig. 3.1 Interconversion of Fischer, Newman, sawhorse, and flying-wedge formulas of *erythro*-$R^1CabCabR^2$

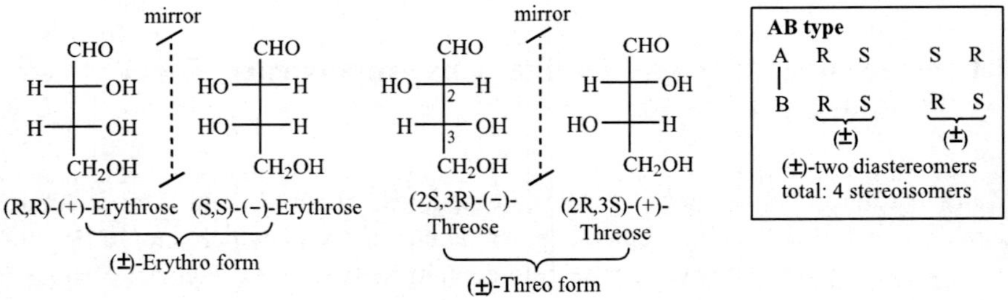

Fig. 3.2 Fischer projection formulas of (±)-erythrose and (±)-threose (AB type)

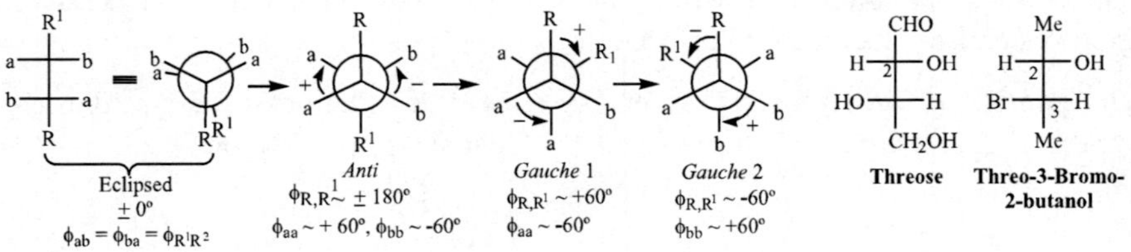

Fig. 3.3 One enantiomer of *threo*-$R^1CabCabR^2$; Fischer and Newman projection formulas; conformers. Examples

***Erythro* isomer**: Examining different two-dimensional (Fischer projection, Newman, and sawhorse) and three-dimensional (flying wedge) formulas of an enantiomer of the *erythro* isomer of $R^1CabCabR^2$ in Fig. 3.1, one can define an *erythro* isomer in one of the following alternative ways:

(i) "At a glance" nomenclature is possible in the Fischer projection (Fp) formula having the main chain vertical in which the *erythro* diastereomer will *have both horizontal pairs of matched like (or similar) groups (or ligands)*, (e.g., two OH groups and two Hs in erythrose) *on the same side, i.e., they are eclipsed* (Torsion angle $\phi \approx \pm 0°$); the other pair of ligands (CHO and CH_2OH in erythrose) will also be eclipsed.
(ii) In the Newman projection, sawhorse, and flying wedge formulas of the eclipsed conformer (≡Fp formula), *at least two pairs of like (or similar) ligands are eclipsed* ($\phi \approx \pm 0°$) (*vide* first column of Fig. 3.1).
(iii) In the *anti*-conformer (second column of Fig. 3.1) *at least two pairs of like (or similar) or identical ligands are antiperiplanar* ($\phi \approx \pm 180°$).
(iv) In any of the two gauche conformers (third and fourth columns of Fig. 3.1), *each pair of the two like (or similar) or identical ligands will be gauche* having $\phi \approx +60°$ (third column) or $-60°$ (fourth column).

***Threo* isomer**: The diastereomer of *erythro* $RCabCabR^1$ is called *threo* isomer. It will have (by default) the following characteristics. Figure 3.3 shows only the Fischer and Newman projection formulas, from which sawhorse and flying formulas can be written (cf. Fig. 3.1).

The different formulas of *threo* isomer results by *exchange of any two ligands* (e.g., a and b) *at any one chiral center* (C1 or C2) of the *erythro* isomer in the Fischer, Newman, and sawhorse projection or flying wedge formulas (cf. Fig. 3.3).

(v) In the Fp formula (having the main chain vertical), the *threo* diastereomer *will have both horizontal pairs of unlike (or dissimilar) ligands* (e.g., the OH and H in threose) *on the same side*; i.e., they are eclipsed ($\phi \approx \pm 0°$). Hence, both pairs of like (or similar) ligands will be on the opposite side.
(vi) In the Newman projection formula (sawhorse and flying wedge formulas are not shown) of the eclipsed conformer (≡Fp formula), at least two pairs of unlike (or dissimilar) ligands at the chiral carbons will be eclipsed, having torsion angle $\phi = \pm 0°$.
(vii) In the *anti*-conformer of Newman projection formula, two pairs of like (or similar) ligands *will not be antiperiplanar* having $\phi = \pm 180°$.
(viii) In either gauche conformer the *torsion angle ϕ between two pairs of like* (or similar) *ligands will be of opposite sign.*

The above statements for recognizing the *threo* isomer are illustrated in Fig. 3.3. The *erythro–threo* nomenclature is possible for compounds of the type $RCabCacR^1$. Of course, the easiest way to identify the *erythro* and *threo* isomers is to examine their Fischer or Newman projection formulas.

***Threo* diastereomer**: (*compare with the first row of* Fig. 3.1).

Examples:

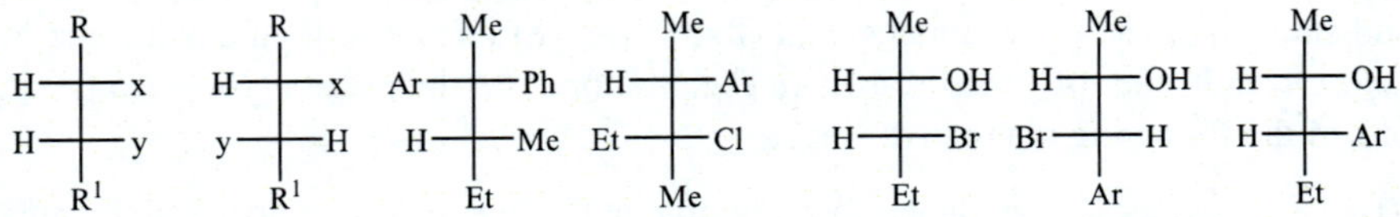

A B C D E F G

Eythro or Threo ? (if R, R′ and x or y are alkyls or aryls) — Erythro or Threo ? — Erythro, Threo, Erythro: two heteroatom ligands

Fig. 3.4 Some cases where ambiguity arises (**A–D**) or does not arise (**E, F, G**)

The ambiguity in threo–erythro nomenclature arises in cases **A** and **B** (Fig. 3.4), when at least all of R, R^1, *x*, and *y* are alkyls or aryls, when matching of the two pairs of ligands at the two chiral centers is not possible; examples are cases **C** and **D**. However, this nomenclature in cases **A** and **B** is still possible if *x* and *y* are ligands with heteroatoms such as OR, NR^1R^2, or halogen, and R and R^1 are alkyl or aryl groups (examples **E**, **F**, **G**).

3.1.2 "Pref" and "Parf" Nomenclature

In cases like **A** and **B** (Fig. 3.4), and also in the general case **H** (Fig. 3.5), when only one or no pair of ligands on the two chiral carbons can be matched the ***pref–parf*** nomenclature, developed in 1982 by Carey and Kuehne [1], can be used conveniently. The original version of this nomenclature has been simplified in the following way: Three ligands at each chiral center (case **H**) are to be ordered by CIP priority sequence rules as $a > b > c$ and $x > y > z$. One does this by viewing the ligands from any one side of the Fp formula joining the two chiral centers, say, *from bottom/down side*. If the three ligands on each chiral center appear in the clockwise (*Re*), or anticlockwise (*Si*) order, when viewed from any end, the relative configuration may then be specified as *Re/Re* (or *Si/Si*) for one diastereomer and as *Re/Si* (or *Si/Re*) for the other. The former case is denoted by Carey and Kuchne system as ***pref*** (***p****riority* ***ref****lective*), and the latter case is denoted by ***parf*** (***p****riority* ***a****nti****ref****lective*) (Fig. 3.5).

For an AB system usually *erythro* isomer corresponds to the *pref* isomer, but the *erythro–threo* nomenclature has no correspondence with the *pref–parf* nomenclature. For example, compound K (Fig. 3.6) appears to be *erythro* isomer by convention, but *parf* according to *pref–parf* system (Fig. 3.6). In fact, *pref–parf* nomenclature is used only when *threo–erythro* nomenclature is not possible. For *pref–parf* nomenclature of compounds having more than two chiral centers, e.g., (**E**) and (**F**), see Fig. 3.9.

An ***advantage*** of the ***pref–parf*** system is that it can be used for specifying relative configuration of two noncontiguous chiral centers, assuming that they are directly linked, disregarding the intervening achiral centers (Fig. 3.7).

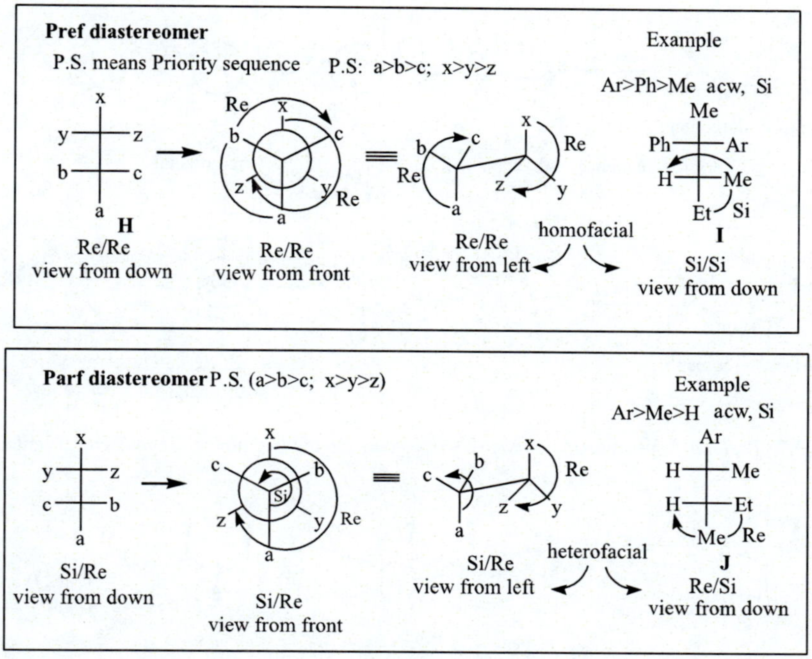

Fig. 3.5 Delineation of *pref* and *parf* diastereomers (one enantiomer of each)

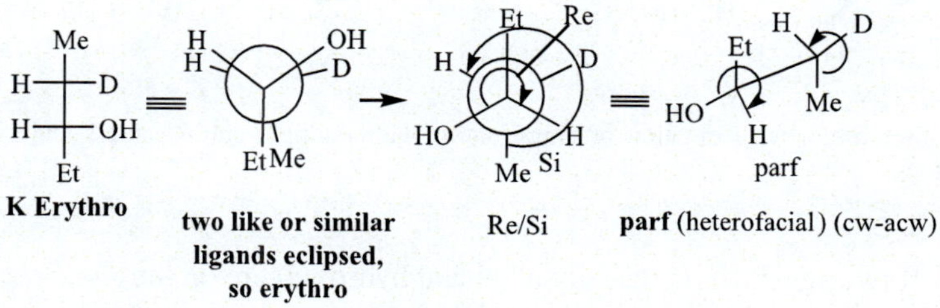

Fig. 3.6 No correspondence between *erythro–threo* and *pref–parf* nomenclatures

3.1.3 Syn *and* Anti *System*

This simple system of nomenclature of relative configuration has been introduced by Masamune [2] for aldol-type compounds containing multiple chiral centers. Thus, the chiral aldehyde (**A**) was transformed stereoselectively into the diastereomers 3,4-*anti*-aldol (**C**) and into 3,4-*syn*-aldol (**D**) in different ratios using Li-enolates of different ketones (**B**) (Fig. 3.8). The longest carbon chain is written in a zigzag

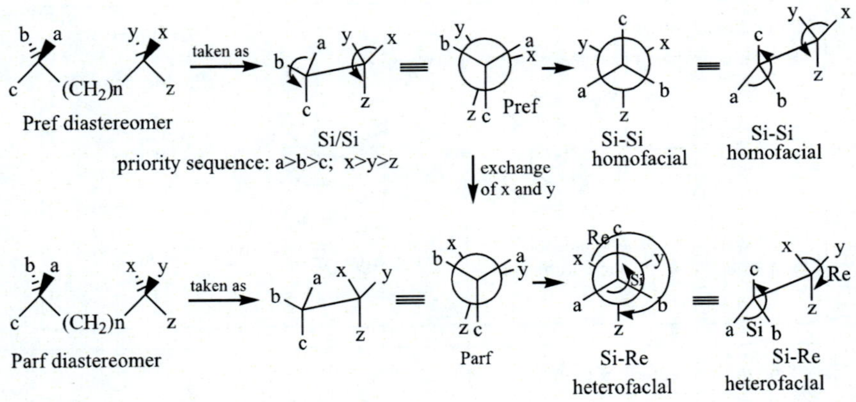

Fig. 3.7 *Pref–parf* specification of two diastereomers having two noncontiguous chiral centers

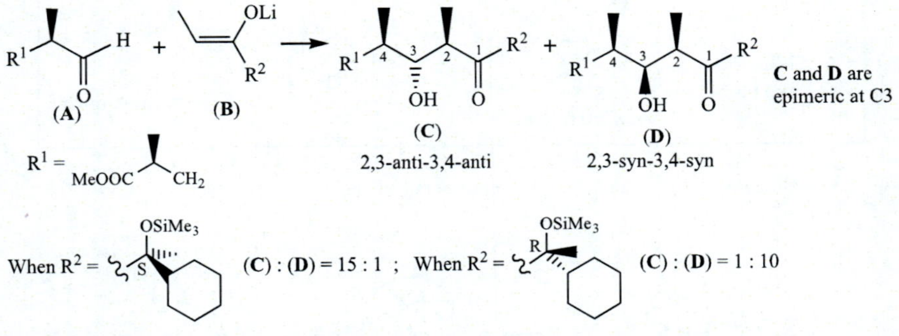

Fig. 3.8 Stereoselective formation of chiral aldols with multiple chiral centers and *syn, anti*-designations

fashion. If two substituents (usually alkyl and hydroxyl) are on the *same side of the carbon chain plane*, the prefix *syn* is used. If they are on the *opposite side*, the prefix *anti* is used, as illustrated in Masamune's stereoselective aldol formation (Fig. 3.8).

3.1.4 Like (l) *and* Unlike (U) *System*

Prelog and Helmchen [3] have proposed a similar system of designation for the relative configuration of acyclic molecules with multiple adjacent chiral centers, in terms of the *R, S* designation of the adjacent chiral centers. If the adjacent chiral centers of the diastereomer are of like chirality (R^*R^*) or of unlike chirality (R^*S^*), they are termed *l* (like) or *u* (unlike) respectively, starting from the lowest-numbered chiral center (locant). Here R^*R^* and R^*S^* refer to relative stereochemistry (Sect. 2.2.4).

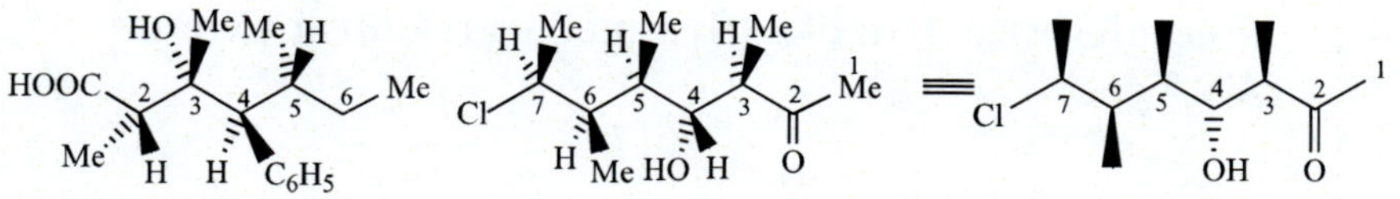

(E) **(F)** **(F)**

2S, 3R, 4S, 5R
u u u (Prelog)
Represents (**E**) or its enantiomer (**E'**).
S (2l, 3u, 4l, 5u) Represents (**E**) (Brewster).
R(2l, 3u, 4l, 5u) Represemts (E') Brewster).
2,3-pref, 3,4-pref, 4,5-parf {Si/Si, Si/Si, Si/Re}
Represents (**E**) or (**E'**).

3R, 4R, 5R, 6R, 7S (CIP)
l l l u (Prelog)
Represents (**F**) or its enantiomer (**F'**).
R (3l, 4l, 5l, 6l, 7u) Represents (**F**) (Brewster).
S (3l, 4l, 5l, 6l, 7u) Represents (**F'**).
3,4-parf, 4,5-parf, 5,6-parf, 6,7-parf (Re/Si, Si/Re, Re/Si, Si/ Re) Represents (**F**) or (**F'**).

Fig. 3.9 Prelog's and Brewster's stereochemical notations for acyclic compounds having multiple chiral centers

This is illustrated by compounds (**E**) and (**F**) containing four and five consecutive chiral centers (Fig. 3.9) which are represented by prefixes *uuu* and *lllu*, respectively.

3.1.5 *Brewster's System of Nomenclature*

A general stereochemical notation, balanced in its emphasis on geometry (here relative configuration) and topography (here absolute configuration), has been developed in 1986 by Brewster [4] from the *R-S* (*R***S**) system "by use of *ul* notation" of Prelog et al. and "the concept of external referencing." Thus, the absolute configuration of the compounds (**E**) (Fig. 3.9) may be designated as *S* (2*l*, 3*u*, 4*l*, 5*u*), where *S* represents the absolute configuration of the lowest locant (here C2), which is written outside the parenthesis (external reference), while *l* or *u* notation (being like or unlike the external reference) for each chiral center is placed inside the parenthesis. The chirality of each chiral center of the enantiomer of (**E**), designated (**E**′) is defined by *R* (2*l*, 3*u*, 4*l*, 5*u*), which means that the nomenclature depicts the chirality as a property of the whole molecule. Thus, topography (absolute configuration) at each center is readily recoverable: *RS* (2*l*, 3*u*, 4*l*, 5*u*) represents a racemic mixture (**E**) and (**E**′). A partially resolved mixture consisting of 80% (**E**) and 20% (**E**′) is represented by [80*S*, 20*R*) (2*l*, 3*u*, 4*l*, 5*u*). Any unknown chirality (say at C3) in compound (**E**) may be indicated by putting an *x* as in *S* (2*l*, 3*x*, 4*l*, 5*u*), which means that the chirality (absolute configuration) of C3 is not known.

3.2 Molecules with Two like (Symmetrical) Chiral Centers (AA Type)

Molecules with two like chiral centers are of AA type and are *constitutionally symmetrical* having two diastereomers: one (±)-pair and one *meso*-isomer (Fig. 3.10). It may be noted that if anyone ligand on any chiral carbon (C2 or C3) is substituted by a new ligand (different from the existing ones), *meso*-isomer becomes *erythro*, and active isomer becomes *threo*. The *anti*-forms of (−)-tartaric acid and (+)-tartaric acid, each with **C2** point group, are shown in the figure.

Optical inactivity of meso-tartaric acid. The optical inactivity of the meso-tartaric acid may be explained in terms of its three staggered conformers, viz. the *anti*-conformer and two gauche conformers which constitute over 99% of the total number of molecules present in a sample of *meso*-tartaric acid. The eclipsed form shown in Fig. 2.50 along with the two other eclipsed forms obtained by rotation of any chiral carbon of the eclipsed form (say C2) around C2–C3 by 120° and 240° (*cw* or *acw*) constitute much less than 1% of the total number of molecules present. The *anti*-conformer with $\mathbf{S_2}$ ($\equiv \mathbf{C_i}$ point group) is optically inactive. The gauche 1 and gauche 2 conformers although optically active (each with $\mathbf{C_1}$ point group) are *equally populated* (statistically symmetrical) *enantiomers*, as evident from the same but opposite torsion angle signs between the like ligands in each of them—negative in gauche **1** and positive in gauche **2**. *Thus, the combined effect of all molecules is to make meso-tartaric acid optically inactive.*

The terminology *homochiral* (*RR* or *SS*) for optically active stereoisomers and *heterochiral* (*RS* or *SR*) for *meso*-stereoisomers has been adopted [5].

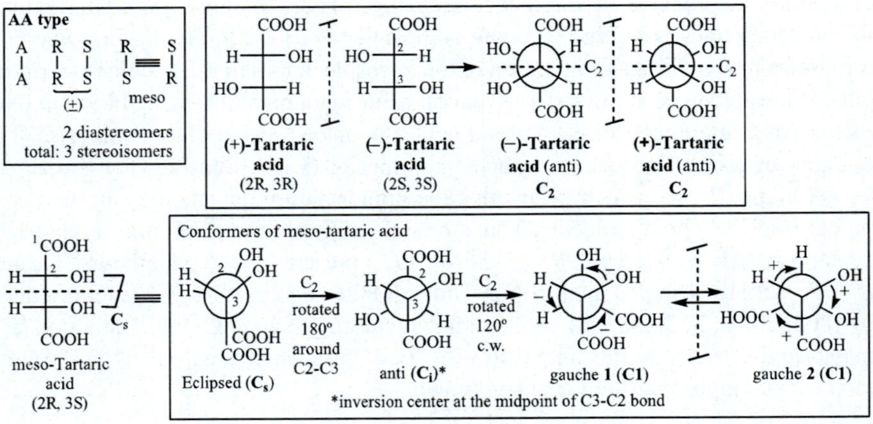

Fig. 3.10 Stereoisomers of tartaric acid (AA type). Optical inactivity of *meso*-tartaric acid

Stereoisomer	1 ; 2	3 ; 4	5 ; 6	7 ; 8
C2 A	R ; S	S ; R	R ; S	S ; R
C3 B	R ; S	R ; S	S ; R	S ; R
C4 C	R ; S	R ; S	R ; S	R ; S
	(±)	(±)	(±)	(±)

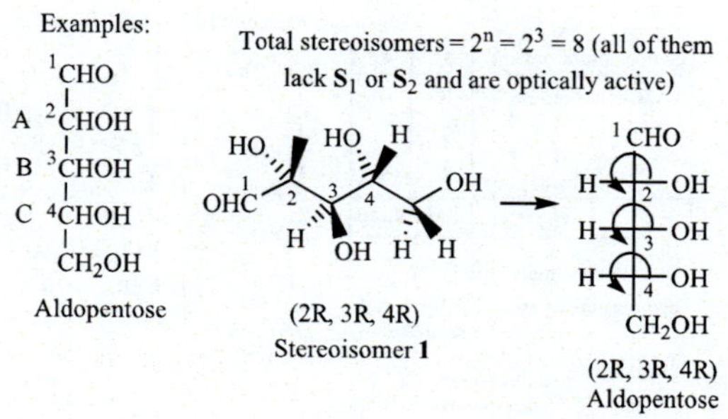

Fig. 3.11 Stereoisomers of ABA type with an example

3.3 Molecules with Three Unlike Chiral Centers (ABC Type)

An aldopentose, CHO–$(CHOH)_3CH_2OH$, a constitutionally unsymmetrical molecule belonging to ABC type, has three chiral centers and exists as eight (2^n, where n = number of chiral centers) stereoisomers involving four diastereomers, each one having its enantiomer (Fig. 3.11). Of the eight stereoisomers, the one with (2*R*, 3*R*, 4*R*) absolute configuration is shown in its zigzag and Fischer projection formulas in Fig. 3.11.

3.4 Constitutionally Symmetrical Molecules Having Three Chiral Centers (ABA Type)

An acyclic molecule containing three or more chiral centers is constitutionally symmetrical if the chiral atoms equidistant from the geometrical center of the molecule are identically substituted. The two end groups of such a molecule are necessarily equivalent. Let us take n as the number of chiral centers in such a molecule. Thus, a constitutionally symmetrical molecule, say 2,3,4-trihydroxyglutaric acid, belongs to ABA type (Fig. 3.12) and exists as $2^{n-1} = 2^{3-1} = 4$ stereoisomers. Two of them (stereoisomers **1** and **2**) are enantiomers, and the remaining two, i.e., stereoisomers **3** and **4** are optically inactive *meso*-forms. The latter two (**3** and **4**) are diastereomeric with each other and also with the two enantiomers **1** and **2**.

Monomethyl ester (**B**) [half ester of (**A**)] (Fig. 3.12) becomes constitutionally unsymmetrical and belongs to ABC type like an aldopentose (Fig. 3.11) and exists as four diastereomers, each having a (±)-pair, making eight stereoisomers in total (Fig. 3.12).

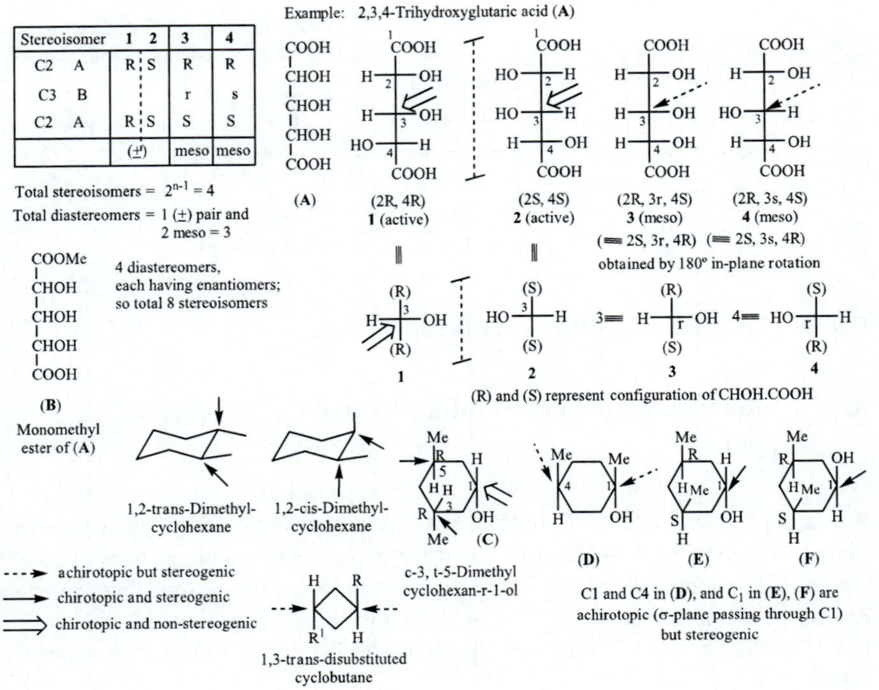

Fig. 3.12 Stereoisomers of compounds of ABA type. Stereogenecity and chirotopicity

3.5 Stereogenecity and Chirotopicity (Fig. 3.12)

The status of the C2, C3, and C4 centers of the stereoisomers **1**, **2** (enantiomers), and **3** and **4** (Fig. 3.12) will now be discussed in terms of the observations of Mislow and Siegel [6] in 1984. Usually, a chiral center has two distinct attributes or characters:

1. *Stereogenecity* refers to bond connectivity—whether permutation or exchange of a pair of ligands gives a stereoisomer. (*R*-*S*)-Designation is associated with stereogenecity.
2. *Chirotopicity* is determined by local symmetry; i.e., whether S_n axis is present locally. A chirotopic atom resides in a chiral environment. Thus, all the five atoms in bromofluorochloromethane (BrCl FCH) are chirotopic and belong to $\mathbf{C_1}$ point group, although only the C atom is a stereogen. Three categories of centers are possible:

   (i) *Chirotopic and stereogenic*. In most cases, the chiral centers, e.g., C2 and C4 of isomers **1–4**, are *chirotopic* as well as *stereogenic*, i.e., the two attributes overlap. The C1 and C2 of *cis*- and *trans*-1,2-dimethylcyclohexanes are stereogenic as well as chirotopic.

   (ii) *Chirotopic but nonstereogenic*. The C3 centers of enantiomers **1** and **2** are chirotopic since they are devoid of any Sn axis (plane or center of

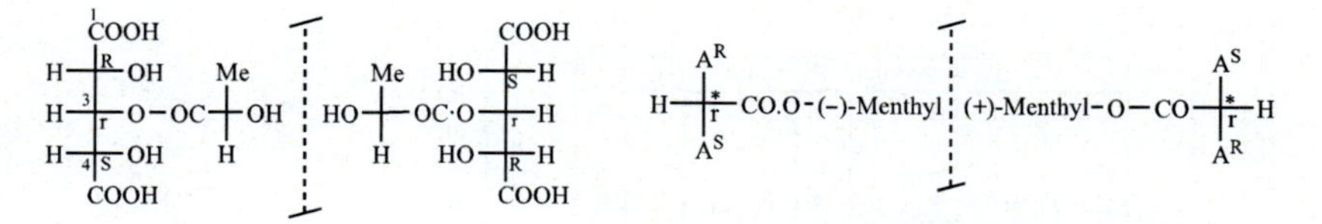

3r
3-Hydroxy gr. esterified by
R-(–)-lactic acid
C3 is reflection invariant

3r
Enantiomers

if –CO.O > A^R > A^S both enantiomers have the
r-configuration of the * asterisked C – and hence reflection invariant

Fig. 3.13 Examples of reflection invariant carbon, which is *chirotopic* as well as *stereogenic*

symmetry), and no local symmetry is present. But they are nonstereogenic since two ligands of C3 in each are identical in both structure and chirality (*R, R*) or (*S, S*). Consequently, exchange of H and OH, for that matter, any two different ligands, does not give any isomer. Another example is the C1 atom of *c*-3,*t*-5-dimethyl-cyclohexan-*r*-1-ol (**C**) (Fig. 3.12), since the two faces (α-) and (β-) are exchangeable by *C2* operation, and exchange of H and OH does not give any new stereoisomer.

(iii) *Achirotopic but stereogenic* [6]. In the meso-isomers **3** and **4**, C3 is achirotopic but stereogenic—thus again showing that stereogenecity and chirotopicity are two distinct properties and can be delinked. Exchange of H and OH at C3 of **3** gives the other *meso*-isomer **4**. C3 in both **3** and **4** can be given configurational descriptors: *r* to **3** and *s* to **4**, since the priority sequence *R* > *S*. C3 of the two molecules **3** and **4** are invariant to reflection; their configuration is denoted by lower case symbols, *r* and *s*. Even if C3 is made chiral by esterifying OH with *R*-lactic acid, the configurational specification *r* or *s* remains invariant to reflection (Fig. 3.13). In classical stereochemistry, such an achirotopic but stereogenic center is called *pseudoasymmetric* and may be designated as C*RS*bc where *R* and *S* represent two enantiomorphous ligands [6].

Some more examples of molecules belonging to the category (iii) are 1,3-disubstituted cyclobutane and the three dimethylcyclohexanols (**D**), (**E**), and (**F**) (Fig. 3.13).

Some examples of compounds with *reflection invariant carbon being chirotopic as well as stereogenic* are depicted in Fig. 3.13.

3.6 Molecules with Four (ABCD Type) or More Unlike Chiral Centers in a Chain

An aldohexose, OHC–$(CHOH)_4$–CH_2OH, a constitutionally unsymmetrical molecule belonging to ABCD type, has four chiral centers and exists as $2^n = 2^4$ =

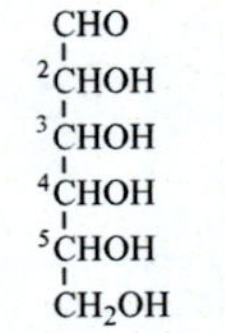

D-Aldohexose (ABCD type)

	Diastereomers (**D**-enantiomers)							
C2 A	R	S	R	S	R	S	R	S
C3 B	R	R	S	S	R	R	S	S
C4 C	R	R	R	R	S	S	S	S
C5 D	R	R	R	R	R	R	R	R
Diaste-reomer #	**1**	**2**	**3**	**4**	**5**	**6**	**7**	**8**

Example:

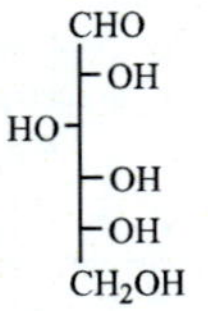

D-(+)-Glucose (3)
(2R, 3S, 4R, 5R)

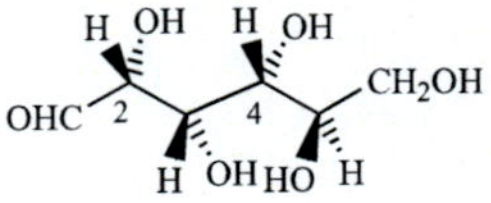

D-(+)-Glucose
(2R, 3S, 4R, 5R)
Zigzag formula
(Diastereomer no. 3)

Fig. 3.14 Eight diastereomers of ABCD type (only D-(+)-Glucose shown)

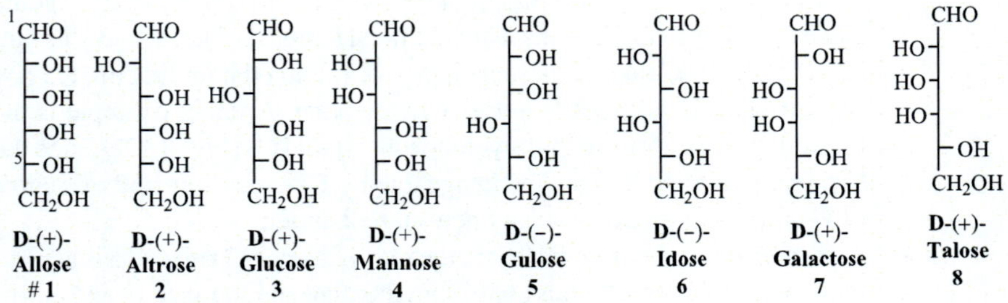

Compare **3** with that of the **Figure 2.54**

Fig. 3.15 Diastereomeric D-aldohexoses

16 (where *n* is the number of chiral centers) stereoisomers involving *eight diastereomers*; each one is optically active and has its enantiomer. Figure 2.54 depicts the D-enantiomer of each of the eight pairs of diastereomeric aldohexoses (ABCD type) (Fig. 3.14).

Much of the early development of stereochemistry was stimulated by investigations of various sugars. It is, therefore, pertinent to depict in Fig. 3.15 the aldohexoses, only the ***D***-enantiomers (C5 OH on the right side possessing *R*-configuration) of the eight diastereomers. Of these aldohexoses **D**-(+)-glucose occurs mostly as plant natural product glycosides. **D**-Gulose and **D**-idose are levorotatory; all other **D**-aldohexoses are dextrorotatory.

3.7 Constitutionally Symmetrical Molecules with Four or More like Chiral Centers in a Chain (ABBA, ABCBA, Etc., Types)

A constitutionally symmetrical molecule has been defined in Sect. 3.4. Such a molecule with four chiral centers (the first and second being similar to the 4th and 3rd chiral centers, respectively) belongs to ABBA type, e.g., tetrahydroxyadipic acids

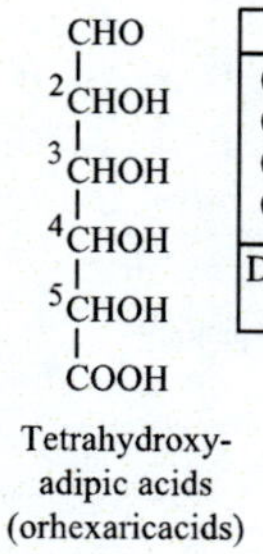

	Diasteromers (**D**-Enantiomer)							
C2 A	R	S	R	S	R	S	R	S
C3 B	R	R	S	S	R	R	S	S
C4 B	R	R	R	R	S	S	S	S
C5 A	R	R	R	R	R	R	R	R
Diastero-mer #	**1**	**2**	**3**	**4**	**5**	**6**	**7**	**8**
	a	a	a	meso		meso	a	a

same (3, 5); enantiomers (2, 8)

a = optically active

Isomer 8 is enantiomeric with isomer 2.
Isomer 5 is same as isomer 3.
So hexaric acid (like perhydrophenanthrene) has four (±) pairs and two meso forms, i.e., 6 diastereomers, and in total 10 stereoisomers, as evident from the Table.

Fig. 3.16 Stereoisomers of hexaric acid (ABBA type)

(or hexaric acids), which exist as six diastereomers (four racemic pairs and two *meso*-forms) and hence ten stereoisomers (Fig. 3.16).

Stereoisomers of compounds with even number of constitutionally symmetrical chiral centers in a chain. In general, in the series of the type Cabd $(Cab)_{n-2}$ Cabd, where n is the number of chiral centers and n is **even**, there are 2^{n-1} optically active stereoisomers and $2^{(n-2)/2}$ *meso*-forms. So, in case of hexaric acid, $n = 4$ and total number of stereoisomers $= 2^{n-1} + 2^{(n-2)/2} = 2^3 + 2^{2/2} = 8 + 2 = 10$. Perhydrophenanthrene (Sect. 9.2) also belongs to ABBA.

Stereoisomers of compounds with odd number of constitutionally symmetrical chiral centers in a chain. In the series of the type Cabd $(Cab)_{n\text{-}2}$ Cabd where n is the number of chiral centers and n is **odd**, total number of stereoisomers $= 2^{n-1}$, of which number of *meso*-forms $= 2^{(n-1)/2}$.

Thus, when $n = 3$, total number of stereoisomers $= 2^{n-1} = 2^2 = 4$, of which number of meso-forms $= 2^{(n-1)/2} = 2^{(3-1)/2} = 2$; so number of optically active isomers = 4–2 = 2 (1± pair) (see Fig. 3.17).

The next higher homologue with n odd, HOOC–$(CHOH)_5$COOH (heptaric acid) is of the ABCBA type. Here $n = 5$, hence total number of stereoisomers $= 2^{5-1} = 2^4 = 16$, of which number of the meso-forms $= 2^{(5-1)/2} = 2^2 = 4$. Hence, number of optically active isomers = 16–4 = 12. So there will be 6 racemic pairs and 4 meso-forms, i.e., 10 diastereomers. All $2^n/2 = 2^5/2 = 16$ diastereomers of ABCDE system can be written down following the system in Fig. 3.14. The six racemic and four meso-isomers of ABCBA system can also be ascertained from the former system (ABCDE) (cf. Fig. 3.16).

3.8 Chiral Compounds with Asymmetric Carbon Atoms in Branched Chains

On rare occasions, one encounters molecules in which chiral carbon atoms cannot be aligned in one chain. We depict the cases (**1**) and (**2**) in Fig. 2.57. Cases (**2**)–(**5**),

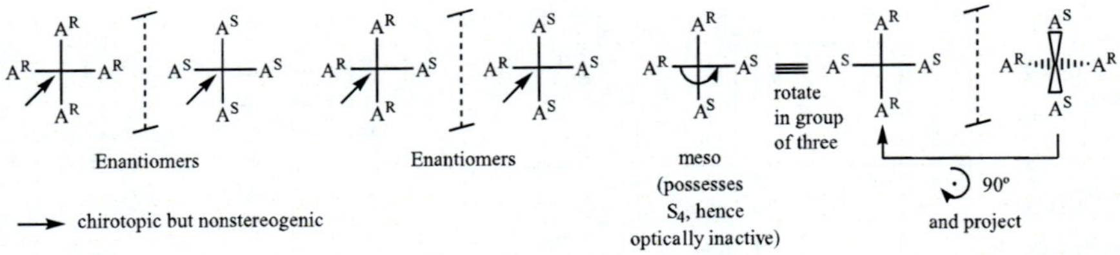

An example of such a meso compound possessing a four-fold alternating axis of symmetry (S_4) and no other element of symmetry, prepared and shown to be optically inactive [5], is given below

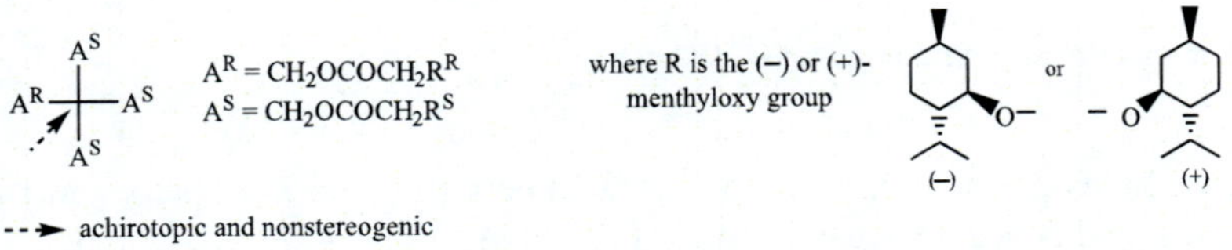

Fig. 3.17 Stereoisomers of cases CA*B*D*E* and CA*$_4$. A molecule possessing S_4

where two or more of the chiral ligands attached to the central carbon atom are alike, are more complicated as delineated in Figs. 3.17 and 3.18.

(3) Case CH(A*)$_3$.

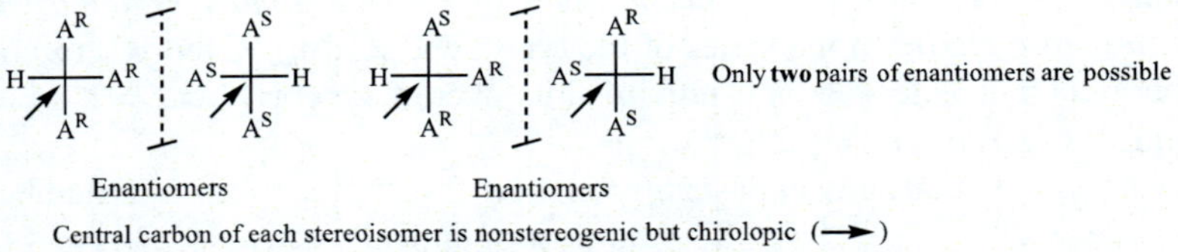

(4) Case CH(A*)$_2$B*. Here **four** enantiomeric pairs (four diastereomers) and **eight** stereoisomers are possible as shown below

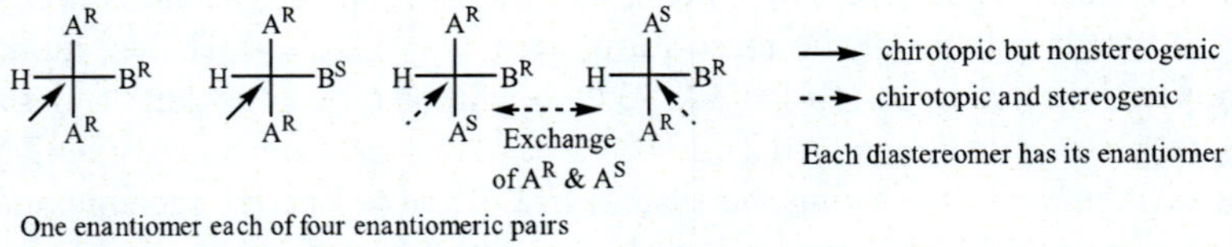

(5) Case C(A*)$_2$(B*)$_2$. Here **five** enantiomeric pairs and **ten** stereoisomers are possible as shown below:

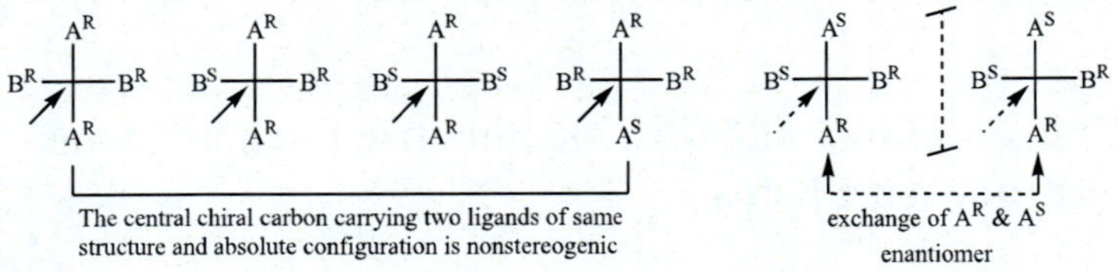

Fig. 3.18 Streoisomers of cases CH(A*)$_3$, CH(A*)$_2$B*, and C(A*)$_2$(B*)$_2$

3.9 Chirality and Dimension. One-, Two-, and Three-Dimensional Chiral Simplexes

Some simple concepts of stereochemistry (in the molecular level) may be conveniently revealed on the basis of "*chiral simplex*" (applicable to ***1D***, ***2D***, or ***3D***), introduced by Prelog and Helmchen [3]. The term means the simplest structure that could be chiral in ***1D***, ***2D***, or ***3D*** world, as illustrated in Fig. 2.59 with examples.

A linear molecule, with two dissimilar points, like A–B (case **A**) is a one-dimensional chiral simplex, i.e., chiral in one dimension (*y*-axis) since its mirror image cannot be superposed on it without the help of a second dimension, i.e., by its rotation around the orthogonal *x*-axis. So a linear molecule A–B is prochiral in a two-dimensional world, since one more dimension (*Z*-axis) is necessary to make it chiral in three dimensions. The molecule A–B is pro-prochiral in the three-dimensional world since two steps are necessary to convert it to a chiral molecule in our *3D* world.

Likewise, a dissimilarly substituted trigonal tricoordinated carbon (case **B**) is chiral in two dimensions (two-dimensional world) and prochiral in tetrahedral tetracoordinated three dimensions.

By the same token, a dissimilarly substituted tetrahedral tetracoordinated carbon is chiral in three dimensions, e.g., any chiral compound with a chiral center. However, one may assume that any chiral compound in three dimensions, which is a three-dimensional chiral simplex, would have been prochiral in an imaginary four-dimensional world, had there been any fourth dimension (cf., time). Some examples of two-dimensional chiral simplex and of pro-, pro^2-, and pro^3-chiral molecules/species are illustrated in Fig. 3.19.

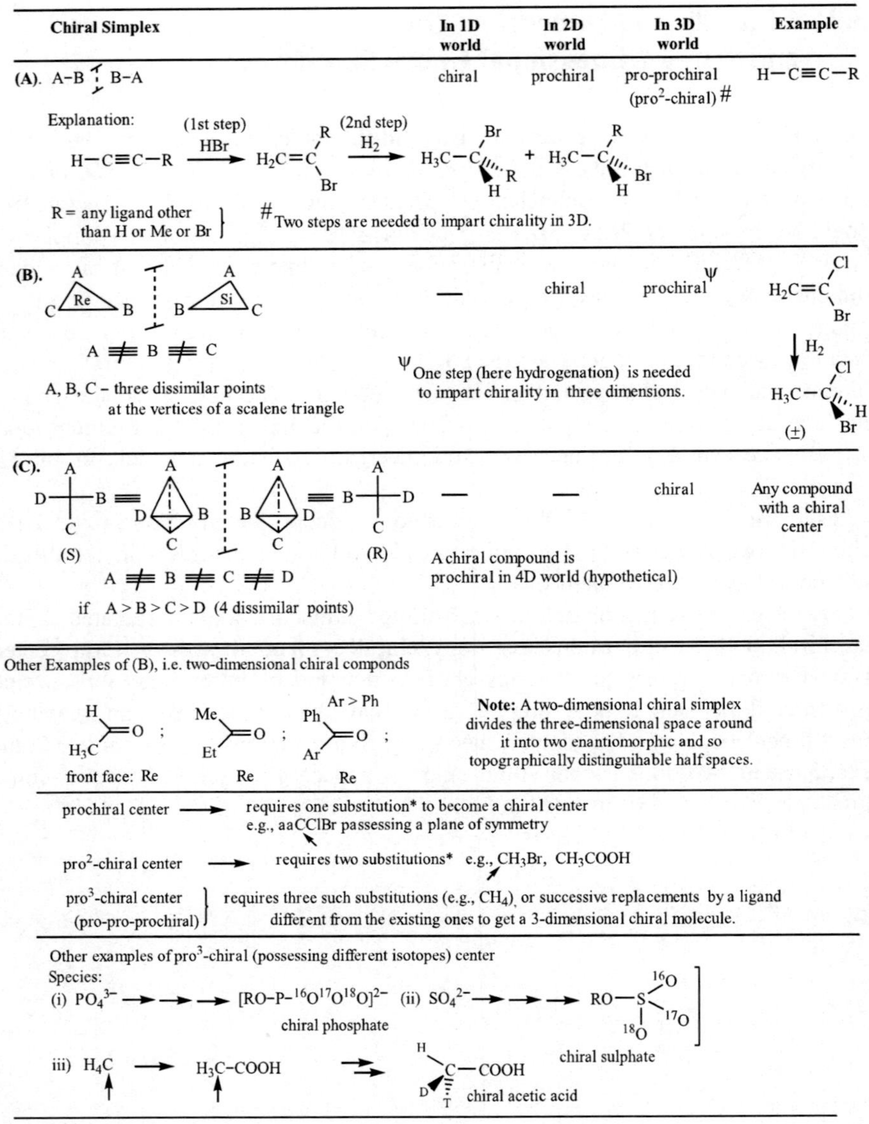

Fig. 3.19 One-, two-, and three-dimensional chiral simplexes. Pro-, pro^2-, and pro^3-chiral molecules

References

1. F.A. Carey, M.E. Kuehne, Beyond erythro and theo. A proposal for specifying relative configuration in molecules with multiple chiral centers. J. Org. Chem. **47**, 3811–3815 (1982)
2. S. Masumune, S.A. Ali, D.L. Snitman, D.S. Garvey, Highly stereoselective aldol condensation using an enantioselective chiral enolate. Angew. Chem. Int. Ed. Engl. **19**, 557–558 (1980)

3. V. Prelog, G. Helmchen, Basic principles of the CIP-system and proposals for a revision. Angew. Chem. Int. Edn. Engl. **21**, 567–583 (1982)
4. H. James, Brewster, on the distinction of diastereomers in the Cahn-Ingold Prelog (RS) notation. J. Org. Chem. **51**, 4751–4753 (1986)
5. V. Schurig, Internal versus external diastereotopism in homochiral *cis* versus *trans* olefin metal π–complexes. Tetrahedron Lett. 3977–3980 (1977)
6. K. Mislow, J. Siegel, Stereoisomerism and local chirality. J. Am. Chem. Soc. **106**, 3319–3328 (1984)

Chapter 4
Prochirality and Prostereoisomerism. Topicity of Ligands and Faces Nomenclature [1–5]

Abstract In this chapter at first a suitable brief introduction to the various terms related to the principle of stereoselective synthesis (vide Chap. 5) is given. This is followed by a useful treatment with examples of homotopic, enantiotopic, and diastereotopic ligands and faces, their nomenclatures and topicities, topic relationship of ligands and faces, stereoheterotopic ligands and NMR spectroscopy, and anisochrony arising out of diastereotopic faces.

4.1 Introduction

In this Chapter at first a suitable brief introduction to the various terms related to the principle of stereoselective synthesis (vide Chap. 5) is given. This is followed by a useful treatment with examples of homotopic, enantiotopic, and diastereotopic ligands and faces, their nomenclatures and topicities, topic relationship of ligands and faces, stereoheterotopic ligands and NMR spectroscopy, and anisochrony arising out of diastereotopic faces.

A nonstereogenic center can be changed to a stereogenic center by replacement of any one of the two *apparently identical ligands* by a different one. Two examples, propionic acid (**1**) or phenylacetic acid (**2**), are given in Fig. 4.1 In each case, the prostereogenic (also prochiral) C2 center is converted to a stereogenic (also chiral) center by replacement of one *homomorphic* ligand (here a hydrogen atom) by a ligand other than those present in the molecule. The term *homomorphic* originates from Greek *homos* meaning "same" and *morphe* meaning "form" [3, 4]. Homomorphic ligands are identical only when separated from the rest of the molecule. Replacement of the other homomorphic ligand by the same new ligand will lead to the corresponding enantiomer. In such cases, the topicity (Greek *topos* meaning "place" or "physical surrounding") of the two *homomorphic* ligands (here H's at C2), in other words, their special relation with the rest of the molecule is different; they reside in enantiomeric environments. Such ligands are called *stereoheterotopic*, in this case more specifically *enantiotopic*.

S. K. Talapatra and B. Talapatra, *Basic Concepts in Organic Stereochemistry*,
https://doi.org/10.1007/978-3-030-95990-6_4

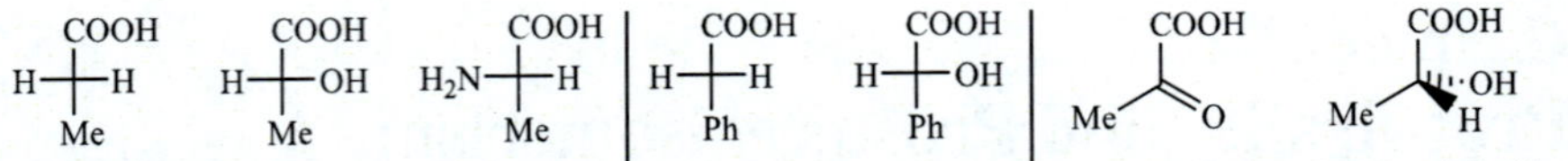

Propionic acid (**1**) (prochiral) | R-**D**-Lactic acid (chiral) | S-**L**-Alanine (chiral) | Phenyl-acetic acid (**2**) (prochiral) | R-**D**-Mandelic acid (chiral) | Pyruvic acid (**3**) (prochiral) | R-**D**-Lactic Acid (R-**D**-2-Hydroxy-propionic acid)

Fig. 4.1 Prochiral (prostereogenic) [(**1**), (**2**), and (**3**)] and chiral (stereogenic) molecules

By the same token, the two faces (front and rear of the molecular plane) of pyruvic acid (**3**) (Fig. 4.1) are enantiotopic, since the two faces are in enantiomeric environments; thus, addition of hydride (or any nucleophile) to the two faces in turn gives rise to enantiomers, viz., (*R*)- and (*S*)-2-hydroxypropionic acid (*R*- and *S*-lactic acid).

Molecules having stereoheterotopic ligands or faces exhibit *prostereoisomerism* (if they can be reacted upon). Prostereoisomerism is attributed to such a center carrying stereoheterotopic ligands, for example, to C2 of *prostereogenic* and *prochiral* compounds (**1**) and (**2**), or to a center carrying stereoheterotopic faces, *e.g.*, pyruvic acid (**3**).

The principle of stereoselective synthesis (Sect. 5.1) is based on the different behavior of stereoheterotopic ligands and faces in the chemical reactions. The concept of stereoheterotopicity has been discussed in 1967 [2], that of prochirality in 1966 [1], and the topic has been reviewed in detail by Eliel in 1982 [5].

Two criteria, viz., (i) substitution or derivatization (in case of ligands) or addition (in case of faces) and (ii) a symmetry criterion, are used to determine the topic relationship of homomorphic ligands and faces, as illustrated by several examples of each of homotopic ligands (Fig. 4.2), homotopic faces (Fig. 4.3), enantiotopic ligands (Fig. 4.4), enantotopic faces (Fig. 4.5), diastereotopic ligands (Fig. 4.7), and diastereotopic faces (Fig. 4.8). Any of the two criteria is sufficient to determine the topicity in each case.

4.2 Homotopic Ligands

The first example of homotopic ligands is of Za_2b_2 type, belonging to the $\mathbf{C_{2V}}$ point group (Fig. 2.61). The homotopic ligands are recognized by either of the two criteria already stated. Two examples of homotopic ligands are illustrated, recognized by both substitution and symmetry criteria, in this figure. Four more examples of homotopic ligands of simple molecules like chloromethane, ethylene, allene, and acetic acid are also included indicating the homotopicity of the homomorphic ligands by symmetry criteria.

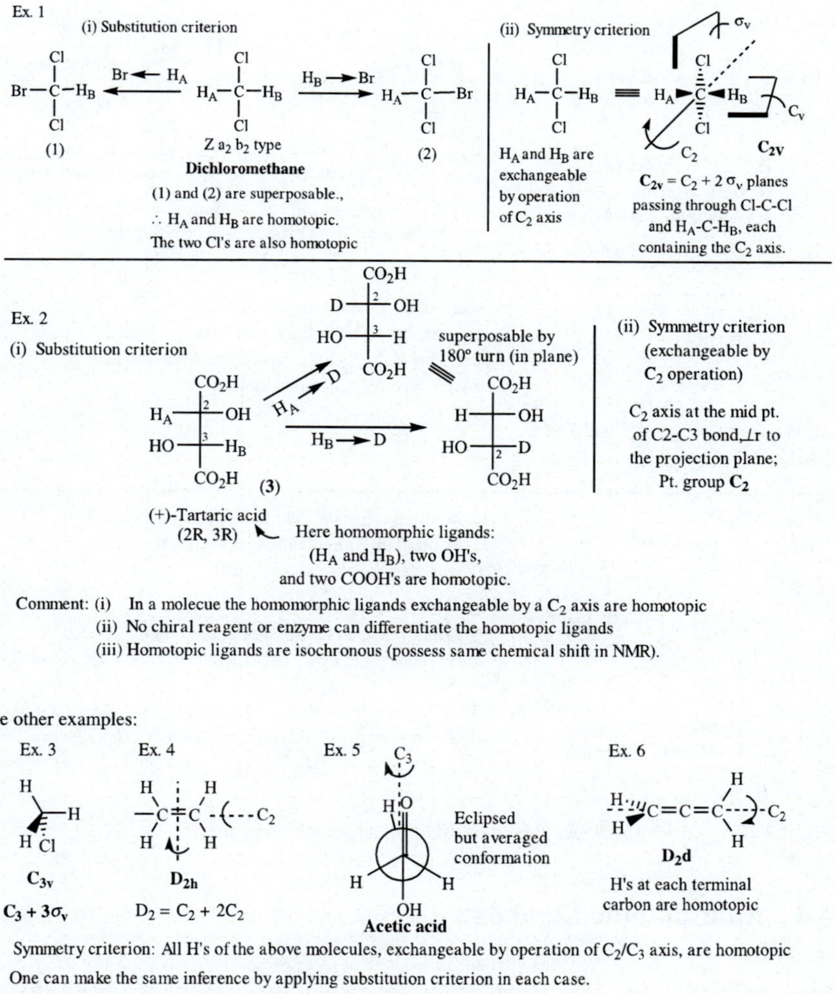

Fig. 4.2 Homotopic ligands recognized by substitution and symmetry criteria

4.3 Homotopic Faces

Homotopic faces are recognized by the two criteria as illustrated in Fig. 4.4.with the help of four examples.

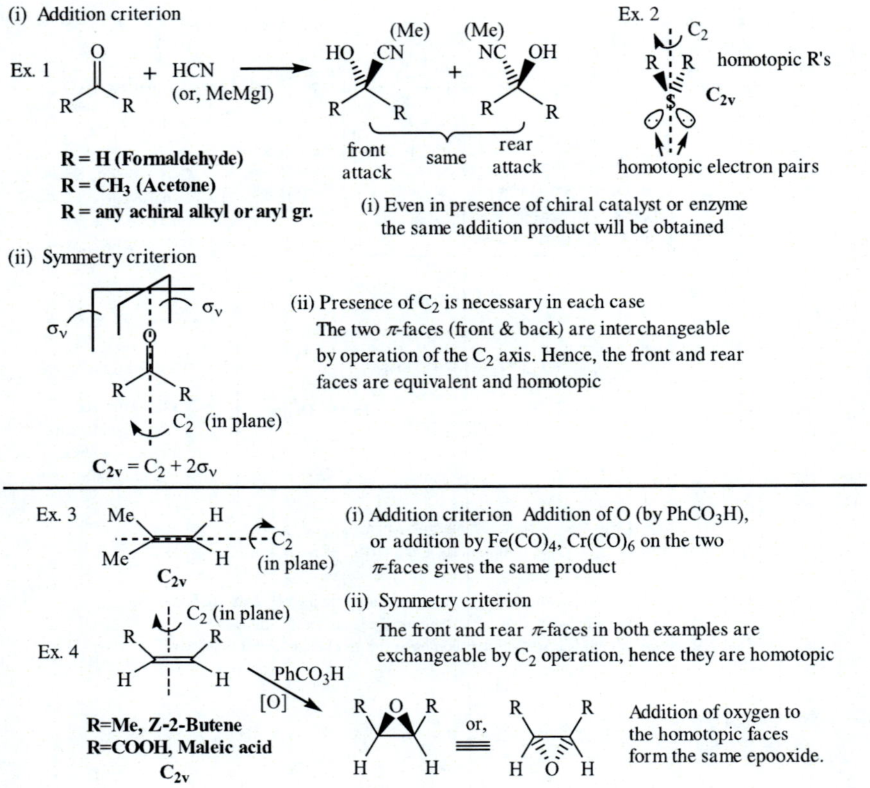

Fig. 4.3 Homotopic faces recognized by addition and symmetry criteria

4.4 Enantiotopic Ligands

Enantiotopic ligands are recognized by any one of the two criteria as illustrated in Fig. 4.4 with six examples.

4.5 Nomenclature of Geminal Enantiotopic Ligands. Pro-R and Pro-S

In compounds **1–3** of Fig. 2.63, of the two homomorphic ligands H_AH_B or H_CH_D (attached to the same carbon) if replacement of any one, say H_A, by an atom of higher priority, but of lower priority than the other two ligands (in such case by D), gives rise to (*S*) configuration of the new chiral molecule, H_A is then termed as ***pro-S*** and is expressed as H_S. H_B in that case, by default, will be termed as ***pro-R*** and is expressed as H_R. H_A and H_B may be replaced by D to give the (*S*) and (*R*)

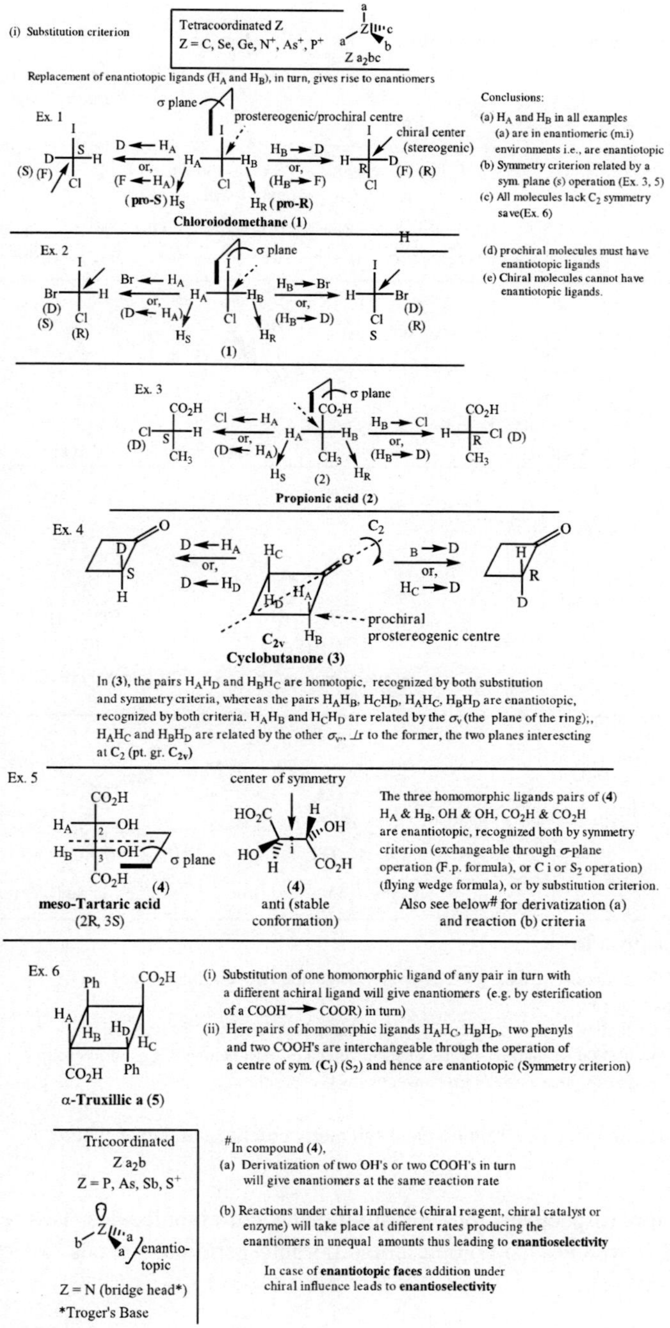

Fig. 4.4 Enantiotopic ligands recognized by substitution and symmetry criteria

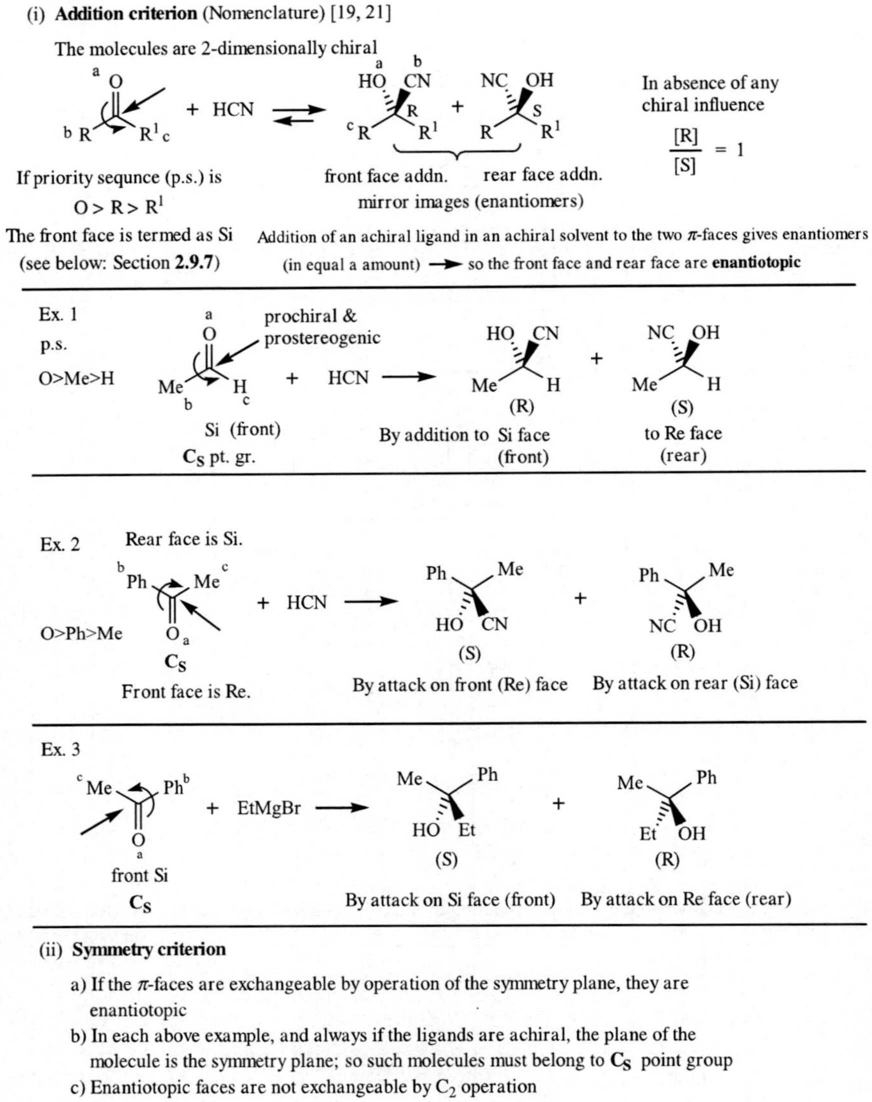

Fig. 4.5 Enantiotopic faces. Addition and symmetry criteria, and nomenclature

configuration, respectively, of the resulting new chiral molecules. This terminology is applicable to any enantiotopic homomorphic germinal ligands (attached to the same carbon) [1].

4.6 Enantiotopic Faces

Enantiotopic faces are recognized by any one of the two criteria as illustrated in Fig. 4.5 with three examples.

4.7 Nomenclature of Enantiotopic Faces

The π-faces of a trigonal atom (olefinic/carbonyl carbon) having three different ligands will be enantiotopic and are two-dimensionally chiral (Sect. 3.9).

If one looks at the plane of a π-face and finds that the CIP priority sequence of the three dissimilar ligands is clockwise, the face is called *Re*. If the sequence is anticlockwise, the face is called *Si*, these being the first two letters of ***Rectus*** and ***Sinister***, respectively [1]. Examples 1–3 are shown in Fig. 4.5. If one face is *Re*, the other face will be *Si*.

In case of a double bond containing two specifiable trigonal atoms, each face may be uniquely defined by two symbols, one for each specifiable atom, as illustrated by the compounds in Fig. 4.6. The π-faces of an oxime are also specified in this Figure. The α- and β-faces of each olefin excepting maleic acid being exchangeable only by a σ-plane (*xy* or paper plane) operation and not by any C_2 operation, each olefin

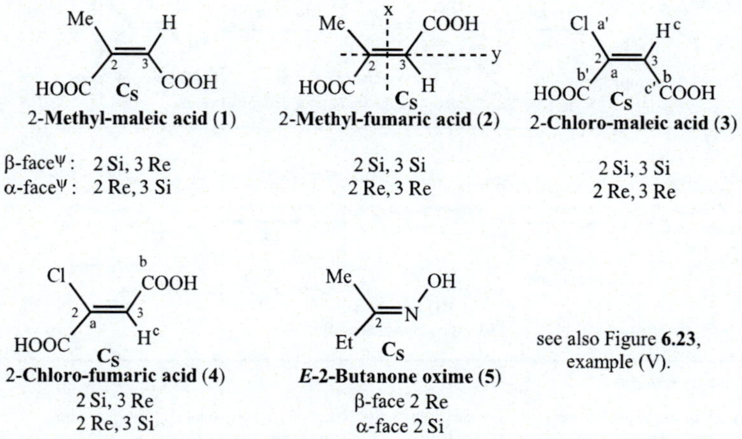

Notes: $^{\Psi}\beta$- and α-faces specify front and rear faces respectively when the structures are written in the xy (paper) plane, as shown. After rotation around in-plane x-axis (vertical) or, around in-plane y-axis (horizontal) the β face becomes α and vice-versa but (Re, Si) specification of a specified face remains unaltered.

Fig. 4.6 Nomenclature of enantiotopic π-faces of some unsymmetrically substituted olefins produces enantiomeric epoxides or complexes with the double bond in equal amount in absence of any chiral influence (see also Fig. 4.14).

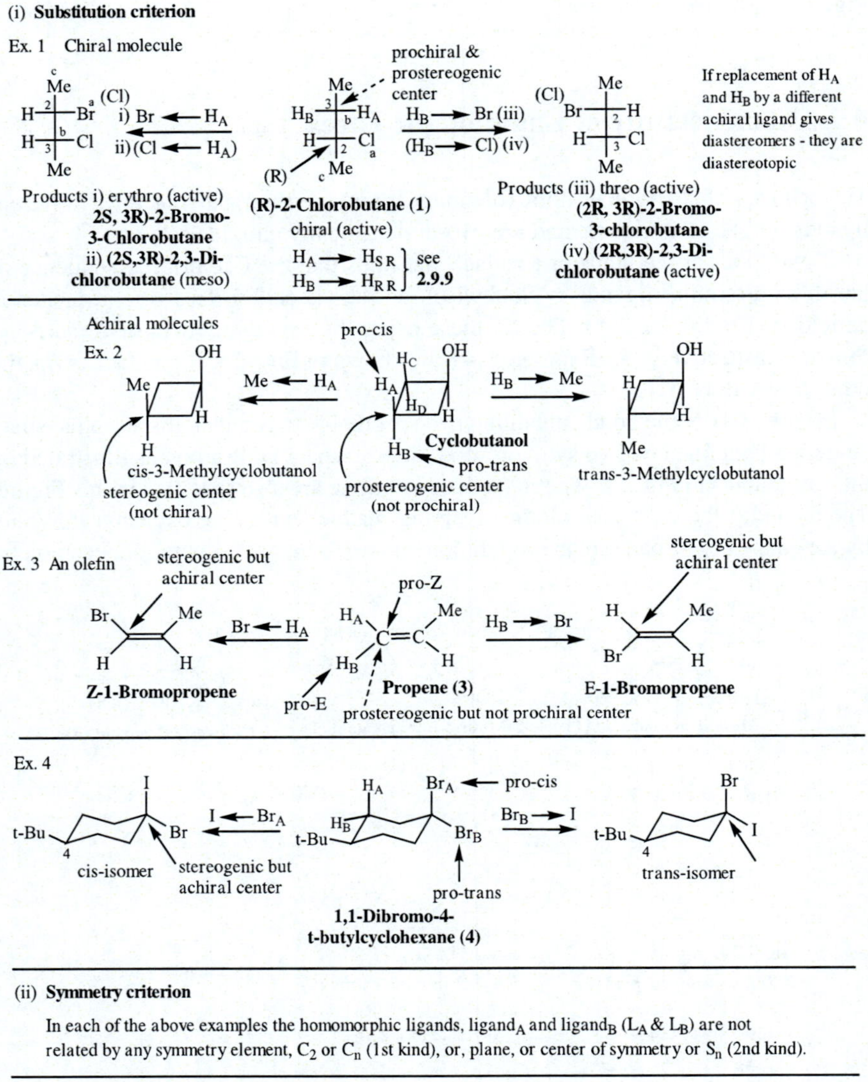

Fig. 4.7 Diastereotopic ligands. Nomenclature

i) **Addition criterion** Addition to diastereotopic π-faces gives diastereomers

Ex. 1

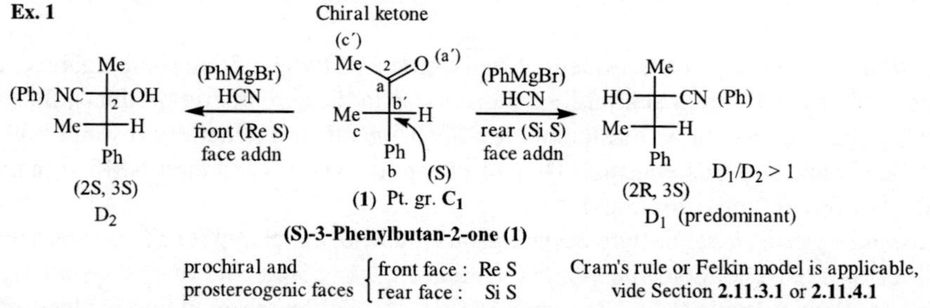

Ex. 2

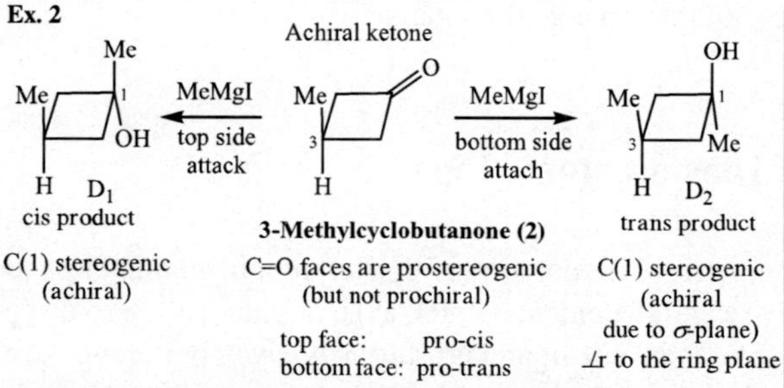

Ex. 3

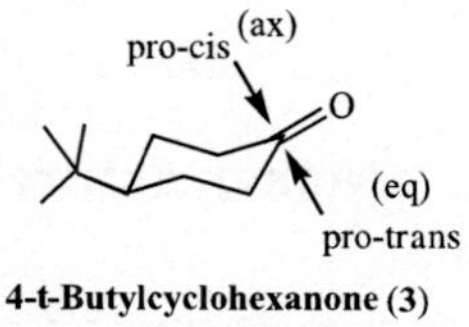

Ex. 4

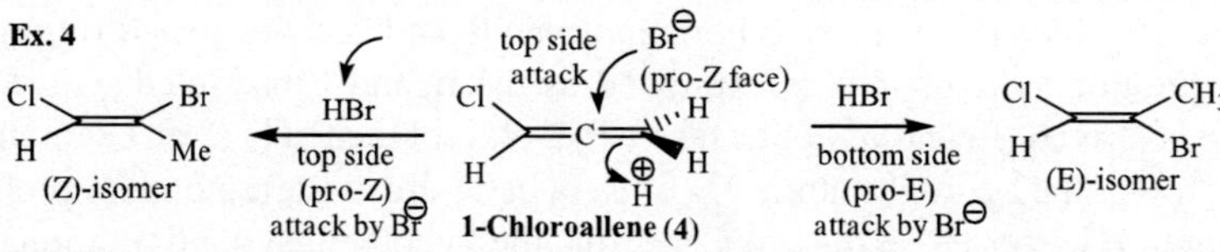

Ex. 5

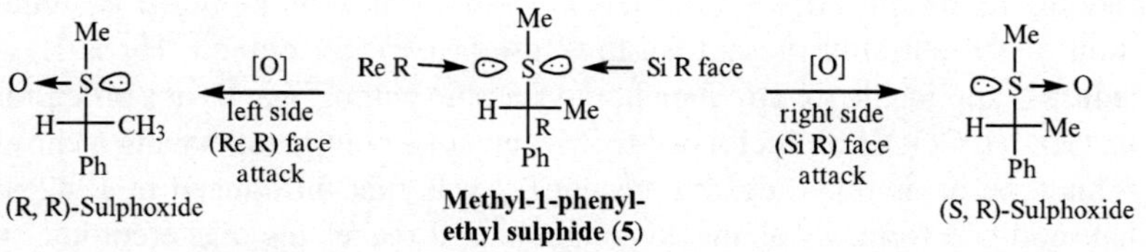

The two diastereomeric sulphoxides are formed in different proportions

(ii) **Symmetry criterion**: The two π-faces of compounds (**1**) to (**4**) are not exchangeable by operation of any C_2 or σ plane or S_n axis, so they are diastereotopic. Same is the case with the two lone pairs of (**5**), which are also diastereotopic.

In each case, diastereoselectivity is observed because of the involvement of diastereomeric transition states.

Fig. 4.8 Diastereotopic faces. Nomenclature

4.8 Diastereotopic Ligands

Substitution criterion. If the sequential replacement of two homomorphic ligands in a molecule by a different achiral ligand gives rise to diastereomeric products, those two ligands are said to be diastereotopic. Such ligands are generally distinct both chemically and spectroscopically. They react at unequal rates and their NMR signals will be different (anisochronous).

Symmetry criterion. Diastereotopic ligands must not be related (exchangeable) by a C_n or S_n axis. The molecule (**1**), depicted in Fig. 4.7, is devoid of any symmetry element; the molecules (**2**), (**3**), and (**4**) (Fig. 4.7) although contains a plane of symmetry, its operation does not interchange the ligands.

4.9 Nomenclature of Diastereotopic Ligands

A *chiral* molecule containing a pair of homomorphic diastereotopic geminal ligands at a prochiral center must have a chiral element (center, axis, or plane) also. To designate the diastereotopic pair of ligands, one ligand is arbitrarily given preference over the other to treat the prochiral center as a chiral one. Now applying CIP rules if the center becomes (*R*), then that particular homomorphic ligand is called ***pro-R*** (hypothetical configuration symbol) and is expressed as the ligand subscripted with *R*, *i.e.*, the ligand is termed $(\text{ligand})_R$. Let us take (*R*)-2-chlorobutane (**1**) as an example (Fig. 4.7, first row). The hydrogen atoms H_A and H_B at C3 are diastereotopic. If preference is given to H_A over H_B in the sequence rule, the hypothetical configurational symbol for C3 of (**1**) would be *S*. The configurational symbol of H_A thus becomes H_{SR} after adding to the subscript *S* a further subscript *R*, the symbol of the absolute configuration of the chirality present (in this case at C2). By default, H_B is H_{RR}. This procedure is applicable for any such homomorphic pair of ligands at a prochiral center, *e.g.*, COOH, COOH or Ph, Ph or Me, Me, etc. of a compound having a chiral element. This type of double indexing system [7] was first introduced in 1982 in a somewhat modified form. It has the advantage that it makes the diastereotopicity immediately obvious.

Ex. 2, Ex. 3, and Ex. 4 of Fig. 4.7 illustrate the nomenclature of different types of *achiral* molecules [*viz.*, cyclic compounds or olefins or cumulenes with odd number of double bonds (having successive planes of π bonds orthogonal to each other)] having diastereotopic homomorphic geminal ligands. The substitution products shown are also achiral since the original plane of symmetry is retained.

The substitution products of homomorphic geminal diastereotopic ligands may become chiral if the original symmetry plane is destroyed, *e.g.*, H_C and H_D at C2 of (**2**) and H_A and H_B at C3 of (**4**) in Fig. 4.7. Thus, C2 of (**2**) and C3 of (**4**) are prostereogenic as well as prochiral centers.

4.10 Diastereotopic Faces. Nomenclature

The diastereotopic faces are designated as follows. First, the face is designated as *Re* or *Si* as stated earlier in Sect. 4.7. To this symbol, the specified absolute configuration of the chiral element present in the molecule is added. Thus, in case of a chiral compound (**1**) (Fig. 4.8), the front face becomes *ReS* by adding *S*, the absolute configuration of the chiral center present, to the *Re* face. Thus, the diastereotopic rear face is designated *SiS*.

Diastereotopic nomenclature in cases of compounds with diastereotopic faces has been illustrated in Fig. 4.8 with examples of an achiral ketone 3-Methylcyclobutanone (**2**), achiral allene with diastereotopic faces 1-chloroallene (**3**), and a chiral sulfide (**4**). In each case, the nomenclature of the diastereotopic faces is self-explanatory (*cf*. Ex. 1).

Diastereotopic faces are recognized by application of any one of the usual two criteria as illustrated in Fig. 4.8.

4.11 Interesting Examples of Topicities of Homomorphic Ligands [8]

In Fig. 4.9, homotopicity, enantiotopicity, and the diastereotopicity of the CH_2OH groups in the *cases 1*, *2*, and *3*, respectively, are illustrated by selective oxidations as well as by symmetry criteria.

In *case 1*, like the CH_2OH groups 2-OH and 5-OH, as well as 3-OH and 4-OH, exchangeable by C_2 operation, are homotopic, and hence the homotopic OHs when derivatized would lead to the same derivative, and the corresponding acetates will also be homotopic.

In *case 2*, like the CH_2OH groups 2-OH and 5-OH, as well as 3-OH and 4-OH are enantiotopic, and hence, the corresponding acetates of the enantiotopic OH groups will lead to enantiomers.

In *case 3*, like the CH_2OH groups, any pair of OH groups are diastereotopic, and hence, the corresponding acetates of any pair of diastereotopic OH groups will also be diastereomers.

4.12 Interrelation of Topicity of Ligands with Isomerism

Topicity of ligands is interrelated with isomerism in general. A classification diagram for topicity is drawn (Fig. 4.10), which may be compared with that drawn for isomerism (Fig. 1.24).

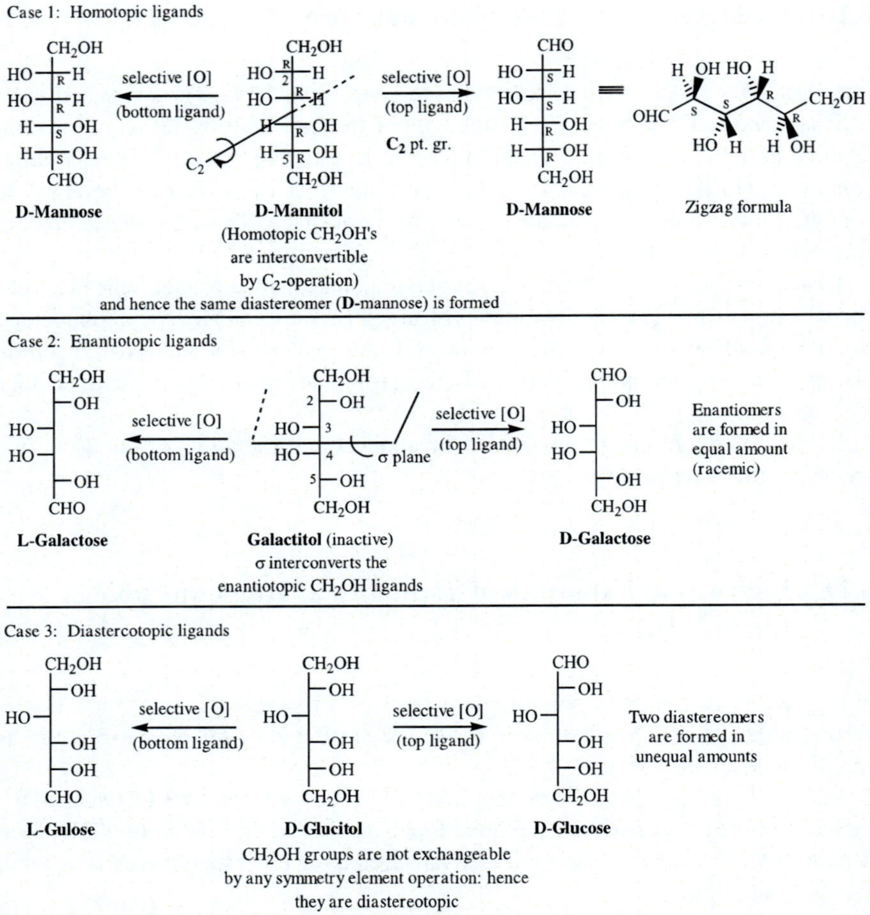

Fig. 4.9 Determination of topicity of CH_2OH groups of a few hexoses by selective oxidation

4.13 Molecules with Prostereogenic but Proachirotopic Center and Multiprochiral Centers

Some interesting examples of topicity and nomenclature are delineated in Fig. 4.11.

4.14 Topic Relationship of Ligands and Faces

Topic relationship of ligands and faces, so far discussed, are summarized in Fig. 4.12. This figure has much in common with the table given by Mislow and Raban in 1967 [2] and by Eliel in 1980 [9].

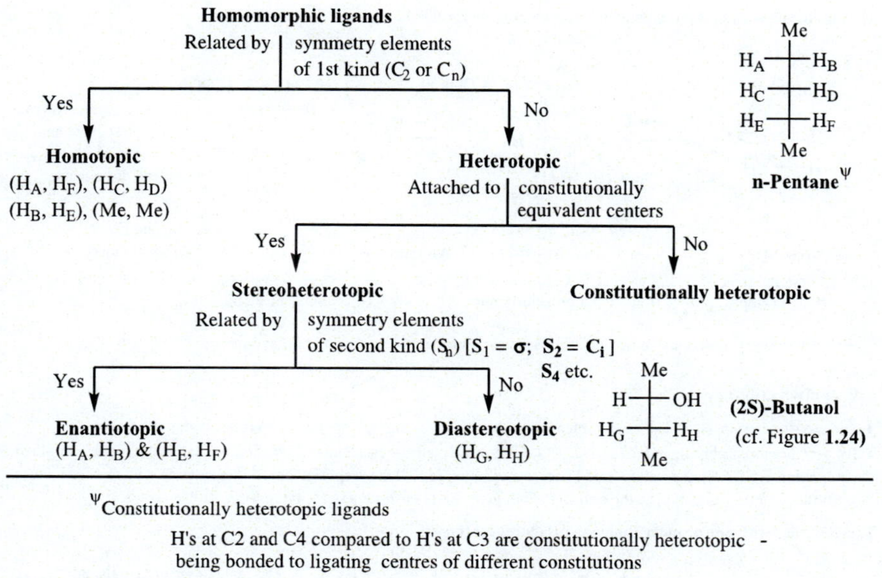

Fig. 4.10 Classification of homomorphic ligands based on topicity

4.15 Stereoheterotopic Ligands and NMR Spectroscopy

NMR (like IR and UV) is an achiral probe. Nuclei that reside in different environments can be distinguished by NMR spectroscopy showing different chemical shifts. Such nuclei exhibiting nonequivalent chemical shifts are called *anisochronous*; they split. Nuclei showing same chemical shift are called isochronous, which do not split. The *isochrony* or *anisochrony* of nuclei is dependent on their topicity. The relationship between topicity, isochrony/anisochrony, and signal multiplicity are discussed in the form of a chart in Fig. 4.13. Anisochrony is often very small (1 ppm or less); a case, when it is undetectable, is called *accidental isochrony*. Use of higher field or different solvents in ^{1}H NMR or ^{13}C NMR may be helpful in increasing anisochrony.

A very instructive example is illustrated in Fig. 4.14.

4.15.1 Anisochrony Arising Out of Diastereotopic Faces

Isochrony arising out of the homotopic faces of the maleic acid (−)-menthyl ester (**1**) and anisochrony arising out of the diastereotopic faces of fumaric acid (−)-menthyl ester (**2**) are much instructive [10] and are illustrated in Fig. 4.14.

The two olefinic protons H_A and H_B of the maleate (**1**), being exchangeable by the C_2 axis (in the molecular plane) operation, are *homotopic* and hence are *isochronous* and appear as a 2H singlet. Again, its front and rear π-faces are also interchangeable

1) **A molecule with prostereogenic but proachirotopic centers (1)**

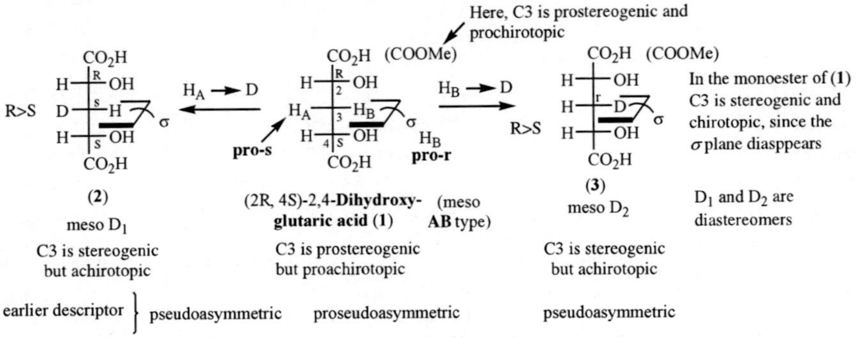

Substitution criteria

- Symmetry criterion: H_A & H_B are not interchangeable by any symmetry operation, so they are diastereotopic
- Pairs of homomorphic ligands (H, H & HO, OH) at C2 and C4 in (**1**) are, however, exchangeable with a σ-plane (containing H_A-C3-H_B) operation, and hence are enantiotopic. Hence replacement of these two H's by D or any other substituent, or derivatization of the two OH's will give rise to enantiomers.

2) Molecules with one or more than one prochiral centers

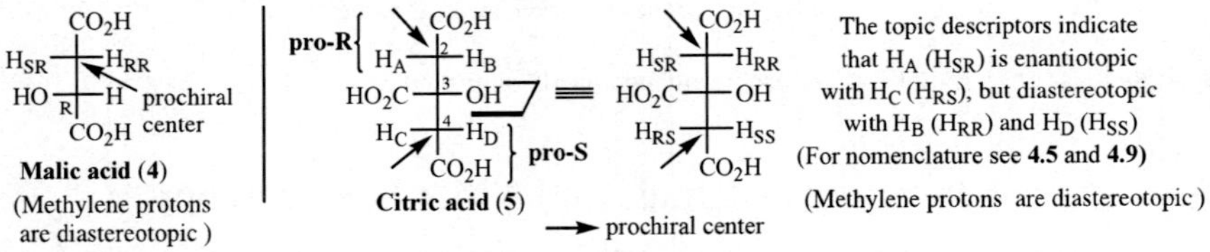

Note: • Citric acid contains three prochiral centres, C2, C3, C4, and both enantiotopic and diastereotopic H's H_A is **pro-S,** H_B **(pro-R)**, H_C **(pro-S)** and H_D **(pro-R)**.

• C3 is a prochiral center since it carries two homomorphic group ligands – CH_2COOH, to which topic descriptors, **pro-R** and **pro-S** may be assigned, as usual (cf. **2.9.5**).

Fig. 4.11 Topicity nomenclature of methylene protons in two interesting examples

Topic Relationship of Ligands and faces

Topicity	Substition/addition/ reaction criterion (by achiral groups)	Symmetry criterion	Behavioral difference
Homotopic	Identical product	**Ligands** related through C_n and **faces** by C_2 axis	No difference by any method
Enantiotopic	Enantiomeric products	**Ligands** related through σ, i or S_n **Faces** related by σ	Usually not distinguishable Distinguishable, in principle, in chiral media (NMR), by chiral reagents and enzyme
Diastereotopic	Diastereomeric products	Ligands and faces not related by any symmetry element	Distinguishable in principle by all physical and chemical methods

Fig. 4.12 Summary of topic relationship of ligands and faces

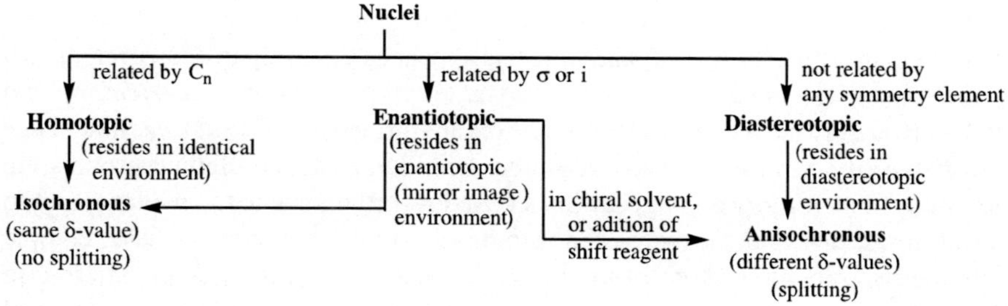

Fig. 4.13 Topicity. NMR chemical shift. Signal multiplicity

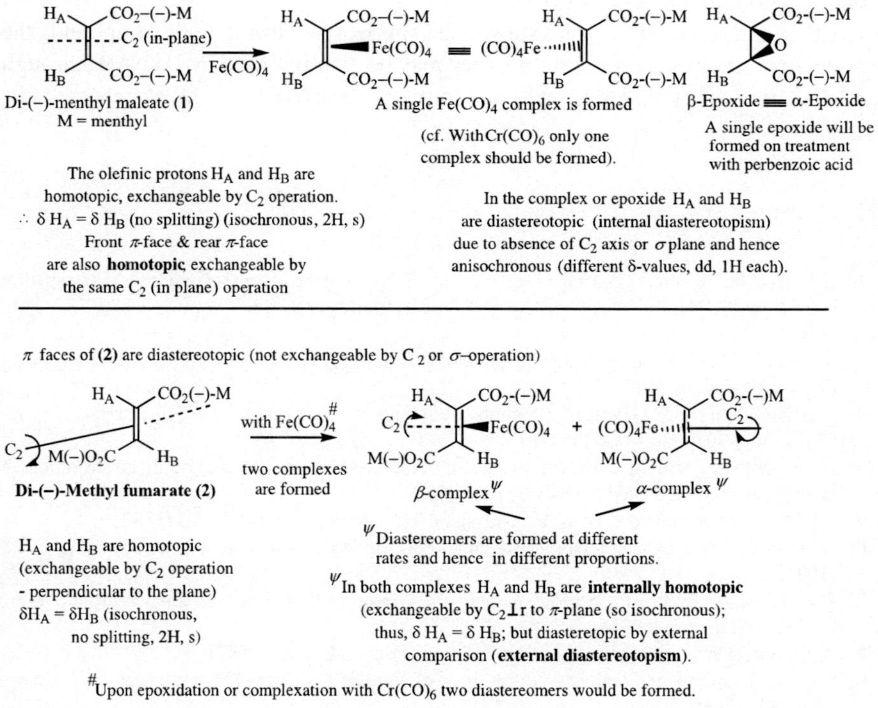

Fig. 4.14 Anisochrony arising out of diastereotopic faces. An instructive example

through the same C_2 axis operation and hence are homotopic. So the complexation by $Fe(CO)_4$ from the homotopic α- or β-face leads to the same complex (*cf.* epoxidation). But H_A and H_B of the $Fe(CO)_4$ complex being *diastereotopic* are *anisochronous* and have different chemical shift values. They split and appear as a double doublet (dd) (1H each) in the NMR. Thus, the maleate (**1**) should form, on treatment with perbenzoic acid, a single epoxide of which H_A and H_B being diastereotopic should appear as a double doublet in the NMR.

In case of the fumarate (**2**) also, H_A and H_B, being exchangeable by C_2 (orthogonal to the π-plane) operation, are *homotopic* and appear as a 2H singlet. However, its two π-faces being not exchangeable by any symmetry operation are *diastereotopic*. So with $Fe(CO)_4$, two diastereomeric complexes are formed *at different rates and hence in different proportions*. In each diastereomeric complex, the olefinic protons are *internally homotopic* (exchangeable by a C_2 axis—orthogonal to the π-plane) and so isochronous. However, the two olefinic protons in one diastereomer are diastereotopic with the two olefinic protons of the other diastereomer *by external comparison*, so they are termed ***externally diastereotopic***. Consequently, the two olefinic protons of the two diastereomeric complexes show two 2H singlet peaks of different intensities.

The two olefinic esters (**1**) and (**2**) thus may be distinguished by NMR through complexation with faces of different topicities.

Although each of (**1**) and (**2**) show a 2H singlet (because in both compounds the two olefinic protons are homotopic), they may be distinguished by ^{1}H NMR through complexation (or epoxidation) with homotopic or diastereotopic faces, respectively.

References

1. K.R. Hanson, Applications of the sequence rule. I. Naming the paired ligands g,g at a tetrahedral atom Xggij. II. Naming the two faces of a trigonal atom Yghi. J. Am. Chem. Soc. **88**, 2731–2742 (1966)
2. K. Mislow, M. Raban, in *Topics in Stereochemistry*, ed. by N.L. Allinger, E.L. Eliel, vol 1 (Wiley, New York, 1967)
3. A. Hirschmann, K.R. Hanson, Elements of stereoisomerism and prostereoisomerism. J. Org. Chem. **36**, 3293–3306 (1971)
4. A. Hirschmann, K.R. Hanson, Prochiral and pseudoasymmetric centers: implications of recent definitions. Tetrahedron **30**, 3649–3656 (1974)
5. E.L. Eliel, Prostereoisomerism (Prochirality) Top. Curr. Chem. **105**, 1–76 (1982)
6. K. Mislow, J. Siegel, Stereoisomerism and local chirality. J. Am. Chem. Soc. **106**, 3319–3328 (1984)
7. J. Retey, J.A. Robinson, *Stereospecificity in Organic Chemistry and Enzymology* (Verlag Chemie, Weinheim, 1982)
8. A. Ghosh, A quiz on stereochemistry. J. Chem. Educ. **64**, 1015 (1987)
9. E.L. Eliel, Stereochemical non-equivalence of ligands and faces (Heterotopicity). J. Chem. Educ. **57**, 52–55 (1980)
10. V. Schurig, Internal versus external diastereotopism in homochiral *cis* versus *trans* olefin metal π−complexes. Tetrahedron Lett. 3977–3980 (**1977**)

Chapter 5
Asymmetric Synthesis

Abstract This Chapter starts with a relevant introduction dealing with principles of enantioselection and diastereoselection and is followed by brief descriptions with examples of asymmetric synthesis—definition, stereoselective and stereospecific reactions, product/substrate stereoselectivity, regioselectivity, optical purity; Cram's rule, Cram's open chain, chelate or cyclic, and dipolar models; Felkin–Anh open chain and dipolar models; Prelog's rule—rationalization and anomalies; and sharpless enantioselective epoxidation, kinetic resolution of racemic allyl alcohols, and mechanism.

5.1 Introduction. Principles of Stereoselection: Enantioselection. Diastereoselection

5.1.1 Lack of Stereoselection

In the absence of chiral influence by way of reactants, reagents, or solvents, a reaction of a starting material having a prochiral center generating a chiral center will produce equal amounts of enantiomers, i.e., a racemic product, since such a reaction proceeds through enantiomeric transition states (TS) of equal energy. Thus, the activation energies being same the reactions take place at the same rate. The principles of the *lack of stereooselection* and of *enantioselection* are illustrated with the help of energy diagrams (Figs. 5.1 and 5.2, respectively).

5.1.2 Enantioselection

In 1950, Mosher and La Combe reported two papers [1, 2] on the first asymmetric Grignard reactions (Fig. 5.3) which are examples of enantioselection (Fig. 5.3). The existing chiral center (or centers) is said to bring about *asymmetric induction.* Thus, addition of a chiral nucleophile to the two enantiotopic faces of a prochiral carbonyl compound leads to enantioselectivity.

S. K. Talapatra and B. Talapatra, *Basic Concepts in Organic Stereochemistry*,
https://doi.org/10.1007/978-3-030-95990-6_5

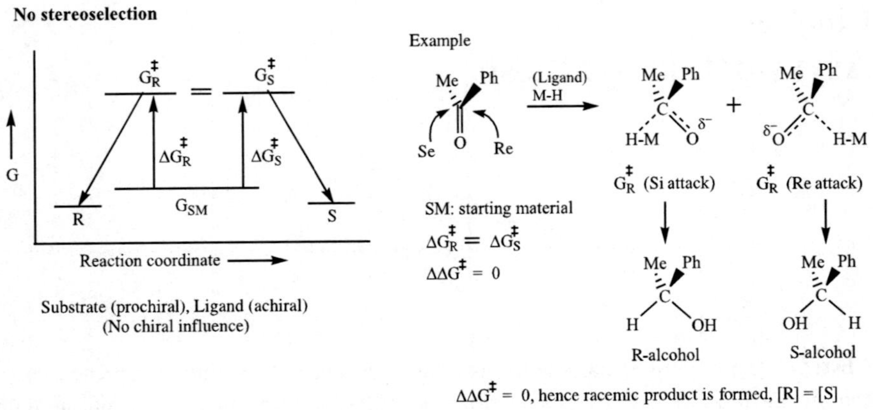

Fig. 5.1 Principle of absence of stereoselection. energetics

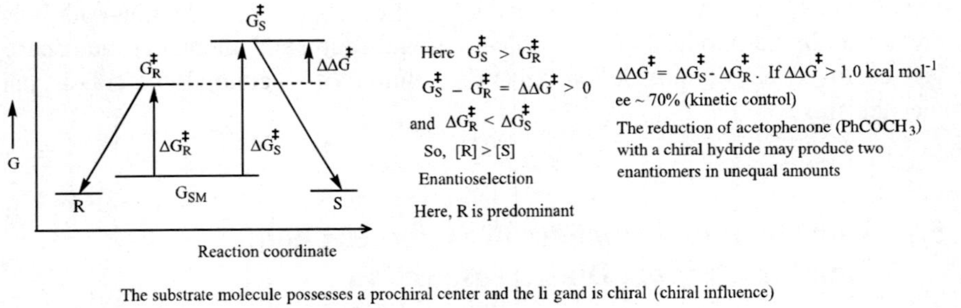

Fig. 5.2 Principle of enantioselection. energetics

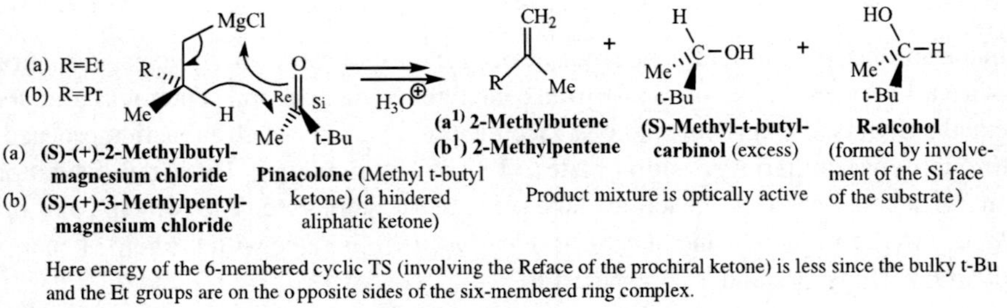

Fig. 5.3 Examples of first enantioselective grignard reactions

5.1.3 *Diastereoselection*

The concept of diastereoselection was first clearly outlined by Emil Fischer in 1894, based upon the conversion of one sugar to the next higher homologue via the cyanohydrin reaction. Thus, the substrate molecule should possess a prochiral center as well as at least one chiral center, and the reagent may be achiral, as in the reaction of a sugar with HCN. The principle of diastereoselection is illustrated with the help of an energy diagram (Fig. 5.4), which is self-explanatory. The expression for $\Delta G°$ in thermodynamic control and that for $\Delta\Delta G^{\neq}$ in kinetic control are shown. In case *the reaction is reversible*, the product which is formed at a slower rate but at the same time if it is of lower free energy (more stable) will be major, and the reaction is said to be of thermodynamic control. Like Fischer's cyanohydrin reaction, reactions to which Cram's rule and Prelog's rule are applicable are also examples of diastereoselection. Moreover, enantiomers have widely different reactivities especially in biological systems. The enantiomer may have different biological activities, and only a particular enantiomer may have the desired drug effect (Fig. 32.2 of Ref. [3]) It is well known that only L-amino acids can participate in protein synthesis. Thus, diastereoselection is a result of involvement of diastereomeric transition states (with different free energies); consequently, the diastereomer, which is formed irreversibly via less activation energy and hence at a faster rate, is the predominant product of kinetic control.

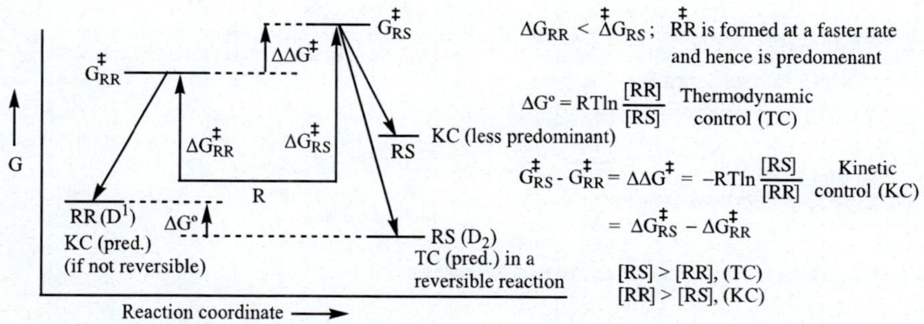

If ΔG or $\Delta\Delta G^{\ddagger} \geq 1.0$ kcal/mole, d.e. would be $\geq 70\% = \%D^1 - \%D^2$ where D^1 is the predominant diastereomer

Fig. 5.4 Principle of diastereoselection. Energetics

5.2 Asymmetric Synthesis. Definition. Stereoselective and Stereospecific Reactions. Product/Substrate Stereoselectivity. Regioselectivity

An *asymmetric synthesis* is a reaction in which a *prochiral unit* in a substrate molecule is converted by a reactant into a *chiral unit* in such a manner that the stereoisomeric products are produced in unequal amounts.

The prochiral unit in a substrate molecule must have enantiotopic or diastereotopic ligands or faces. In rare cases, only one stereoisomer is formed with the complete exclusion of the other stereoisomer, and the reaction is termed *100% stereoselective*. One stereoisomer may be optically inactive also (Ex. 1 of Fig. 5.5). This is an example of *product stereoselectivity*.

Example 2 provides another simple case of *product stereoselectivity* (Fig. 5.5).

When two diastereomers or two enantiomers react at different rates to give two stereoisomers (Fig. 5.6)—the reaction is termed *stereospecific*. Enzymes usually react with one enantiomer only and sometimes with both enantiomers in different ways (Fig. 32.2 of Ref. [3]) and thus show total substrate enantioselectivity.

Product stereoselectivity

Ex. 1

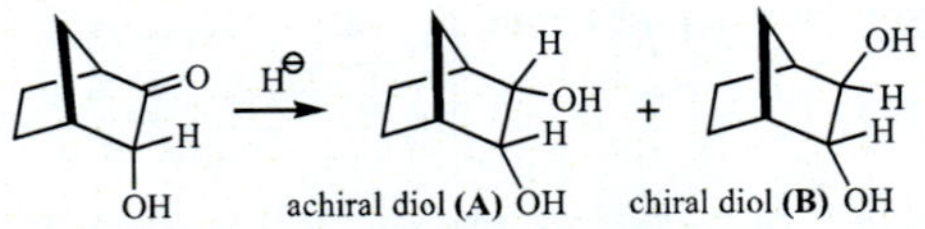

By use of an adequately bulky hydride reagent if the achiral diastereomeric diol (**A**) is formed exclusively, still it is an asymmetric synthesis, and the reaction is said to be 100% stereoselective as a result of 100% diastereoselection. This is an example of product stereoselectivity.

Ex. 2

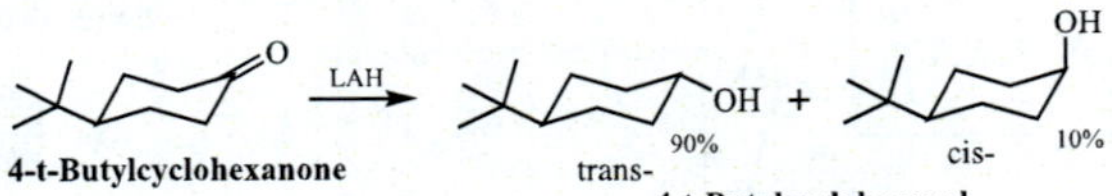

4-t-Butylcyclohexanone **4-t-Butylcyclohexanol**

Fig. 5.5 Stereoselective reactions. Product stereoselectivity

Substrate diastereoselectivity

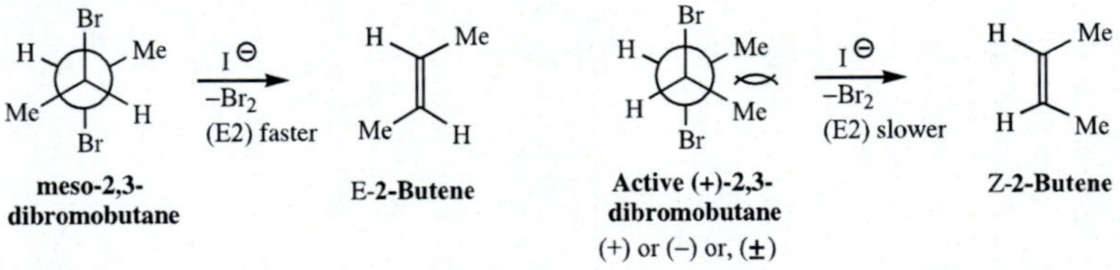

This is a case of substrate selectivity or, better substrate diastereoselectivity. In some particular case one diastereomer may not react at all under the same condition.

Fig. 5.6 Stereospecific reactions. Substrate selectivity

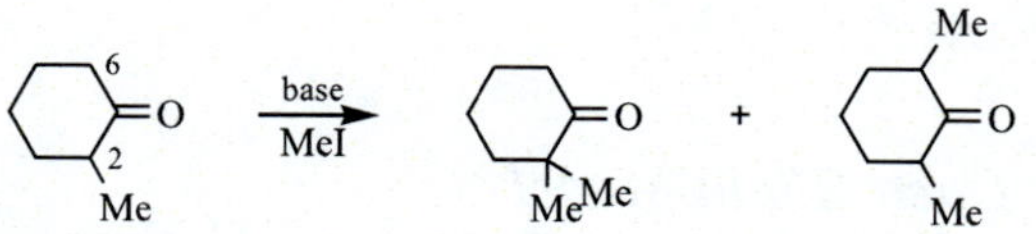

2-Methylcyclohexanone **2,2-Dimethyl-cyclohexanone** (mostly) **2,6-Dimethyl-cyclohexanone** (a little)

The generated C2 anion is more rea ctive than that generated at C6, because of the inductive effect ofthe methyl group.

Fig. 5.7 Example of regioselectivity

Regioselectivity. If a substrate is capable of reacting at more than one center (polydent molecules) and reacts at one center with a higher rate than the other centers, this is known as *regioselectivity* (Fig. 5.7).

5.2.1 Enantiomeric Excess. Diastereomeric Excess. Optical Purity

In a mixture of a pure enantiomer (*R* or *S*) and a racemate (*RS*), *enantiomeric excess* (*ee*) is the percent excess of the enantiomer over the racemate. Thus, *percent enantioselective excess* (*% ee*) or *percent diastereoselective excess* (*% de*) of *R* is illustrated by the following expression:

$$\%\ \text{ee or}\ \%\ \text{de of R} = \frac{[R]-[S]}{[R]+[S]} \times 100 = \%\ \text{R} - \%\ \text{S} = \%\ \text{stereoselectivity of R},$$

where R and *S represent* two enantiomers or configurations of the chiral center created in the two diastereomers.

Enantiomeric excess usually corresponds to the older expression, ***optical purity***, *op*, $([\alpha]_{obs.}/[\alpha]_{max}) \times 100\%$, *i.e.*, the absolute value of the ratio of the observed specific rotation of a sample made up of two enantiomers, which is otherwise chemically pure, to the corresponding specific rotation of one pure enantiomer, expressed as percentage. The term $[\alpha]_{max}$ signifies the specific rotation of one enantiomerically pure sample. Currently, *ee* or *de* is usually measured by NMR and by chromatographic analysis, and hence, the term optical purity is gradually becoming outmoded.

1,2-Addition of Achiral Nucleophiles to Chiral Ketones and Aldehydes (Diastereoselection) (Sects. 5.3–5.5).

5.3 Cram's Rule

5.3.1 *Cram's Open Chain Model*

A carbonyl group attached to a chiral center (*e.g.*, RCOCLMS) (L, M, and S represent large, medium sized, and small ligands, respectively) undergoes nucleophilic addition with metal hydride or achiral organometallic reagents (*e.g.*, R^1MgX or R^1M) to produce two diastereomeric products of which one predominates. The relative configuration of the asymmetric center, created in the predominant diastereomer, and its absolute configuration—if the configuration of the chiral carbon already present in the substrate is known—is predicted by *Cram's rule* based on some arbitrary TS models [4–7]. The rule is essentially empirical but has excellent predictive value.

In the open chain model illustrated in Fig. 5.8, the substrate molecule is complexed with the reagent from the right side of the carbonyl π-plane to give (**A**), from which R^1 is transferred to the trigonal carbon from the right diastereotopic face (from the side of the small ligand S) (route **a**), in preference to the complex ($\mathbf{A^1}$) which transfers R^1 from the left diastereotopic face (route b). Thus, 1,2-asymmetric induction takes place by the chiral center already present. Reagents R^1M: EtLi, MeLi R^1Li, etc.; $LiAlH_4$, $NaBH_4$; R^1MgX: R^1 = Me, Et, Ph, iPr, $iPrCH_2$, etc., X = Br, I.

One other possible open chain TS model (**B**) (Fig. 5.8) with carbonyl group staggered between L and M and the nucleophile a attacking the carbonyl carbon from the side of M (smaller than L) correctly predicts the stereochemical course to give the predominant product, but it involves stronger steric interactions of the bulkiest complexed O (of CO) with gauche L and M groups making the TS energy much higher and hence is not tenable. The third possible TS model (**C**) (Fig. 5.8) with CO staggered between S and L involving the preferential nucleophilic attack

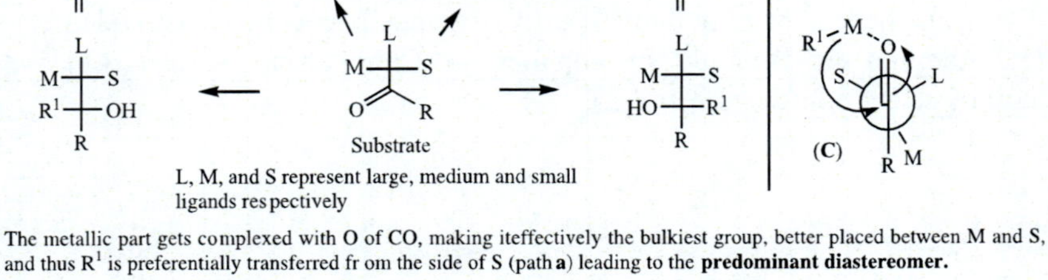

Fig. 5.8 Cram's open chain model

on the left diastereotopic face (the side of S) leads to the less predominant product and hence is eliminated.

Cram's rule is equally applicable to racemic substrate when the products are also racemic.

Limitations of Cram's Open Chain Model

(i) It applies only to kinetically controlled reactions; for Meerwein–Pondorf–Varley (MPV) reaction Cram's model may hold good for a short reaction time.
(ii) It does not apply to catalytic reduction. These two limitations hold good for other models also, to be discussed later.
(iii) It does not apply when the small ligand S is OH, OR, NH_2—capable of complexing with the reagent R^1M or R^1MgX. In such cases cyclic model will be applicable.

Some examples of the applications of Cram's rule are shown in Fig. 5.9.

Stereoselectivity through the open chain nonrigid model (Fig. 5.9) is not usually high except for a case in which the difference of bulkiness of the ligands M and S is high.

5.3.2 Cram's Chelate or Cyclic Model

If the substrate chiral ketone contains at the α-position an OH, NH_2, or OMe group which is capable of coordinating with the reagent, Cram's rule based on a rigid *chelate or cyclic model* predicts the stereochemistry of the predominant product, as delineated in Fig. 5.10. The metallic part of the reagent is doubly coordinated, as shown, to form a 5-membered ring. The nucleophile preferentially approaches the electrophilic carbon from the side of the ligand S. If the chelating group is M (usual case), the cyclic model predicts the same stereochemistry as the open chain model (case 1). If the chelating group is the ligand S or L, the cyclic model predicts the correct stereochemistry of the predominant product, while the open chain model predicts the opposite stereochemistry (case 2). For such substrates, the chelate model should always be applied, irrespective of the bulkiness of the chelating ligand.

5.3.3 Cram's Dipolar Model

Cram suggested a *dipolar model* for prediction of stereochemistry when the substrate carries a strongly electronegative group, *e.g.*, a halogen atom at C_α position [5, 7]. To minimize the dipole repulsion and to increase the electrophilic character of the carbonyl carbon, the $C = O$ bond and the C–X bond (having opposite dipoles) are placed anti in the dipolar model (Fig. 5.11). The nucleophile adds from the side of the

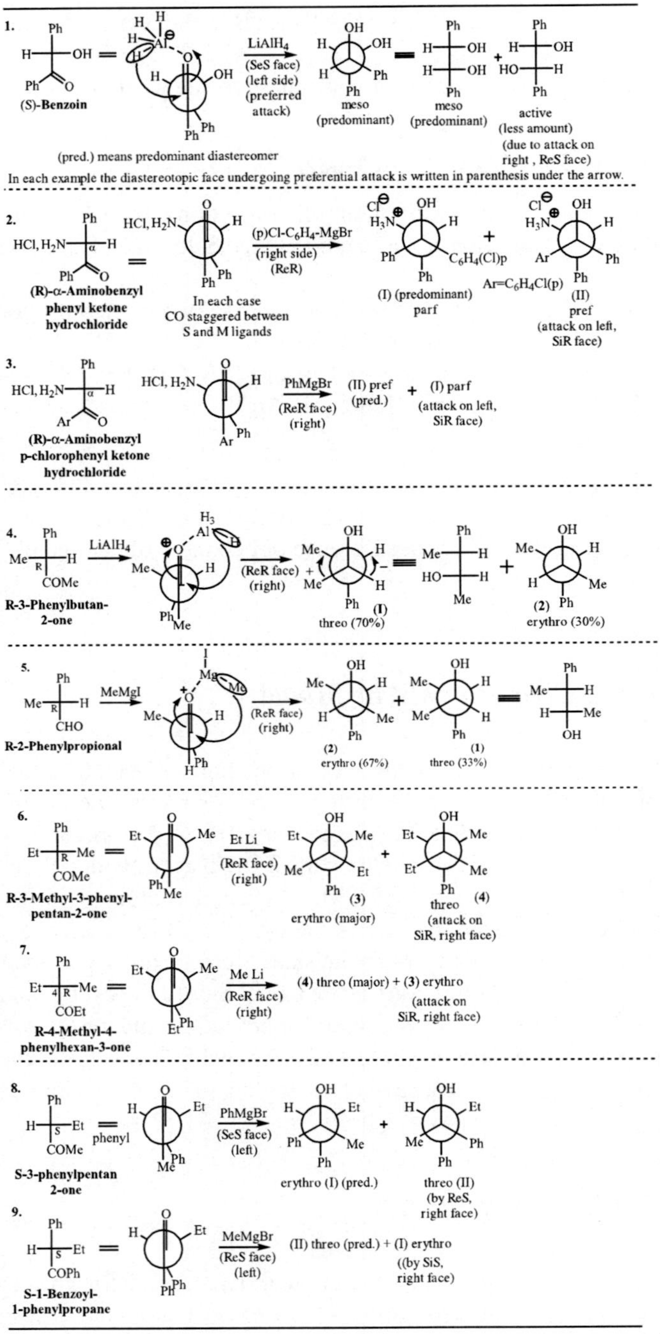

Fig. 5.9 Cram's rule. Open chain model, nine examples

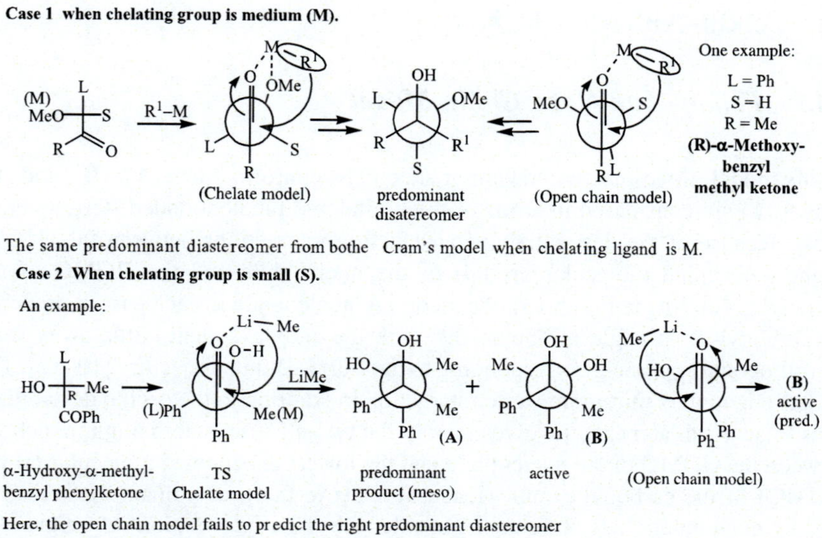

Fig. 5.10 Cram's chelate (cyclic) model. Examples

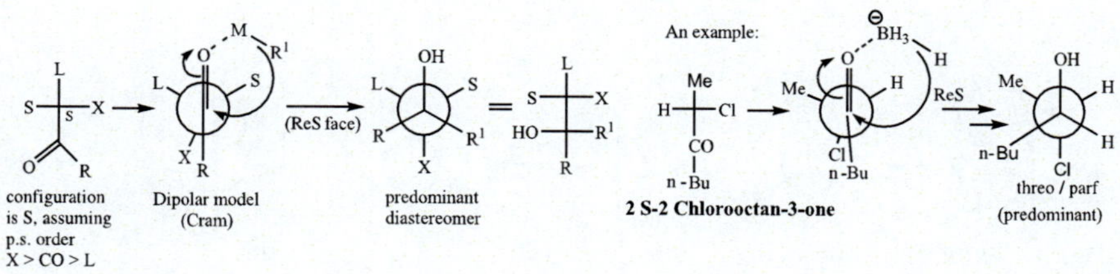

Fig. 5.11 Cram's dipolar model. An example

small ligand S, giving the predominant product, as shown. Cram's model correctly predicts the stereochemical course of the reactions but does not always succeed to give a quantitative assessment of the asymmetric induction based on steric interactions. A few alternative models have been suggested [6] of which the Felkin–Anh model [8, 9] has been widely accepted.

5.4 Felkin–Anh Models [8, 9]

5.4.1 Felkin–Anh Open Chain Model

In this model, three reactive transition state (TS) conformations (**A**), (**B**), and (**C**) (Fig. 5.12) are considered in terms of the orbital and the nonbonded steric interactions. Because of the destabilizing four-electron interactions between the HOMO (highest occupied molecular orbital) of the nucleophile and the HOMO of the carbonyl group [(**a**) in Fig. 5.13], the incoming nucleophile in each case approaches the carbonyl at an angle of about 109° with the carbonyl plane, little away from the orthogonal approach, corresponding to the Bürgi–Dunitz trajectory [10–12]. The TS (**A**) encounters minimum steric interaction in addition to the orbital destabilizations (discussed later) and involves mainly the two-electron stabilizing interactions between the HOMO of the nucleophile and the lowest unoccupied molecular orbital (LUMO) of the carbonyl group. Hence, it leads to the predominant diastereomer (**D**^{1}) by nucleophilic attack on the right side. The TS conformation (**C**) might also lead to the same predominant diastereomer (**D**^{1}), but its contribution should be *very little* because of its high TS energy due to nonbonded interactions R ↔ L and L ↔ R^1 (approaching nucleophile). The TS conformation (**B**) involves the nonbonded

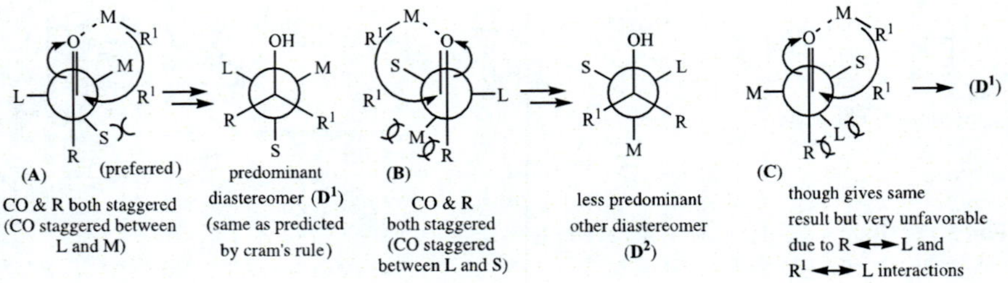

Fig. 5.12 The Felkin–Anh model for the nucleophilic addition to the carbonyl group

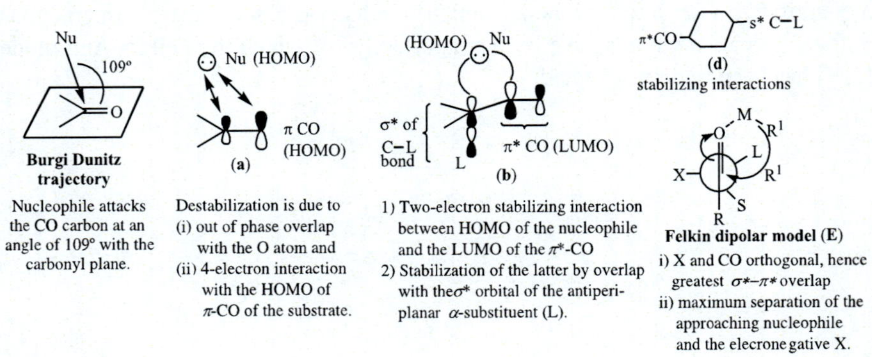

Fig. 5.13 Felkin dipolar model. Orbital interactions in the Felkin–Anh model

interactions M ↔ R and R^1 ↔ M; hence, it possesses higher free energy and leads to the less predominant diastereomer (**D**^{2}).

Moreover, the antibonding π* orbital is stabilized by its overlap with the σ*orbital.

Thus, consideration of the *molecular orbitals involved* reveal that the *orthogonal approach* of the nucleophile leads to its destabilization due to (i) out of phase overlap with the oxygen atom and (ii) from the four-electron interaction with the HOMO of the π-CO of the substrate [(**a**) in Fig. 5.13], and hence the approach of the nucleophile invokes the *Bürgi–Dunitz trajectory* to minimize these destabilizing factors. The approach of the nucleophile involves further stabilization if the largest ligand L is placed orthogonal to the C = O double bond, thus providing greatest overlap between the antibonding π* orbital (lowest unoccupied molecular orbital, LUMO) of the carbonyl group and the antibonding orbital σ* of the antiperiplanar α-substituent, in addition to the stabilizing two electron *n*–σ* interaction with the electron pair of the nucleophile as shown in (**b**).

In fact, any example of the Cram's open chain model (see Fig. 5.10) can be explained in a more quantitative manner by Felkin–Anh's model (**A**) (Fig. 5.12).

5.4.2 *Felkin–Anh Dipolar Model*

Cram's dipolar model for the prediction of the predominant diastereomeric product may also be replaced by a different Felkin–Ahn model (**E**) (Fig. 5.13), based on the tenets already mentioned, with the additional proviso that in the TS the halogen or any strongly electronegative ligand is placed orthogonal to the C = O bond, thus providing greatest overlap between σ* orbital of the C–X bond and the π* orbital of the C = O, as well as allowing maximum separation of the electronegative α-substituent (X) and the negatively charged nucleophilic reagent (Fig. 5.13). Of course, Felkin–Anh dipolar model also leads to the same predominant diastereomer as the Cram's dipolar model (Fig. 5.11).

5.5 Prelog's Rule

Prelog's rule, though empirical, allows the prediction of the steric course of an asymmetric synthesis carried out with a chiral α-ketoester (usually a glyoxylate or pyruvate) of a chiral secondary or tertiary alcohol, SMLC-OH (S, M, and L stand for small, medium sized, and large substituents, respectively). The prediction has been confirmed in the majority of the cases.

It is the outcome of generalization of the results of the asymmetric synthesis carried out by McKenzic group in early twentieth century. In McKenzie's first example [13], the phenylglyoxalate ester of (−)-menthol was reduced with aluminum

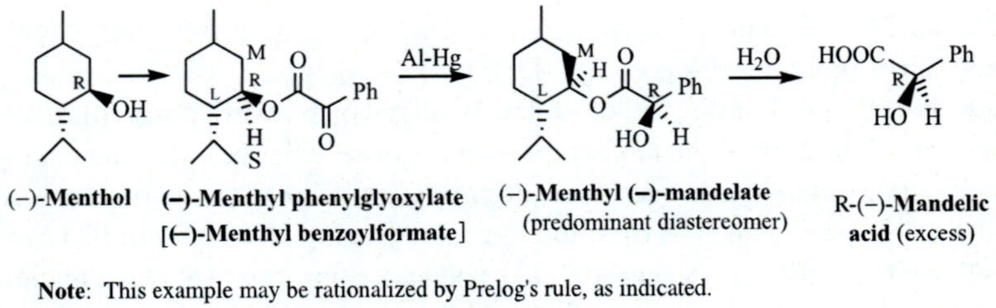

Fig. 5.14 Asymmetric synthesis of (−)-mandelic acid

amalgam to give the (−)-mandalate ester of (−)-menthol as the predominant product over the corresponding (+)-mandalate ester (Fig. 5.14).

For application of Prelog's rule [14, 15], by convention, in the transition state, the α-ketoester (**1A**) (Fig. 5.15) is so oriented that the two carbonyl groups are antiperiplanar (to minimize the dipolar repulsion), and the L group occupies the same

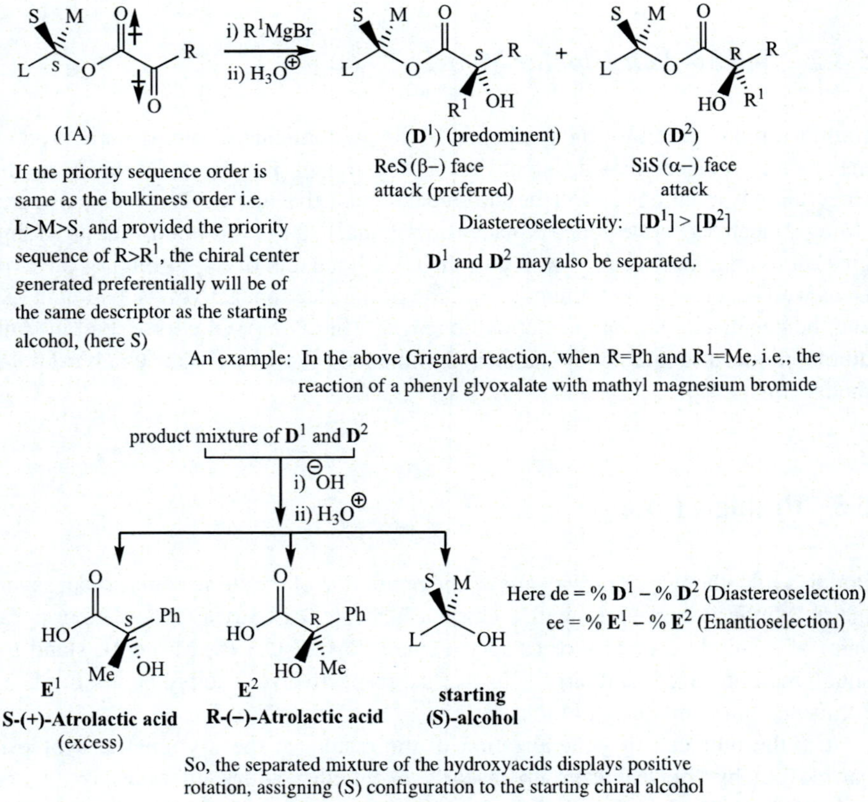

Fig. 5.15 Prelog's rule. Diastereoselection. Enantioselection

plane as the two carbonyl groups and the alkyl oxygen bond; thus, the bulky ester carbonyl group (complexed with the reagent) becomes staggered between two smaller groups S and M (Fig. 5.15). In this TS conformation, the nucleophile (usually the Grignard reagent R^1MgX) will predominantly approach the ketocarbonyl function from the side of the small group S. This rule thus correlates the configuration of the starting chiral alcohol SMLC-OH with the predominant α-hydroxyester formed, and therefore, of the α-hydroxy acid obtained in excess by hydrolysis along with the starting alcohol. Prelog's rule holds good since such nucleophilic addition is kinetically controlled.

5.5.1 Attempted Rationalization of Prelog's Model

The generalization for predicting the stereochemistry of the products from the α-ketoester asymmetric synthesis was put forward as an empirical model; yet its reasonable success implies some correlation of the model with the conformation of the TS (Fig. 5.16).

It is pertinent to note that trans-coplanar conformation does not represent the most stable ground state; it may represent the most stable TS, because of maximum separation of charge between negative centers is possible in this form. Thus, it represents better distribution of charge density under the influence of the reagent in the TS. X-ray crystal studies revealed that the ϕ between two CO's of (**2**) (Fig. 5.16) is 104°. This conformation may exist in solution also.

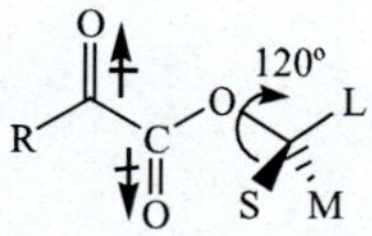

(**1A**) (cf. Figure **2.88**)

Choosing this TS conformation (**1A**) was largely intuitive, but there is a rationale as follows:
The ester CO being complexed with the reagent is staggered between the smaller groups M and S.

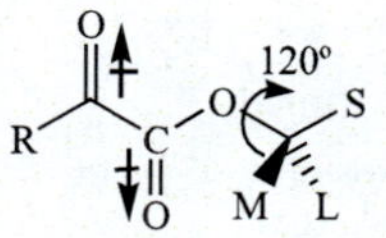

(**1B**)

Although this TS(**1B**) gives the right diastereomer **D**[1], in excess, the complexed ester C=O is staggered between **M** and **L**, so (**1B**) becomes of higher energy and may have minor contribution.

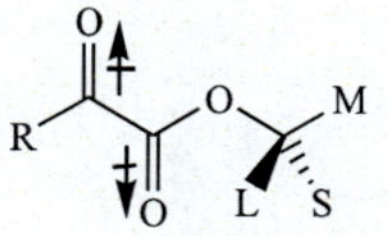

(**1C**)

This TS conformer gives the wrong diastereomer as the predominant one, hence is discarded.

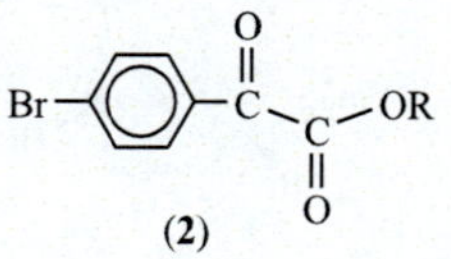

(**2**)

Fig. 5.16 Rationalization of Prelog's rule

5.5.2 More Examples of the Application of Prelog's Rule

The atrolactic acid synthesis and the Prelog's rule have been widely used for the determination (or confirmation) of the absolute configuration of secondary alcohols, *e.g.*, alicyclic alcohols, monoterpene alcohols, sesquiterpene alcohols, triterpene alcohols, and steroid alcohols [16]. Some examples of Prelog's rule are shown in Fig. 5.17. For the asymmetric atrolactic acid synthesis, the chiral secondary alcohol is converted to its phenylglyoxylate ester. The latter is then reacted with MeMgBr to give the α-hydroxy ester in excess predicted by Prelog's rule. The product mixture upon hydrolysis gives the predicted atrolactic acid in excess (as shown in Fig. 5.15). In case the *priority sequence order is the same as the bulkiness order and (R) (here Ph) has higher priority than the incoming nucleophile (R)[1]* (here Me), *(R)-alcohol gives the (R)-hydroxy acid (here (R)-atrolactic acid) in excess* and the *(S)-alcohol gives (S)-hydroxy acid* (here (*S*)-atrolactic acid). The other two possibilities of getting

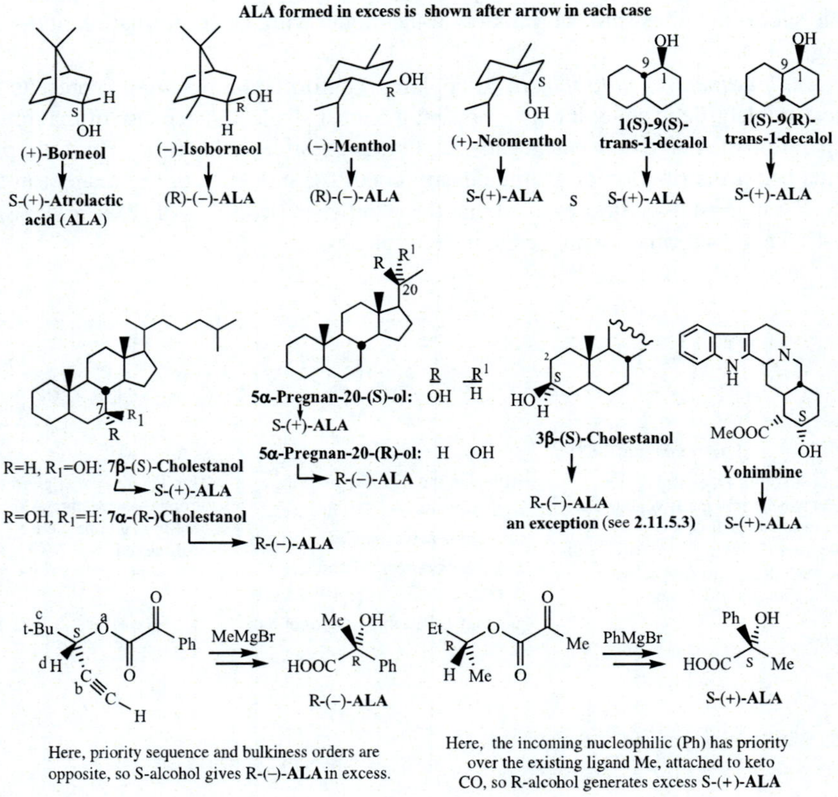

Fig. 5.17 More examples of application of Prelog's rule. Atrolactic acid synthesis

(*R*)-hydroxy acid from (*S*)-alcohol and vice versa are illustrated in the last row of Fig. 5.17.

5.5.3 Exception to and Anomalies of Prelog's Rule

In atrolactic acid synthesis with (*S*)-3β-cholestanol (see Fig. 5.17), *R*-(−)-atrolactic acid is obtained in only 1.7% enantiomeric excess [16]. Thus, *S*-alcohol is providing the *R*-atrolactic acid. This may be taken as an exception to Prelog's rule. Perhaps, here bulkiness order is not the same as the CIP priority sequence. Examination of models reveals that the steric crowding at C2 (with its environment) is greater than that at C4. However, asymmetric induction being so low, no definite conclusion can be drawn from this result.

Atrolactic acid synthesis of the tetramethyl ethers of both (−)-epicatechin (*R*-3-ol) and (+)-catechin (*S*-3-ol) leads to the preferential formation of *R*-(−)-atrolactic acid. These results must be considered as an anomaly rather than an exception to the rule [17].

5.6 Horeau's Rule

Horeau developed an empirical method [18, 19] for the correlation of configuration of secondary alcohols based on the principle of *kinetic resolution.* During the esterification of an optically active secondary alcohol, say, (−)-RR^1CHOH with *an excess of racemic* (±)-R^2COOH the transition states, [(+)-RR^1CHOH.(+)-R^2COOH]$^{\#}$ and [(+)-RR^1CHOH.(−)-R^2COOH]$^{\#}$ are diastereomeric and thus are unequal in enthalpy (*cf*. Fig. 5.4). Hence, the activation energies and the reaction rates of esterification with the enantiomeric acids will be unequal. Thus, the enantiomeric acid involving lower activation will be esterified at a faster rate, leaving the other enantiomeric acid in excess being unreacted (kinetic resolution) (Fig. 5.18).

An enantiomerically pure (or enriched) secondary alcohol is treated with an excess of (±)-α-phenylbutyric anhydride (Fig. 5.18) (or sometimes the corresponding chloride). The residual anhydride is hydrolyzed and the optical rotation of the resultant α-phenylbutyric acid is measured. The Horeau's rule states that (provided, the CIP priority $L > M$ holds good) an alcohol with *R*-configuration gives an excess of *S*-(+)-phenylbutyric acid and vice versa (Fig. 5.18). Brewster [20] and Horeau [21] have published detailed reviews on this method and its limitations.

Horeau's method has the advantage that in case the optically active alcohol is not available, the racemic alcohol itself may be treated with optically active α-phenylbutyric anhydride taken in much excess. The rotation of the unreacted alcohol is measured. Based on kinetic resolution now it may be concluded that if *S*-(+)-phenylbutyric anhydride is used, the unreacted alcohol would have *R*-configuration (provided L gets priority over M) and *vice verse* (Fig. 5.18). The sign of rotation of the

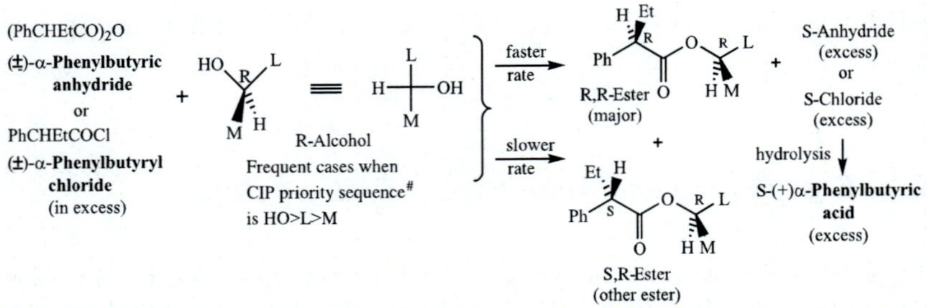

Thus, excess of S-(+)-α-phenylbutyric acid implies R-configuration of the unknown alcohol.
Likewise, excess of R(–)-α-phenylbutyric acid implies S-configuration of the unknown alcohol.

#In rare cases when priority sequence is HO>M>L, the conclusion will be reverse, i.e. excess S-acid implies S-alcohol and excess R-acid implies R-alcohol. Examples: M= –C≡CH and L= t-Bu; M= –CH=CH$_2$, L= –CHMe$_2$

Fig. 5.18 Horeau's rule illustrated

remaining unreacted alcohol needs to be determined. The rule is strictly empirical, and no rationalization has yet appeared.

The Horeau's procedure can be used [22, 23] for microscale determination if a highly sensitive method (GC, HPLC, MS, or CD) is available to determine the composition of the diastereomeric esters formed.

5.7 Sharpless Enantioselective Epoxidation

Sharpless in 1980 discovered a new metal-catalyzed asymmetric epoxidation process [24] which is by far more selective than any previously described methods for such asymmetric transformation. The strained epoxide ring serves as a valuable synthon for many natural products including insect pheromones. One of the most attractive aspects of this new process is the use of its readily available simple components.

In this method a prochiral primary allyl alcohol (mono-, di-, or tri-substituted, R^1, R^2, and R^3 are achiral) is treated with *t*-butylhydroperoxide (TBHP) in the presence of titanium (IV) isopropoxide and optically active diethyl tartrate (DET) (or di-isopropyl tartrate, DIPT) to produce the corresponding epoxide in a highly enantioselective manner, as shown in Fig. 5.19. The system provides consistent enantioselection.

With Sharpless reagent using (+)-DET geraniol undergoes regiospecific epoxidation of the allylic 2,3-double bond to form 2*S*,3*S*-epoxygeraniol in 90% *ee* in accordance with the Sharpless rule [24]; here attack takes place almost entirely at the 2*Re*-,3*Si* (α-face) [25]. Extensive application of the Sharpless epoxidation method to numerous multistep syntheses demonstrates its utility and reliability [26, 27].

The catalyst is sensitive to preexisting chirality in the substrate. Thus, the epoxidation of racemic secondary allylic alcohols with Sharpless reagent proceeds rapidly with only one enantiomeric alcohol, leaving behind the other slower reacting enantiomer.

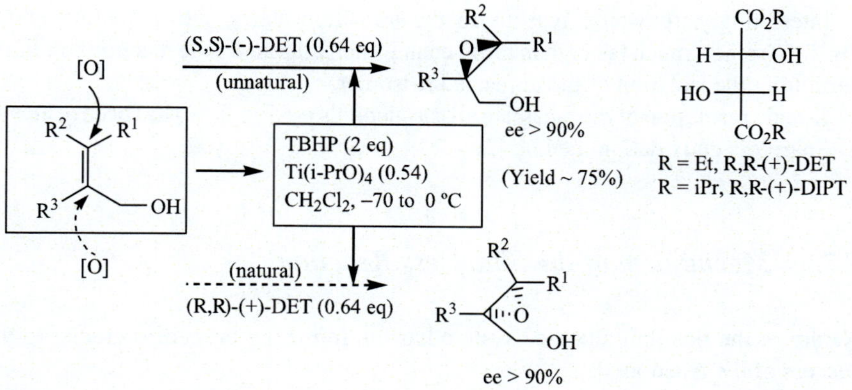

Fig. 5.19 Prediction of the chiral epoxides formed enantioselectively from prochiral allylic alcohols by sharpless rule

5.7.1 *Kinetic Resolution of Racemate Allyl Alcohols*

Resolution of racemate allyl alcohols may be accomplished *by treating with half equivalent of TBHP in the presence of the chiral catalyst like Ti-(i-Pro)$_4$ complexed with an active DET (or DIPT).*

The epoxidation pattern of *E*-cyclohexylpropenylcarbinol using (+)-Diisopropyltartrate is shown in Fig. 5.20 [28].

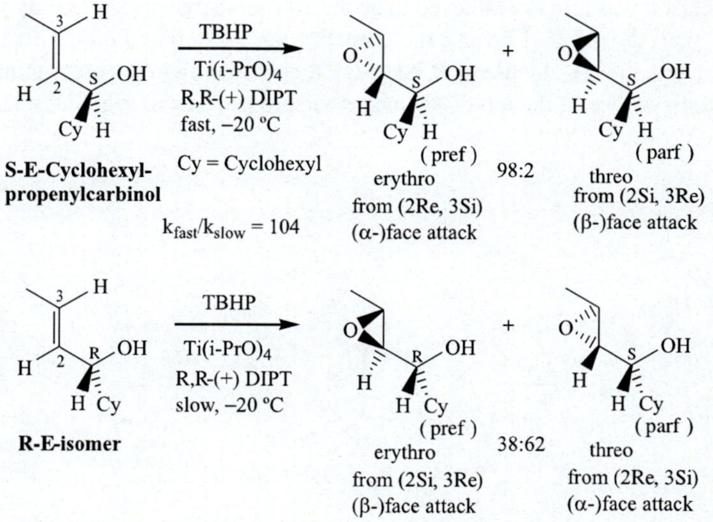

Fig. 5.20 Kinetic resolution of racemate allyl alcohol by sharpless epoxidation method

Interestingly, for steric reasons in the transition states, the *S*-enantiomer (in Fig. 5.20) reacts much faster than the *R*-enantiomer, thus under the reaction condition permitting the isolation of the unreacted *R*-isomer.

Kinetic resolution of the racemate allylic alcohols (*cf.* Fig. 5.20) has been achieved by Sharpless epoxidation method [29].

5.7.2 Mechanism of the Sharpless Reaction

Nature of the titanium alkoxide system has the following properties crucial to the success of the reaction:

(i) Exchange of monodentate alkoxide ligands is rapid in solution.
(ii) Of the four covalent bonds of Ti^{IV} two participate for the divalent chiral auxiliary (tartrate) and one each for TBHP and the allyl alcohol.
(iii) The coordination chemistry of Ti^{IV} (d°) alkoxide system is flexible since this system displays a range of coordination numbers and geometries. This property may be responsible for the catalysts' ability to accommodate sterically widely differing substrates.
(iv) Ti^{IV} alkoxides are weak Lewis acids and thus activate a coordinated alkylperoxo ligand toward nucleophilic attack by olefin of a bound allylic alcohol.

The mechanism of the reaction has been extensively studied by NMR, and numerous crystal structures for the complexes involved have been determined. The active catalyst was initially believed to be the 10-membered structure [30]. Comprehensive analyses [31, 33] favor a mechanism (Fig. 5.21) based on the structure (**I**) to be converted to (**II**), by use of *R,R*-(+)-DET, epoxidizing much predominantly the enantiotopic α-face of the allylic segment (when written as in Fig. 5.19).

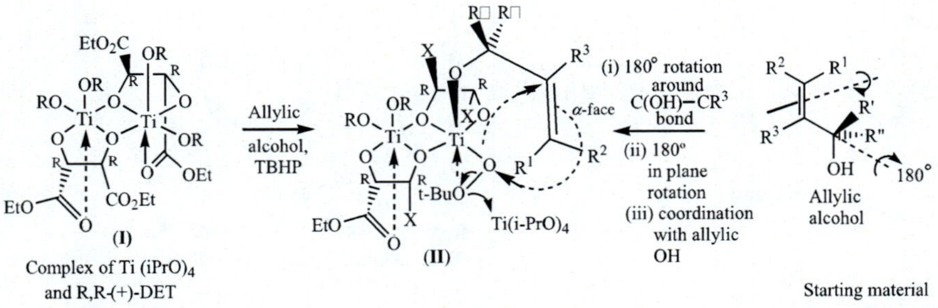

Fig. 5.21 Mechanism of the sharpless epoxidation of an allylic alcohol by the active catalyst (**I**) from *R,R*-(+)-DET

Most importantly, the species containing equimolar amount of titanium tetraalkoxide and tartrate is shown to be the most active catalyst in the reaction mixture, mediating the reaction at a much faster rate than titanium tetraalkoxide alone [31].

In (**I**) due to weak coordination the carbonyl oxygen atoms readily dissociate and recoordinate to the metal centers, providing a means of exchanging the alkoxide ligands for the substrate molecules [*R*,*R*,-(+)-DET], TBHP, and allylic alcohol as shown in (**II**). The lower energy conformation (here oxygen is delivered from the α-face of the double bond of the β-substituted allylic alcohol) of the allylic segment dictates which enantiotopic face is epoxidized. Many allylic alcohols with different substituents have been highly enantioselectively epoxidized following the Sharpless rule.

References

1. H.S. Mosher, E.L. Combe, Asymmetric reductions. I. the action of (+)-2-Methylbutylmagnesium chloride on Methyl *t*-Butyl Ketone. J. Am. Chem. Soc. **72**, 3994–3999 (1950)
2. H.S. Mosher, E.L. Combe, Asymmetric reductions. II. the action of (+)-3-Methylpentylagnesium Chloride on Methyl *t*-Butyl Ketone. J. Am. Chem. Soc. **72**, 4991–4994 (1950)
3. S.K. Talapatra, B. Talapatra, in *Chemistry of Plant Natural Products: Stereochemistry, Conformation, Synthesis, Biology, and Medicine,* vol. 2 (Foreword by Professor K. C. Nicolaou) (Springer, Berlin, Heidelberg, 2015), Sect. 32.2, pp. 955–957
4. D.J. Gram, F.A.A. Elhafez, Studies in stereochemistry. X. the rule of steric control of asymmetric induction. In the synthesis of acyclic systems. J. Am. Chem. Soc. **74**, 5828–5835 (1952)
5. D.J. Gram, D.R.Wilson, Studies in stereochemistry, XXXII. models for 1,2-asymmetric induction. J. Am. Chem. Soc. **85**, 1245–1249 (1963)
6. E.L. Eliel, in *Asymmetric Synthesis* vol 1. ed. by J.D. Morrison, (Academic Press, New York, 1983)
7. D.R. Boyd, M.A. McKervey, Asymmetric synthesis. Quart. Rev. **22**, 95–122 (1968)
8. M. Chérest, H. Felkin, Torsional strain involving partial bonds. The steric course of the reaction between allyl magnesium bromide and 4-t-butylcyclohexanone. Tetrahedron Lett. 2205–2208 (1968)
9. N.T. Anh, Regio-and stereo-selectivities in some nucleophilic reactions. Top. Curr. Chem. **88**, 145–162 (1980)
10. H.B. Bürgi, J.D. Dunitz, E. Shefter, Geometrical reaction coordinates. II. nucleophilic, addition to carbonyl group. J. Am. Chem. Soc. **95**, 5065–5067 (1973)
11. H.B. Bürgi, J.D. Dunitz, J.M. Lehn, G. Wipff, Stereochemistry of reaction paths at carbonyl centers. Tetrahedron **30**, 1563–1572 (1974)
12. H.B. Bürgi, J.M. Lehn, G. Wipff, An *ab initio* study of nucleophilic addition to a carbonyl group. J. Am. Chem. Soc. **96**, 1956–1957 (1974)
13. A. McKenzie, Studies in asymmetric synthesis. I. reduction of Menthyl Benzoylformate. II. action of magnesium alkyl haloids on menthyl benzoylformate. J. Chem. Soc. **85**, 1249–1262 (1904)
14. V. Prelog, Untersuchungen über asymmetriche Synthesen 1. Űber den sterischen Verlauf der Reaction von α-Keto-säure-estern optisch aktiver Alcohol emit Grignard'schen Verbindungen. Helv. Chim. Acta **36**, 308–319 (1953)

15. D. Arigoni, J.D. Dunitz, A. Eschenmoser, in *Biographical Memoirs of Fellowsof the Royal Society,* vol 46. (2000). pp. 444–464. on Vladimer Prelog, *Bibliography*, pp.Prelog 1- Prelog 15.
16. J.C. Fiaud, Prelog's methods. in *Stereochemistry, Fundamentals and Methods*, ed. by H.B. Kagan, vol 3. (Determination of Configurations by Chemical Methods", George Thieme Publishers, Stuttgart, 1977). pp. 19–49. (Chapter 2 Pertinent pages 20–24; and pertinent references cited
17. S.K. Talapatra, B. Talapatra, in *Chemistry of Plant Natural Products: Stereochemistry, Conformation, Synthesis, Biology, and Medicine,* vol 2. (Foreword by Professor K. C. Nicolaou) (Springer, Berlin, Heidelberg, 2015), Section 14.5.3.2, pp. 698–699
18. A. Horeau, Principe et applications d'une nouvelle methode de determination des configurations dite "par Dedoublement Partiel". Tetrahedron Lett. 506–512 (1961)
19. A. Horeau, Determination des configurations par "Dedoublement Partiel-II precisions complements. Tetrahedron Lett. 965–969 (1962)
20. J.H. Brewster, Assignment of stereochemical configuration by chemical methods. in *Elucidation of Organic Structures by Physical and Chemical Methods*, ed. by K.W. Bentley, G.W. Kirby, vol IV, Part III, (Wiley-Interscience, New York, 1972). pp. 1–249
21. A. Horeau, Determination of the configuration of secondary alcohols by partial resolution. in *Stereochemistry, Fundamentals and Methods*, ed. by H.B. Kagan, vol 3. (Determination of Configurations by Chemical Methods, George Thieme Publishers, Stuttgart, 1977). Chapter 3, pp. 51–94
22. R. Weidmann, A. Horeau, Determination des configurations d'alcohols secondaire par "Dedoublement Partiel" IX(I)—Semimicromethode non polarimetrique. Tetrahedron Lett. 2979–2982 (1973)
23. A. Horeau, A. Nouaille, Micromethode de determination de la configuration des alcohols secondaires par dedoublement cinetique emploi de la spectrographie de masse. Tetrahedron Lett. **31**, 2707–2710 (1990)
24. T. Katsuki, K. Barry Sharpless, The first practical method for asymmetric epoxidation. J. Am. Chem. Soc. **102**, 5974–5976 (1980)
25. S.K. Talapatra, B. Talapatra, in *Chemistry of Plant Natural Products: Stereochemistry, Conformation, Synthesis, Biology, and Medicine,* vol 1. (Foreword by Professor K. C. Nicolaou) (Springer, Berlin Heidelberg, 2015) Fig. 6.23 (example **V**), pp. 363
26. B.E. Rossiter, T. Katsuki, K. Barry Sharpless, Asymmetric epoxidation provides shortest routes to four chiral epoxy alcohols which are key intermediates in syntheses of Methymycin, Erythromycin, Leukotriene C-1, and disparlure. J. Am. Chem. Soc. **103**, 464–465 (1981)
27. A. Pfenninger, Asymmetric epoxidation of allylic alcohols—the sharpless epoxidation. Synthesis 89–116 (1986)
28. V.S. Martin, S.S. Woodford, T. Katsuki, Y. Yamada, M. Ikeda, K. Barry Sharpless, Kinetic resolution of racemic allylic alcohols by enantioselective epoxidation, a route to substances of absolute enantiomeric purity. J. Am. Chem. Soc. **103**, 6237–6240 (1981)
29. Y. Gao, R.M. Hanson, J.M. Khender, S.Y. Ko, H. Masamune, K. Barry Sharpless, Catalytic asymmetric epoxidation and kinetic resolution. modified procedures including in situ derivatization. J. Am. Chem. Soc. **109**, 5765–5778 (1987)
30. K. Barry Sharpless, S.S. Woodard, H.G. Finn, On the mechanism of titanium-tartrate catalyzed asymmetric epoxidation. Pure and Appl. Chem. 55, 1821–1836 (1983)
31. S.S. Woodard, M.G. Finn, K. Barry Sharpless, Mechanism of asymmetric epoxidation. 1. kinetics, most importantly, the species containing equimolar amounts of Ti and tartrate is shown to be the most active catalyst in the reaction mixture, mediating reaction at much faster rates than a titanium tetraalkoxide alone. J. Am. Chem. Soc. **113**, 106–113 (1991)
32. M.G. Finn, K. Barry Sharpless, Mechanism of asymmetric epoxidation. 2. catalyst structure. J. Am. Chem. Soc. **113**, 113–126 (1991)

Further Reading

33. R.S. Ward, Selectivity versus specificity. Chemistry in Britain 803–804 (1991)

Chapter 6
Some Other Aspects of Dynamic Stereochemistry: Conformation and Reactivity

Abstract This chapter starts with a relevant introduction followed by the portrayal of the following: Conformational analysis of chemical reactions: stereoelectronic effects and steric effects, conformationally rigid diastereomers, *trans*-2-*trans*-decalol and *cis*-2-*cis*-decalol, conformational energy, anancomeric systems, conformationally locked, biased, and mobile substrate molecules, Winstein–Holness (W–H) equation and Curtin–Hammett (C–H) principle; reactions of 2-bromobutane, benzylphenylcarbinyl triethylbenzoate, 1-chloro-1,2-diphenylethane, 2,3-dibromobutane diastereomers; major conformer leading to less predominant product, free energy diagram, derivation of C–H principle; cope reaction of butane-2-dimethylamine oxide, quaternization of tropane by ^{13}C labeled iodomethane, Chugaev reaction on 3-phenyl-4-t-butylcyclohemane and free energy diagram, Winstein–Holness and Eliel equations, 4-t-butylcyclohexyl tosylate and cyclohexyl tosylate, 4-t-butylcyclohexanol and cyclohexanol, 1-methyl-2-phenylpyrrlidine and 1-methyl-2-arylpyrrolidine, equilibrium constant by ^{1}H NMR method; correlation of axial chirality with centrodissymmetry: (R)-(−)-1-chloro-3-t-butyl-3-methylallene with (R)-(−)-methyl-t-butylethynylcarbinol, (R)- (−)-2,2-dimethyl-3,4-hexadienal with (R)-(+)-3-bulyn-2-ol, (R)-(−)-2-hydroxy-3,5′-dimethoxy-2′-vinyl-6E-ω-phenylvinyl-biphenyl with (−)-thebaine; corrrelation of centrodissymmetry with axial chirality: conversion of (R)-(+)-glutamic acid to (−)-norcamphor, conversion of (S)-(+)-3-phenylallene carboxylic acids to (S)-(+)-bromo-lactones, Baldwin's rules for ring closure examples of different types, limitations of Baldwin's rules, rules for tetrahedral, trigonal and digonal systems.

6.1 Introduction

A knowledge of stereochemistry and conformation of molecules at rest (static stereochemistry) would only have been of limited interest to a chemist if he did not know how to use this information for explaining and predicting the chemical behavior of compounds. This is what dynamic stereochemistry is all about.

Dynamic stereochemistry refers to the stereochemical aspects of chemical reactions, i.e., relationship in space between the atoms and groups in a molecule during

S. K. Talapatra and B. Talapatra, *Basic Concepts in Organic Stereochemistry*,
https://doi.org/10.1007/978-3-030-95990-6_6

Z

the chemical reactions, especially in the transition states and the way in which rates of reactions and chemical equilibria and the product stereochemistry are affected by those relationship. This aspect of stereochemistry borders on the study of reaction mechanism.

Chapter 5, titled "Asymmetric Synthesis" already discussed **(pp. 65–85)**, involving different rules and enantioselective reactions, may thus be regarded as some aspects of *dynamic stereochemisty*.

The present chapter depiction, reactivities of some types of conformations, also belongs to dynamic stereochemistry.

Some other aspects of dynamic stereochemistry namely different types of stereoselective reactions and pericyclic reactions, which have been discussed in detail in some text books (*vide* the relevant references in "Further Reading, **p. 163**), have not been included in the present monograph because of space constraint.

6.2 Conformational Analysis of Chemical Reactions

It has long been recognized that in any type of chemical reaction proceed through diastereomeric transition states, thus involving different activation energies and hence differ significantly in terms of their relative rates, and stereochemistry, whenever applicable, of the products. The stereochemistry of the starting materials and the products formed in kinetically controlled rate processes can be correlated by considering the conformation of the transition states of the rate determining step. Such correlation in conformation of steroid nucleus was first pointed out by Barton [1] in 1950 by citing the difference in chemical behavior between rigid cyclohexane like steroidal compounds, having the reacting substituents in the axial or equational orientation. Such correlation studies are also applicable in case of flexible acyclic systems.

There are two kinds of effects or factors involving the relationship of conformation and reactivity as stated below.

6.2.1 Stereoelectronic Effects and Steric Effects

These are generally important in determining the stability of the transition state of any chemical reaction.

1. 'Stereoelectronic effects' involve effects on reactivity of the spatial disposition in the transition state of particular electron pairs:

(a) bonded: e pair in bonds that are broken, formed, or otherwise dislocated in the reaction in question;

(b) non-bonded: unshared electrons on exocyclic or endocyclic heteroatoms (cf. anomeric effect)

Stereoelectronic effects arise from delocalization of electrons or from overlap of orbitals in transition states and decide the feasibility of the reaction and pertains to its mechanism and stereochemistry.

2. "*Steric effects*" arise due to close approach of two groups in a molecule (or between molecules)—such that appreciable van der Waals forces are involved. In most cases, only repulsive forces matter. Attractive forces are to be considered rarely, since attractive potential becomes very high within van der Waals radii.

Torsional strain, bond angle and bond length distortion strain, H-bonding or electrostatic interaction, if any are also included in steric factor.

Steric effects occur in the ground state of a molecule or in the transition state for a given reaction or both and should be considered to assess the activation energy of the reaction (kinetic energy).

Steric effects are concerned with the conformational requirements of the non-participating groups and decide the rate of the reaction by determining the extent of steric hindrance or steric assistance.

The ground state energy of the product should be considered to determine the thermodynamic energy, if the reaction is reversible under a particular condition.

6.2.2 *Conformationally Rigid Diastereomers*

If the reacting groups are locked into two different orientations, equational and axial, because of the absence of ring inversion of compounds—such compounds become conformationally rigid diastereomers. Compounds with two or more cyclohexane rings linked through *trans*-ring junction/s unable to undergo ring inversion—as found in diterpenes, triterpenes and steroid [2] belong to this category. We may consider the simplest case of *trans*-2-decalol diastereomers (Fig. 6.1). In *trans*-2-*trans*-decalol (**1**), the OH group is equational (*e*), whereas in *trans*-2-*cis*-decalol (**2**), the OH group is axial (*a*).

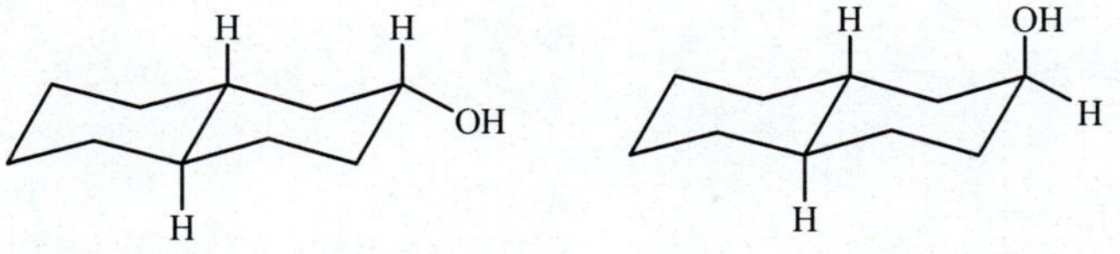

trans-2-trans-Decalol (**1**) trans-2-cis-Decalol (**2**)

Acetylation: k_{rel} 7.69 k_{rel} 1.00

Fig. 6.1 Conformationally rigid diastereomers showing relative reaction rates of equational and axial isomers

6.2.3 *Conformational Energy: Anancomeric Systems*

The difference in potential energies between the most stable conformer of a molecule and a designated less stable one is called *conformational energy* (see also Sects. 1.3.5 and 7.12.4). The equilibrium population of any two conformers is expression in the well-known reaction equation:

$\Delta G^0 = -RT\ln K$ at a constant absolution temperature T, and K is the equilibrium constant being the ratio of the enthalpies of the more stable to that of the less stable conform and hence is negative. ΔG^0 is called than *Gibbs free energy* at temp. T for that system.

The standard free energy change is related to the enthalpy difference through an entropy term, as shown in the following equation

$$\Delta G^0 = \Delta H^0 - T\Delta S^0$$

In most cases, ΔS^0 is too small and is neglected. The free energy difference in case of n-butane conformers, the gauchi conformer being two times populated than the anti-conformer, enhances the entropy by Rln2 at 25 °C. This is same as the entropy of mixing of the gauchi former. With increase of temperature, the population of the less stable conformer increases eventually to become equal to that of the stable conformer. On the contrary with the decrease of temperature, the stable *anti*-conformer prevails to become the only detectable form.

If ΔG exceeds 10 kJ mol^{-1}, the stable conformer constitutes 98% of the total at 25 °C. Such an equilibrium is called *anancomeric* [3, 4] or conformationally biased; however, the rate of interconversion remains facile. The equation $G\Delta^0 = -RT\ln K$ has been used [5] to give the populations of the stable conformer and the corresponding conformational free energies.

Compounds confined to a single conformation in which the substituent is either strictly equational (*e*) (as in **3**) or strictly axial (*a*) (as in **4**) of Fig. 6.2.

X = a substituent

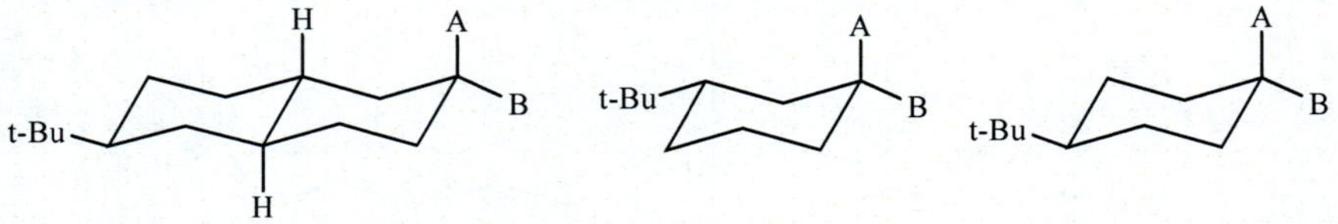

A = H, B = X
e Isomer (trans) (**3**)

A = X, B = H
a Isomer (cis) (**4**)

A = H, B = X
e Isomer (cis) (**5**)

A = X, B = H
a Isomer (trans) (**6**)

A = H, B = X
e Isomer (trans) (**7**)

A = X, B = H
a Isomer (cis) (**8**)

Fig. 6.2 Conformationally locked molecules (**3** and **4**) and conformationally biased bia (**5, 6, 7** and **8**) molecules

Compounds belonging to **3** and **4** are having *conformational blocking*, whereas compounds belonging to **5**, **6**, **7** and **8**) are conformationally biased. Barton's classical paper [1] dealt with the physical and chemical properties of equationally and axially substituted cyclohexyl compounds especially of steroidal compounds. 4-t-Butyl substituted compounds are prototypes of biased model [6]. Compounds of this type are called *anancomeric* meaning "*fixed in one conformation.*" (The word is derived from Greek *anankein* meaning to fix by some fate of law).

Conformational energies, $-\Delta G^0$ or A values [6] (25 to −100 °C) for a large number (108) of common substituents have been presented by Eliel in a tabular form [7]. *Conformational equilibria* have been widely determined by low-temperature NMR studies. Eliel determined [8] the chemical equilibration of the conformational energy of *cis*- and *trans*-4-*t*-butylcyclohexanol from the study of their chemical equilibration in presence of $Al(O\text{–}iPr)_3$ or Raney Ni.

6.2.4 Conformation and Chemical Reactivity for a Conformationally Mobile Substrate Molecule: The Winstein–Holness Equation and the Curtin–Hammett Principle

In conformationally mobile systems, the primary relationship of reactivity is to conformation. Both stereoelectronic and steric factors on reactivity and product composition depend primarily upon conformational factors in such systems (Sect. 6.2.1). Both reaction rates and reaction products may depend upon the conformational composition of starting materials and the conformation of the preferred transition states. The dependence of reaction rate on conformational composition is most aptly described by the Winstein–Holness equation [6, 9–12]; the dependence of the product composition is governed by the Curtin–Hammett principle [11–13]. Both the Winstein–Holness equation and the Curtin–Hammett principle involve the kinetic scheme shown in Fig. 6.3 and may be expressed as in Fig. 6.4.

Conditions for the Validity of Winstein–Holness Equation and Curtin–Hammett Principle

Curtin–Hammett principle stipulates that the activation energy of the reaction is large compared with the barrier to rotation (which is usually quite low), the proportion of the products in no way reflects the relative population of the ground state conformations but depends on the activation energies of the processes leading to these products [13]. This principle is an extension of a more general proposition, according to which the ratio of the products formed from one starting material of two conformers of unequal energy (or different energies) depends only on the difference in the free energy levels of the two transition states and is in no way related to

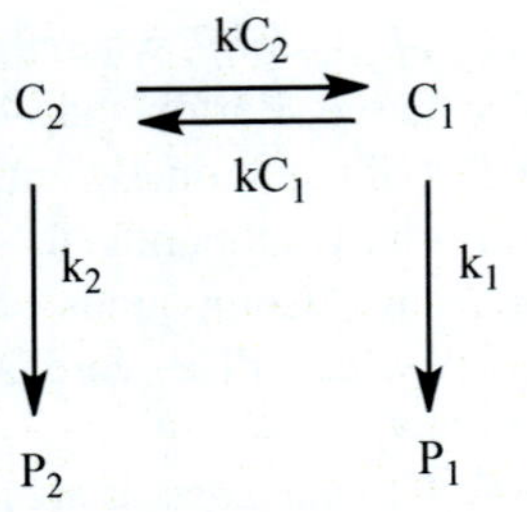

Here, C_2 and C_1 are conformations of a compound in rapid equilibrium (i.e., kC_2, $kC_1 >> k_2$, k_1); such as C_2 and C_1 ae tantomers or conformational isomers.

Products P_2 and P_1 may be different substances not interconvertible, or conformers of the same substance.

Fig. 6.3 Kinetic schemes for both Winstein–Holness equation and the Curtin–Hammett principle

1) Winstein-Holness equation (W-H equation)

$$k_{WH} = \sum_i n_i k_i \qquad i = \text{number of contributing conformers}$$

(observed rate constant) $\qquad n_i$ = mole fraction of the ith conformer

$= n_1k_1 + n_2k_2$, here

2) Curtin-Hammett principle (C-H principle)

$$\frac{[P_1]}{[P_2]} = e^{\Delta G^{\ddagger}/RT} \qquad \text{when } [P_1] > [P_2]$$

$(kC_1, kC_2) >> (k_1, k_2)$ (see Fig. 6.3) — when kC_1 & $kC_2 > 10$ (k_1 or k_2) and C_2C_1 equilibrium is not highly one-sided, i.e., kC_1 and kC_2 are of the same order of magnitude.

Both W-H equation and C-H principle apply to the case of Fig. 6.3.

Fig. 6.4 Conditions for the validity of Winstein–Holness equation and Curtin–Hammett principle

the free energy level/s or population/s of the ground state that the reactions follow the same rate equation (homo-competitive) (Figs. 6.4 and 6.5).

Energetics of the formation of two products P_1 and P_2 from two conformers C_1 and C_2 of unequal energy are shown in Fig. 5.6. In this case, one should note that the activation energies $\Delta G_1^{\#}$ and $\Delta G_2^{\#}$ ands the free energy difference between the two conformers ΔG^0 (GC_2–GC_1) are vectors pointing upwards and hence positive. Let us now proceed to derive the Winstein–Holness equation as delineated in this Fig. 5.6.

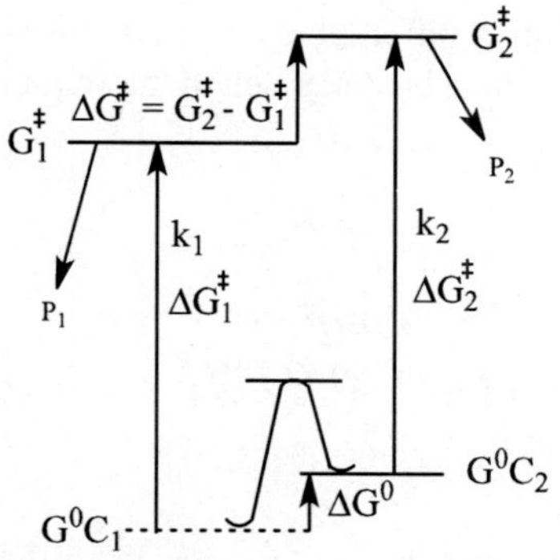

Two conformers C_1 and C_2

<u>Assumptions</u> : i) $G^0C_i < G^0C_2$ $\therefore$ C_1 is more stable than C_2, and $\Delta G^{\ddagger} > \Delta G^0$

$$P_2 \xleftarrow{k_2} C_2 \underset{kC_1}{\overset{kC_2}{\rightleftharpoons}} C_1 \xrightarrow{k_1} P_1$$

Example: for this case $\begin{cases} C_1 \rightarrow \text{equatorial or more stable conformer} \\ C_2 \rightarrow \text{axial or less stable conformer} \end{cases}$

ii) $\Delta G_1^{\ddagger} < \Delta G_2^{\ddagger} \therefore$ C_1 reacts faster
i.e. $k_1 > k_2$ (*cf*. steric hindrance)

<u>Derivation</u> (from first principle)

Here, $G^0P_1 < G^0P_2$ $\therefore$ P_1 is major

$$\frac{[P_1]}{[P_2]} = \frac{dP_1/dt}{dP_2/dt}\ \frac{k_1[C_1]X}{k_2[C_2]X} = \frac{k_1K}{k_2} \quad .....(i)$$

$$= \frac{e^{-\Delta G_1^{\ddagger}/RT} \cdot e^{\Delta G^0/RT}}{e^{-\Delta G_2^{\ddagger}/RT}}$$

$$= e^{(-\Delta G_1^{\ddagger} + \Delta G^0 + \Delta G_2^{\ddagger})/RT} \quad .....(ii)$$

$$= e^{(G_2^{\ddagger} - G_1^{\ddagger})/RT} = e^{\Delta G^{\ddagger}/RT} \quad .....(iii)$$

$\therefore$ The product ratio > 1.

Thus, the product ratio gives $\Delta G^{\ddagger}$
Again, from equation (i)
product ratio may be assessed
in many cases from K, k_1 & k_2

$\because \dfrac{[C_1]}{[C_2]} = K$ (equilibrium constant)

(for case 1); $\dfrac{[P_1]}{[P_2]} = \dfrac{k_1}{k_2}$

X = product of concentration of all other species in the rate equation

Since, <u>using Transition State theory</u>,

$k_1 = Q.T.e^{-\Delta G_1^{\ddagger}/RT}$, where $Q = \dfrac{Kk}{h}$

$k_2 = Q.T.e^{-\Delta G_2^{\ddagger}/RT}$

$K = e^{-(-\Delta G^0)/RT}$

$= e^{\Delta G^0/RT}\ [G^0C_1 - G^0C_2 = \Delta G^0]$

K (kappa) = transmission coefficient
k = observed rate constant

Fig. 6.5 Energy diagram for Winstein–Holness equation and Curtin–Hammett principle for two conformers of unequal energy forming two products. Derivation of the Curtin–Hammett principle from the first principle

Applications of Curtin–Hammett principle. Three different cases of Curtin–Hammett principle with examples in each case will hence be discussed in the sequel and delineated in Figures.

6.2.5 *Curtin–Hammett Principle. Case 1, Predominant Conformer Leading to Predominant Product* ($G^oP_2 > G^oP_1$)

In the energy diagram in this case, (cf. Fig. 6.5) the product P_1, formed from the more stable conformer C_1, will have lower energy than that of the product P_2, i.e., $G^0C_1 < G^0C_2$ and $G^0P_1 < G^0P_2$.

C–H Principle. Case 1. Example 1. Base Catalyzed Dehydrobromination of 2-Bromopropane

Here, the concentration of the product P_1, *E*-2-butene is found to be 6 times that of the product P_2, *Z*-2-butene (Fig. 6.6).

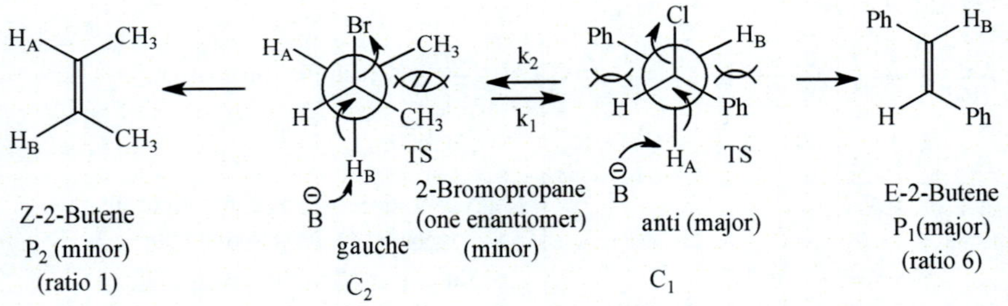

Population/stability: $[C_1] > [C_2]$ (cf. energy diagram in Fig. 6.5)

Here, $[P_1]/[P_2] = 6 = e^{(G_2^\ddagger - G_1^\ddagger)/RT} = e^{\Delta G^\ddagger/RT}$

Thus, $\Delta G^\ddagger$ can be determined.

Comments :

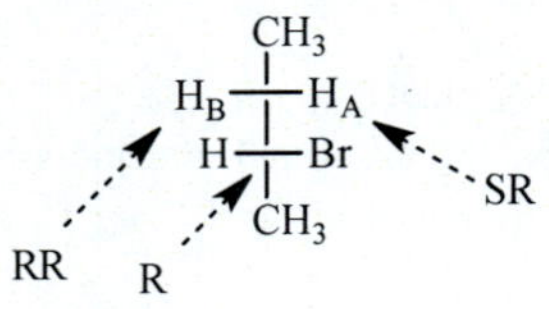

H_A and H_B are diastereotopic.
Thus, elimination of each one involves
a diastereotopic transition state to
yield two diastereomers in unequal amounts.

Fig. 6.6 Curtin–Hammett principle. Case 1. Example 1. Base catalyzed dehydrobromination of 2-bromopropane

Curtin–Hammett Principle. Case 1. Example 2. Base Catalyzed Elimination Reaction of Benzylphenylcarbinyl 2,4,6-Triethylbenzonate

Here, the concentration of the product P_1, *E*-stilbene, formed predominantly from the predominant *anti*-conformer of the substrate molecule is found to be 100 times that of the minor product P_2, *Z*-stilbene, formed from the less predominant *gauche* conformer of the substrate molecule [14]. This type of interaction has been termed [13] the "*cis*-effect."

Curtin–Hammett Principle. Case 1. Example 3. Base Catalyzed Elimination (Dehydrochlorination) of 1-Choro-1,2-Diphenylethane

Here the product P_1 *E*-stilbene formed from the predominant conformer (*anti*) of an enantiomer (*S*) of 1-chloro-1,2-dediphenylethane is formed predominantly. The other product P_2, *Z*-stilbene is formed as the less predominant one.

6.2.6 Elimination Reactions of Two Diastereomers trans- *and* Cis-*2,3-Dibromobutane Leading to Two Diastereomeric Products*

One should note that the above three examples (Figs. 6.6, 6.7, and 6.8) are obviously different from the case of reactions of diastereomers displayed in Fig. 6.9.

In Fig. 6.9, the *elimination reactions of two diastereomers* of 2,3-dibromobutane (*RS*, meso; *RR* one enantiomer of *dl*) give rise to *Z*-2-butene (*cis*-2-butene) and *E*-2-buteme (*trans*-2-butene), proceeding through transition states (A) and (B), respectively. This is obviously different from three foregoing and two following examples of reactions of conformers. (A) and (B), derived from the corresponding 2,3-dibromo *anti*-conformers.

6.2.7 Curtin–Hammett Principle. Case 2. Energy Diagram for Predominant Conformer Leading to the Less Predominant Product. Derivation of C-H Principle

In the energy diagram (Fig. 6.10) (cf. Figure 6.5), the product P_1, formed from the predominant conformer C_1, will be the less predominant one, i.e., $G^0C_1 < G^0C_2$ and $G^0P_1 > G^0P_2$. The ratio $[P_1]/[P_2]$, being equal to $e^{-}\Delta^{G\#}/RT$, would be a fraction (<1); hence, P_2 will be the predominant product.

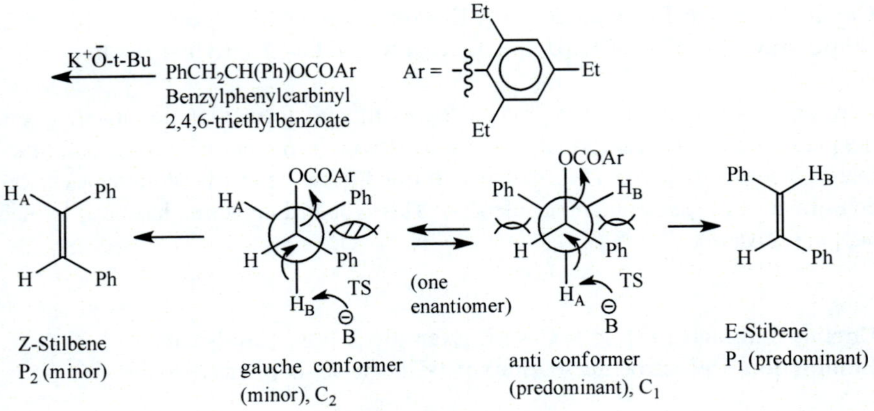

Here, $[C_2] < [C_1]$

$$G^0C_1 < G^0C_2 \text{ and } \Delta G_1^{\ddagger} < \Delta G_2^{\ddagger}$$

(cf. energy diagram in Fig. 6.5)

$$\text{Here, } P_1/P_2 \text{ observed, } 100/1 = e^{(G_2^{\ddagger} - G_1^{\ddagger})/RT} = e^{\Delta G^{\ddagger}/RT}$$

$$\therefore \ \ln 100 = \Delta G^{\ddagger}/RT$$

$$\Longrightarrow \Delta G^{\ddagger} = RT \ln 100 = RT\ 2.303 \log 100$$

$$\text{at } 27^{\circ}C \Longrightarrow \Delta G^{\ddagger} = 2X300\ (2.303)\ 2 \text{ cal/mole}$$

$$= 2.764 \text{ Kcal/mole}$$

Thus, $\Delta G^{\ddagger}$ is reasonable, since E- and Z-stilbenes differ in stability by about 5.7 Kcal/mole.

Hence, $\Delta G^{\ddagger}$ reflects the difference in stability of the products to the extent that the eclipsing interaction (because of cogwheeling) existing in the ground state gouche conformation of Z-stilbene (planar) exists in the transition state (approaching coplanarity) to a somewhat lesser extent, leading to the product P_2.

Here, $\Delta G^{\ddagger} < \Delta G^0$.

Fig. 6.7 Curtin–Hammett principle. Case 1. Example 2. Base catalyzed elimination of benzylphenylcarbinyl 2,4,6-triethylbenzoate

Curtin–Hammett Principle. Case 2. Example 1. Pyrolytic *Cis*-Elimination (Cope Reaction) of Butane-2-Dimethylamine Oxide (Fig. 6.11)

Pyrolysis of butane-2-dimethylamine oxide with the formation of olefins and elimination of dimethylhydroxylamine is believed to involve a cyclic transition state as shown in Fig. 6.11. Here, the predominant conformer through the five-membered cyclic transition state (A) forms the less predominant product P_1, while the less populated conformer through the cyclic transition state (B) forms the predominant product P_2, the latter reaction being *sterically assisted*, as shown. The two staggered conformers of the *R*-enantiomer of the substrate molecule have been involved in the *cis*-elimination. Such *cis*-elimination is known as Cope reaction.

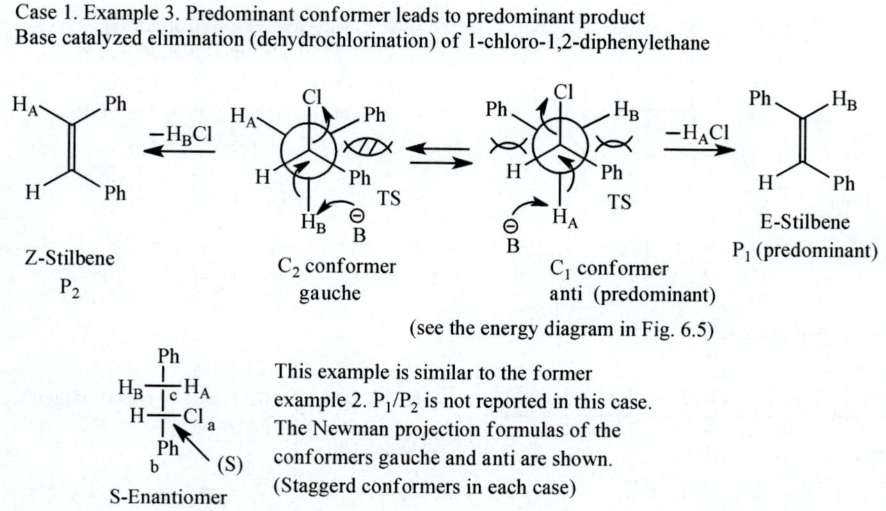

Fig. 6.8 Curtin–Hammett principle. Case 1. Example 3. Base catalyzed dehydrochlorination of 1-chloro-1,2-diphenylethane

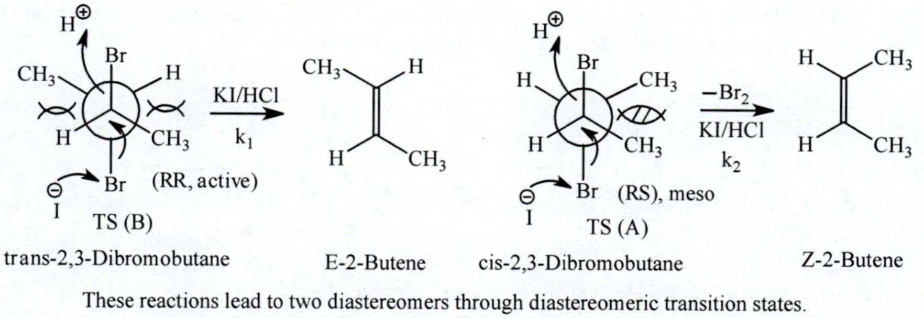

Fig. 6.9 Two diastereomers, *trans*- and *cis*-2,3-dibromobutane leading to two diastereomeric products by elimination reactions

Curtin–Hammett Principle. Case 2. Example 2. Quaternization of Tropane by ^{13}C Labeled Iodomethane ($^{13}CH_3I$) (Fig. 6.12)

Tropane conformers are less sterically hindered on the side of the pyrrolichine ring, than on the side of the piperidine ring; hence the conformer C_1 is more stable than the other conformer C_2. So on methylation with ^{13}C labeled iodomethane ($^{13}CH_3I$), the more stable major conformer C_1 having the Me on the equatorial side reacts with the electron pair on the more hindered axial side at a slower rate (k_{ax}), producing the minor product P_1 (Fig. 6.12).

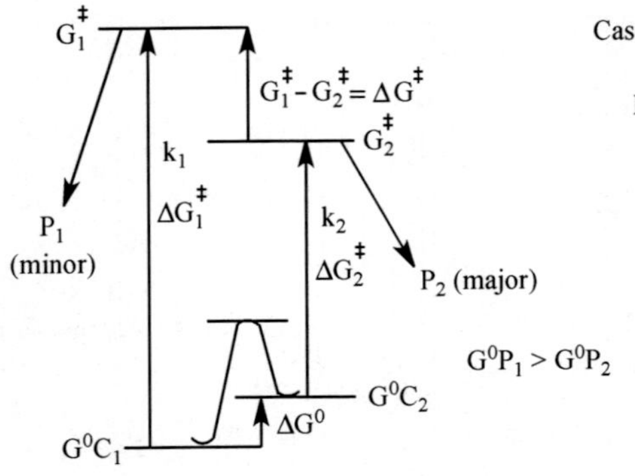

Case 2 : More abundant conformer leads to the minor product

Here also $G^0C_1 < G^0C_2$ (cf. Fig. 6.5)

i.e. C_1 is the more stable conformer.

But $\Delta G_1^{\ddagger} > \Delta G_2^{\ddagger}$, $\therefore k_2 > k_1$

$$C_2 \rightleftharpoons C_1 \quad K = \frac{[C_1]}{[C_2]}$$

$$G^0C_1 - G^0C_2 = -\Delta G^0$$

$$G_1^{\ddagger} - G_2^{\ddagger} = (+G_1^{\ddagger} - G_2^{\ddagger}) = -\Delta G^{\ddagger}$$

$$\text{Here, } \frac{[P_1]}{[P_2]} = \frac{dP_1/dt}{dP_2/dt} = \frac{k_1[C_1]X}{k_2[C_2]X} = \frac{k_1}{k_2}K$$

where X stands for the prodct of concentration of all other species that enter into the rate equation.

$$= \frac{e^{-\Delta G_1^{\ddagger}/RT} \cdot e^{\Delta G^0/RT}}{e^{-\Delta G_2^{\ddagger}/RT}} = e^{(-\Delta G_1^{\ddagger} + \Delta G^0 + \Delta G_2^{\ddagger})/RT} = e^{-(G_1^{\ddagger} - G_2^{\ddagger})/RT} = e^{-\Delta G^{\ddagger}/RT}$$

Thus, the product ratio would be a fraction, hence, P_2 will be the major product

Fig. 6.10 Curtin–Hammett principle. Case 2. Energy diagram for more abundant conformer leading to less abundant product. Derivation of Curtin–Hammett principle

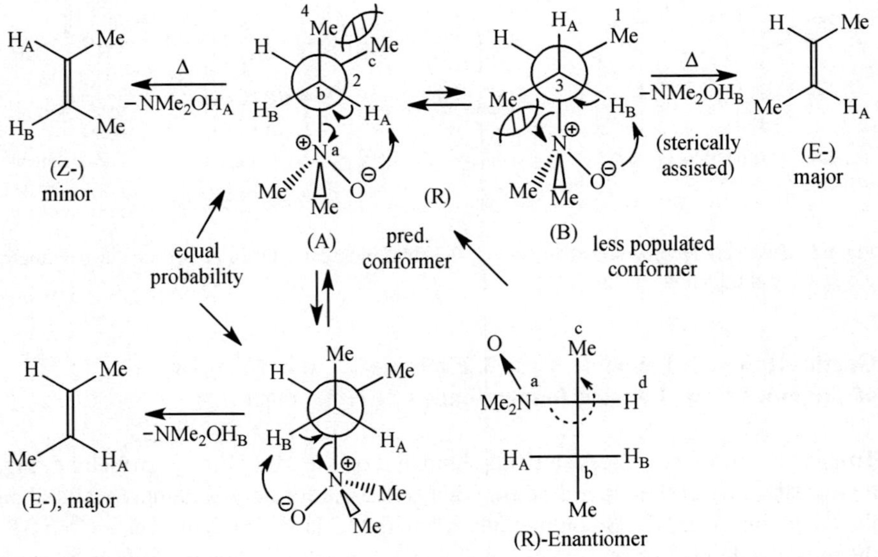

Fig. 6.11 Curtin–Hammett principle. Case 2. Example 1. Pyrolytic *cis*-elimination (Cope reaction) of butane-2-dimethyamine oxide

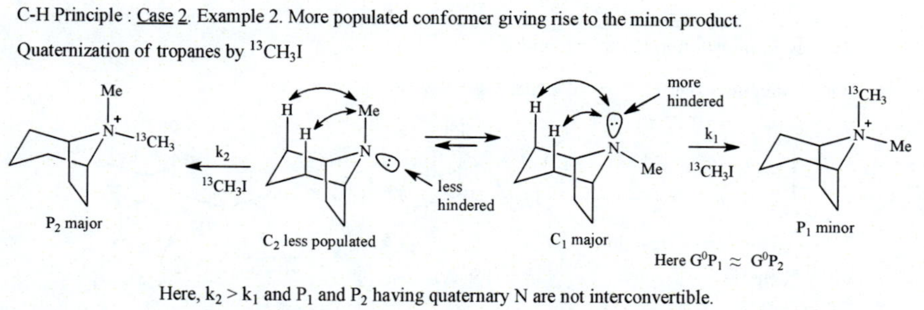

Fig. 6.12 Curtin–Hammett principle. Case 2. Example 2. Quaternization of tropane by $^{13}CH_3I$

On the other hand, the less stable minor conformer C_2 having the electron pair on the less hindered equatorial side of the piperidine ring produces at a faster rate (k_{eq}) the major product P_2.

It is implied that in the transition state, factors such $^{\&+}$N–CH$_3$–I$^{\&}$(−) geometry relative to the tropane geometry are also likely to play a key role in these alkylation selectivities [17].

Here, $k_{eq} > k_{ax}$ and the products P_1 and P_2 are not interconvertible.

6.2.8 *Curtin–Hammett Principle. Case 3. Different Conformers of* Trans*-2-Phenyl-*Trans*-3-*t*-Butylcyclohexan-*r*-1-ol Xanthate Having Same Free Energy Leading to Different Products. Chugaev Reaction*

On pyrolysis, different conformations C_1 and C_2 of *trans*-2-phenyl-*trans*-3-*t*-butylcyclohexan-*r*-1-ol xanthate having same free energy produce different products of unequal energy. The C_1 conformer through transition state (A) produces P_1, while the C_2 conformer through transition state (B) forms the expected major product P_1 (Fig. 6.13). The expected free energy diagram representing such a Chugaev reaction displayed in this figure shows that $G^0C_1 = G^0C_2$ and $[P_1] > [P_2]$.

6.2.9 *Quantitative Treatment of Mobile Systems. Winstein–Holness and Eliel Equations*

The Curtin–Hammett principle was proposed independently of each other by two groups, Winstein and Holness in 1955 [6] and Eliel and Ro in 1956 [9]. Since there is no basic difference between the equation of Winstein–Holness and that derived by

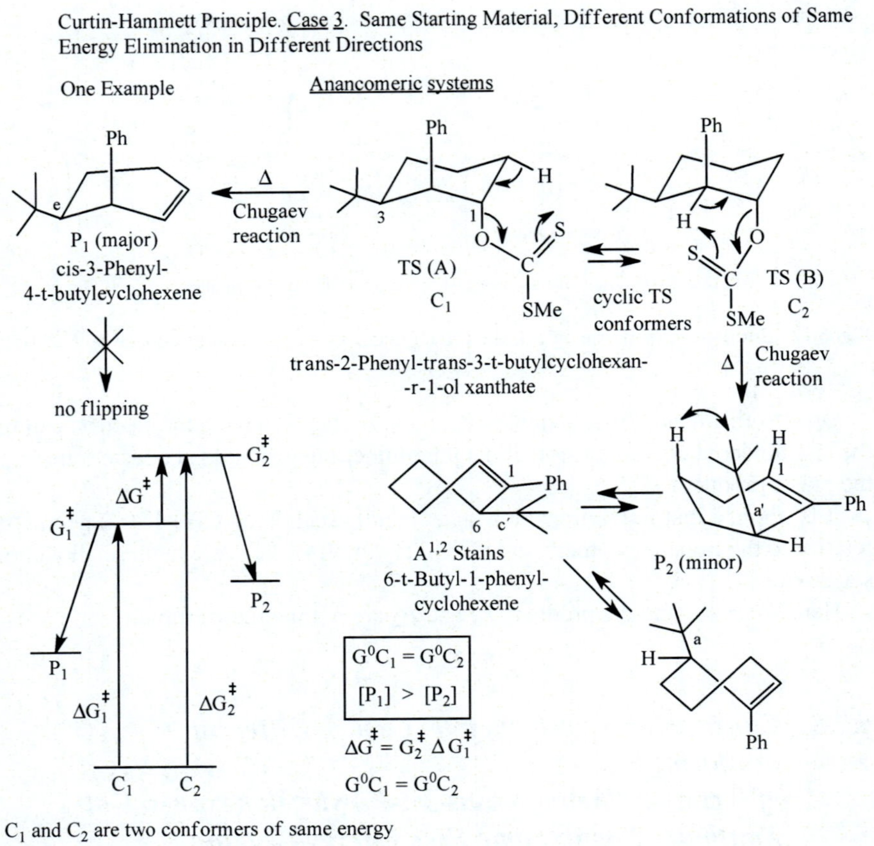

Fig. 6.13 Curtin–Hammett principle. Case 3. Energy diagram. One example (Chugaev reaction)

Eliel's group, the two equations may be appropriately referred to as Winstein–Eliel equations which are delineated in Fig. 6.14.

Equation (4) correlating the overall empirical reaction rate constant (k) of a conformationally mobile substrate of concentration [C] with the specific rate constants k_E and k_A at which the isomers E and A react is delineated in Fig. 6.14. Equation (5) for equilibrium constant K can be derived from the Eq. (4). The Eliel Eq. (4) and the Winstein–Holness Eq. (6) can be easily interconverted as shown in Fig. 6.14.

6.2.10 *Application of Winstein–Holness Equation. Examples 1 and 2*

See Fig. 6.15.

Thermodynamic equilibrium

Here, products ratio is determined by $\Delta G = -RT\ln K$
(K)

Simple monosubstituted cyclohexanes, as well as a number of di- or polysubstituted ones may exist in two conformations. So, the reactivity of both conformations must be considered to assess the reactivity of such mlecules.

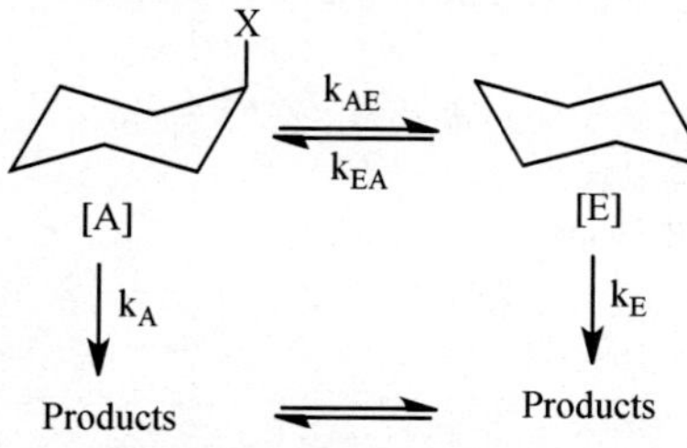

Reaction of mobile cyclohexane systems

Let

[A] be the concentration of the A (a) isomer
[E] be the concentration of the E (e) isomer

and $K = \dfrac{[E]}{[A]}$ equilibrium constant between the isomers

$$= \frac{N_E}{N_A} = \frac{k_{AE}}{k_{EA}}$$

N_E and N_A are mole fractions of E & A

k_E and k_A are the specific rates at which the E and A isomers react and k = normally reported rate constant

Then the overall rate $= k_E\,[E]\,[P] + k_A\,[A]\,[P]$ ....(1)

[P] stands for the product of the concentration of all species other than the substrate that enters into the rate equation.
e.g. i) solvolysis of cyclohexyl tostylate; [P] = 1
ii) saponification of ethylcyclohexyl carboxylate, $[P] = [\bar{O}H]$
[Derived by Eliel & Ro [8]]

But, [C] = [E] + [A] stoichiometric concentration of the substrate

Overall Rate $= k\,[C]\,[P]$, ...(2)
(where k is the specific rate constant.)

$\therefore\; k_E\,[E] + k_A\,[A] = k[C] = k\{[E] + [A]\}$ ....(3)

$\Rightarrow k_E.K[A] + k_A[A] = k\{[E] + [A]\}$

Dividing by [A], $k_EK + k_A = k\,(K+1)$

Regrouping, $k = \dfrac{k_EK + k_A}{K+1}$ ....(4) Eliel equation

Again regrouping, $k_EK - kK = k - k_A$

$\Rightarrow K = \dfrac{k - k_A}{k_E - k} = \dfrac{k_A - k}{k - k_E}$ ....(5) Eliel equation

Thus k & K can be found out from equation (4) and (5) respectively.

From equation (4) $k = \left(k_E\dfrac{N_E}{N_A} + k_A\right) \Big/ \dfrac{1}{N_A}$

$= k_EN_E + k_AN_A$ .....W-H Equation (6)

Since, $K + 1 = \dfrac{N_E}{N_A} + 1 = \dfrac{N_E + N_A}{N_A} = \dfrac{1}{N_A}$

We know, overall rate $= k[C] = k_E[E] + k_A[A]$

Dividing by [C] $\Rightarrow k = k_E\dfrac{[E]}{[C]} + k_A\dfrac{[A]}{[C]} = k_EN_E + k_AN_A$ ....W-H equation (6)

Fig. 6.14 Quantitative treatment of mobile systems. Winstein–Holness and Eliel equations

Example 1.

Simplest situation arises when only one conformer (a) is reactive

Thus,

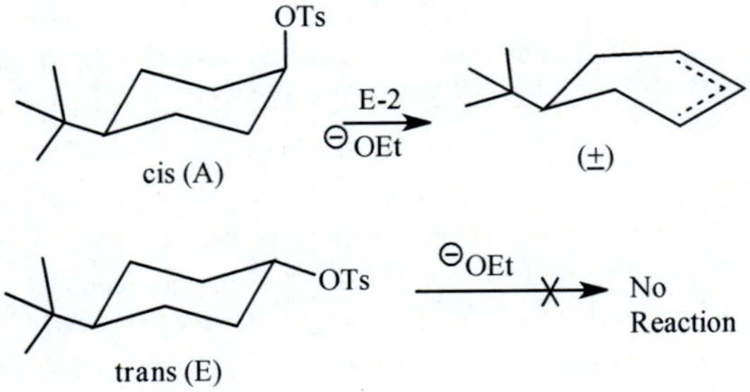

For the cis- isomer
$k = k_A N_A = k_A \because N_A = 1, N_E = O$
observed k
$= 7.1 \times 10^{-3}$ l.mol.$^{-1}$ sec.$^{-1}$
at 75°C (589 nm)

Example 2.

The conformational equilibrium constant K for the tosylate from the following data. The specific rate constant for the ionic elimination of cis-4-t-butylcyclohexyltosylate in presence of base is 7.1×10^{-3} l. mol^{-1}sec^{-1} at 75° (from example 1), and that for cyclohexyl tosylate = 2.4×10^{-3} l. mol^{-1}sec^{-1} at 75°

Now,

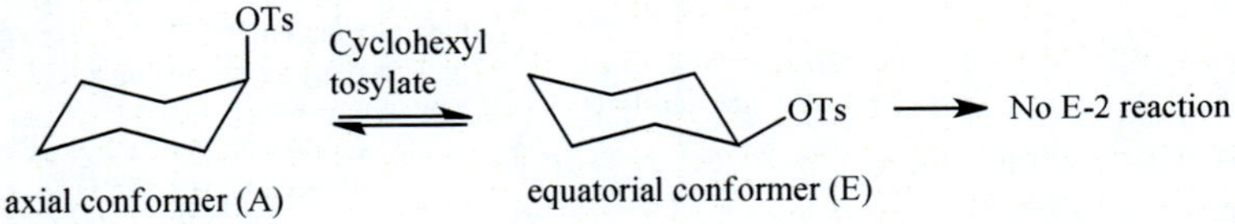

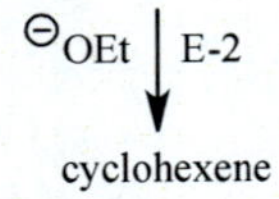

So, here $k = k_A N_A$, $k_E = 0$
observed rate constant, $2.4 = k_A N_A = 7.1 \times N_A$
$\therefore\ N_A = 2.4 / 7.1 = 0.34$

$$K = \frac{N_E}{NA} = \frac{1 - N_A}{N_A} = \left(1 - \frac{2.4}{7.1}\right) \Big/ \frac{2.4}{7.1} = \frac{4.7}{7.1} \div \frac{2.4}{7.1} = \frac{4.7}{2.4} = 1.96 \approx 2$$

Fig. 6.15 Winstein–Holness equation. Elimination of 4-*t*-butylcyclohexyl tosylate and cyclohexyl tosylate

6.2.11 Application of Eliel Equation. Examples 1 and 2

See Fig. 6.16.

Example 1.

The acetylation of cyclohexanol with Ac_2O / Py at 25°C: the specific reaction rate = 8.37×10^{-5} l $mole^{-1}sec^{-1}$. In this case the reactivity of both conformational isomers should be taken into account.

The specific reaction rates in the acetylations (Ac_2O / Py at 25°C) of cis-4-t-butylcyclohexanol (A) trans-4-t-butylcyclohexanol (E), and cyclohexanol are 2.89, 10.65, and 8.37 units respectively, the unit being 10^{-5} l $mole^{-1}sec^{-1}$. The conformational equilibrium constant K for hydroxyl group and its free energy difference ΔF_{OH} in cyclohexanol can be found out as shown below.

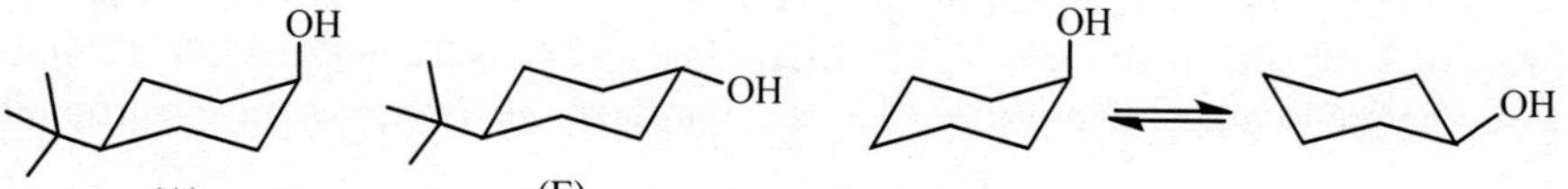

(A) (E)

k is the specific rate contant.

k = 8.37

$N_A = 1, N_E = 0$
$k = k_A = 2.89$ units

$N_E = 1, N_A = 0$
$k = k_E = 10.65$ units

From Equation 5 of Fig. 6.14 : $K = \dfrac{k - k_A}{k_E - k} = \dfrac{8.37 - 2.89}{10.65 - 8.37} = \dfrac{5.48}{2.28} \approx 2.40$ units

$F_A - F_E = \Delta F_{OH} = -RT\ln k \approx 0.52$ Kcal/mole

Example 2.

In case of disubstituted cyclohexanes the ΔF values might be expected to be additive.

4-Methylcyclohexanol

Axial isomer Equatorial isomer

Kcal mol^{-1}
$\Delta F_{OH} \approx 0.5$
$\Delta F_{Me} \approx 1.8$

Here, one may calculate $\Delta F = 1.8 - 0.5 = 1.3$ Kcal mol^{-1}; whereas 0.5 Kcal/mole is gained by changing OH from a to e conformation, 1.8 Kcal/mole is lost by flipping Me form e to a position.

$\Delta F = -RT\ln \underline{K} = 1.24$ Kcal $mole^{-1}$

Experimental value for specific rate constant k of acetylation of cis-4-methylcyclohexanol is 3.76 units.

We know, from **Example 1** above, k_A and k_E in acetylation rection under same condition are 2.89 and 10.65 units respectively.

So, $K = \dfrac{k - k_A}{k_E - k} = \dfrac{3.76 - 2.89}{10.65 - 3.76} = \dfrac{0.87}{6.89} \approx 0.126$

Fig. 6.16 Applications of Eliel Equation. Examples 1 and 2

6.2.12 *Quaternization of 1-Methyl-2-Phenylpyrrolidine and 1-Methyl-2-Arylpyrrolidine. Product Ratio*

Seeman [11] along with his coworkers [17] made a thorough analysis of the combination of the Winstein–Holness equation and Curtin–Hammett kinetic schemes (see Sect. 6.2.4, and Figs. 6.4 and 6.5). They studied the quaternization of N-methyl-2-phenylpyrrolidine ($R = H$) (Fig. 6.17) and N-methyl-2-arylpyrrolidine (Fig. 6.17) and experimentally found that the product ratio $[P_1]/[P_2]$ is more than 1 when the *ortho* or 2-substituent has $R = H$, Me, Et, or iPr); i.e., P_1 is the major product formed from the predominant conformer C_1, since the *steric hindrance* to the electrophilic

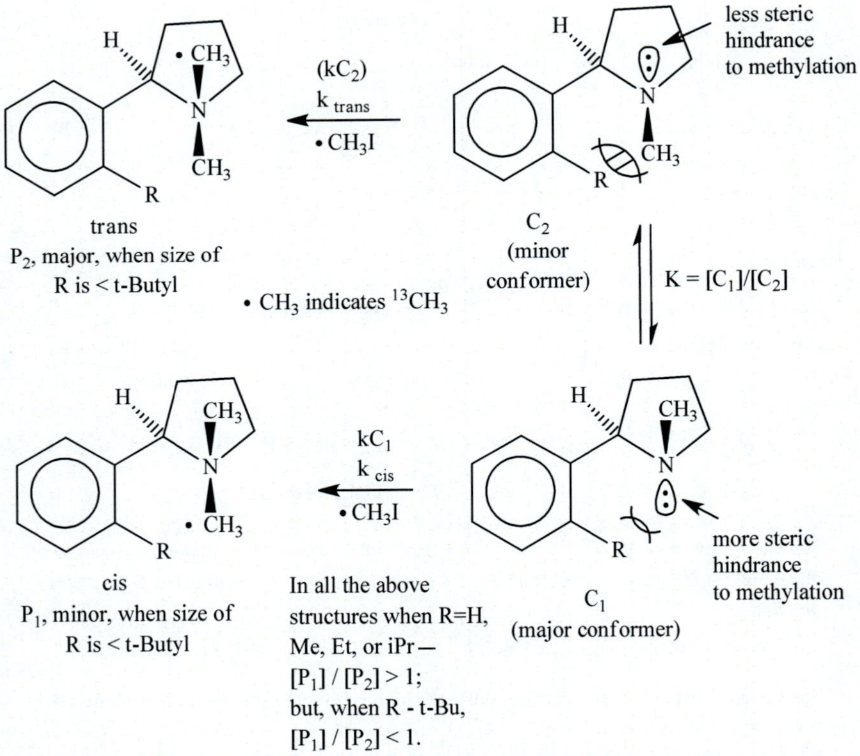

R	$[P_1]/[P_2]$	P_1
H	1.7	major
CH_3	1.4	major
C_2H_5	1.3	major
$(CH_3)_2CH$	1.3	major
$(CH_3)_3C$	0.28	minor

Fig. 6.17 Quaternization of N-methyl-2-arylpyrrolidine and N-methyl-2-phenylpyrrolidine with $^{13}CH_3I$

methylation on the N lone pair in C_1 is less than that in the minor conformer C_2 (Fig. 6.17); P_1 and P_2 having quaternary N are not interconvertible.

Whereas, *the prouct ratio is reversed* by being less than 1, when $R = t$-butyl; i.e., here P_2, formed from the minor conformer C_2, becomes predominant, since the *steric hindrance* to the methylation on the N lone pair in the minor conformer C_2 is less than that in the major conformer C_1 (Fig. 6.17).

6.2.13 *The Conformational Preference of Substituents by ^1H NMR Method. Calculation of Equilibrium Constant K (Cf. Kinetic Method) [18, 19]*

Before discussion of the NMR method, it may be mentioned here that direct observation of the conformers is possible by IR spectroscopy. The *Frank–Condon* principle applies, and each conformer has enough time to give its own spectrum before changing to another conformer (time scale: IR-Raman $10^{-14\,\mathrm{s}}$; NMR 10^{-1}–10^{-9} s.)

NMR has proved to be one of the most powerful techniques for conformational analysis and may be conveniently used at ambient temperature or low temperature depending on the system but requires energy barrier above 20 kJ mol^{-1} for recording of individual signals [18].

In principle, conformational composition can be measured at any temperature (usually at ambient temperature) by use of Winstein–Holness equation $k_{WH} = P = \&P_i n_i$, where P is any property. Application of chemical shifts, &, and coupling constants, J gives $= \&n_E \&_E + n_A \&_A$, and $J = n_E J_E + n_A J_A$, that is both observed chemical shifts and observed coupling constants in an equilibrating system (Fig. 6.18) are the weighted averages of the corresponding constants in the individual conformers. Thus, K can be calculated if &, $\&_A$ and $\&_E$ as well as J, J_A and J_E are known. But J measurements give some problems since J varies with temperature [19].

6.2.14 *Absolute Configuration of Allene Molecules (Figs. 6.18–6.22)*

There are five sub-sections namely 6.2.14.1, 6.2.14.2, 6.2.14.3, 6.2.14.4, and 6.2.14.5. The titles of the subsections are recorded below. The matters given in each figure are self-explanatory.

The use of NMR is quite convenient for studying the conformational equilibrium

$$A \overset{K}{\rightleftharpoons} E \qquad K = \frac{[E]}{[A]} = \frac{[N_E]}{[N_A]}$$

For any flexible cyclohexane system $P = P_E n_E + P_A n_A$

$$\therefore\ N_E = KN_A$$

The equilibrium is more rapid than the NMR method by which the mobile system is observed (like the chemical method)

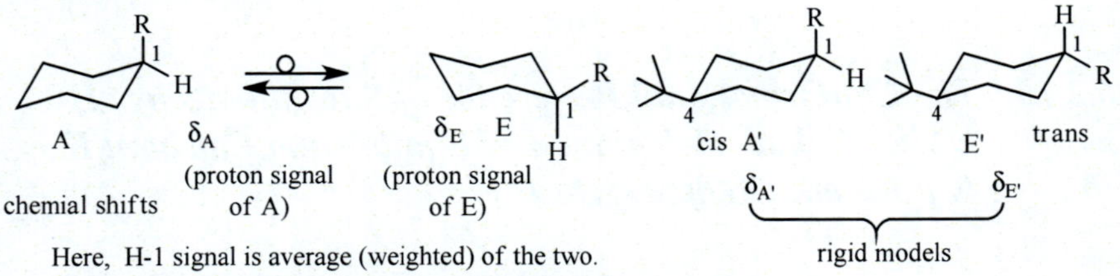

Here, H-1 signal is average (weighted) of the two.
Proton signals are observed when R = EWG like halogen, NO_2, OCH_3 or $COOCH_3$ etc.
When R = Me, the Me doublet signal is observed.
The Me signal of the Axial conformer A is δ_A

Here (by W-H equation), $\delta = N_A\delta_A + N_E\delta_E$ .....(1)

$= NA\,(\delta_A + K\delta_E)$ $\because K = N_E / N_A$

$= \frac{1}{K+1}(\delta_A + K\delta_E)$

Since $N_A = \frac{1}{\frac{1}{N_A}} = \frac{1}{\frac{N_E+N_A}{N_A}} = \frac{1}{K+1}$

$\Longrightarrow K\delta + \delta = \delta_A + K\delta_E$

$\Longrightarrow K\delta - K\delta_E = \delta_A - \delta \qquad \therefore K = \frac{\delta_A - \delta}{\delta - \delta_E}$ .... (2) $\qquad$ [cf. $K = \frac{k_A - k}{k - k_E}$]

Thus K can be calculated if δ, δ_A and δ_E are known.

- Barrier above 20 kJ mol^{-1} is necessary for recording of individual signals

  ν (in cps) = δ.x when measured in x MHz machine.

- The inverse of $\Delta\nu_{AE}$ gives a measure of the NMR time scale and the rate constant (k) of the process is of similar order of magnitude as $\Delta\nu_{AE}$, typically, 10^{-1} to 10^{-5} sec, the system amenable to NMR.

  If there are three conformers, $J = n_1J_1 + n_2J_2 + n_3J_3$ and also $n_1 + n_2 + n_3 = 1$

Fig. 6.18 Conformational preference of substituents by ^{1}H NMR method. Calculation of *K* (cf. kinetic method)

Correlation of Axial Chirality with Centrodissymmetry. Example 1. (*R*)-(−)-1-Chloro-3-*T*-Butyl-3-Methylallene with (*R*)-(−)-Methyl-*T*-Butylethynylcarbinol

See Fig 6.19.

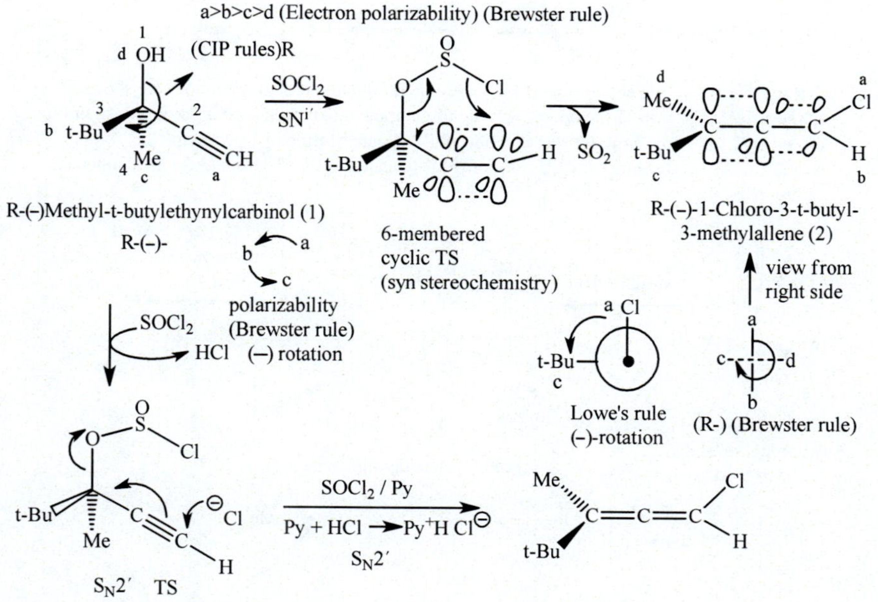

N.B.
- This is a substituted propargyl alcohol rearrangement.
- In the TS of either mechanism, $SN_{i'}$ or $SN_{2'}$ Cl approaches from the side on which the departing group OSOCl (i.e. OH) is located (syn stereochemistry).
- The absolute configuration of (1) was determined by Prelog's method (Atrolactic acid) (see Section 5.5).
- The levo rotation of (1) is in accordance with the Brewster rule (see Section 12.3).
- The levo rotation of (2) is in accordance with the Lowe's rule (see Section 12.4).
- Optical rotation of both (1) and (2) were expermentally found to be (–).
- The absolute configuration of both (1) and (2) are designated (R), according to CIP rules (Section 2.2.2) and Brewster rule (Section 12.3).

Fig. 6.19 Correlation of axial chirality with centrodissymmetry. Example 1. (*R*)-(−)-1-chloro-3-*t*-butyl-3-methylallene (2) with (*R*)-(−)-methyl-*t*-butylethynylcarbinol (1)

Correlation of Axial Chirality with Centrodissymmetry. Example 2. (*R*)-(−)-2,2-Dimethyl-3,4-Hexadienal with (*R*)-(+)-3-Butyn-2-Ol

See Fig 6.20.

Correlation of Axial Chirality with Centrodissymmetry. Example 3. (*R*)-2-Hydroxy-3,5'-Dimethoxy-2'-Vinyl-6-*E*-Ω-Phenylvinyl-Biphenyl with (−)-Thebaine

The reaction sequence (given below) of the conversion of (−)-thebaine to phenyldihydrothebaine, maintaining the stereochemistry of the ethanamine bridge, followed

$$\text{(R)-(+)-3-Butyn-2-ol} \xrightarrow[H^{\oplus}]{Me_2CH.CHO} (4) \xrightarrow[\text{on } SiO_2]{\Delta} \underset{(-)}{(6)}$$

The mechanism of the above reaction sequence involing the π orbital interctions are shown in the sequel. The sign of the optical activity of compound (**3**) reveals the absolute configuration (R) by application of Brewster rule (Section 12.3). By application of Lowe's rule (cf. Section 12.4) the absolute configuration of (**6**) is revealed as (R). These matters are delineated in Figure 6.20.

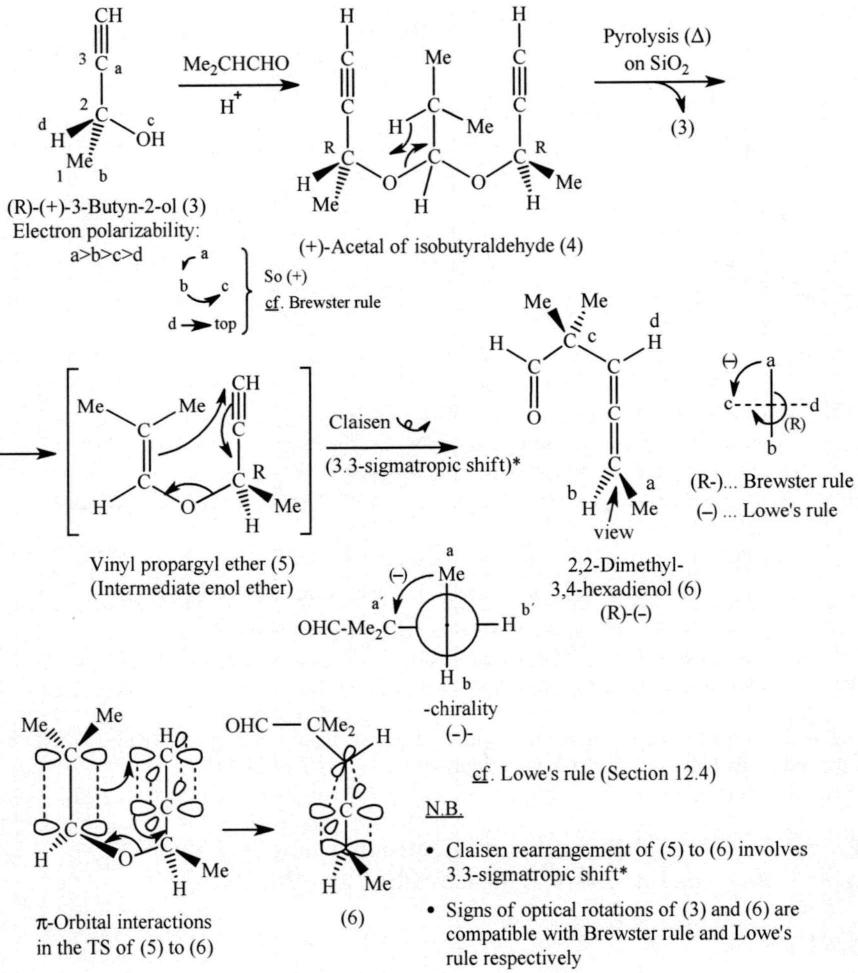

- The absolute configuration of (3) was determined by Prelog's method

* Detachment of O and attachment of C occur in the same side of the produced chiral allene

Fig. 6.20 Correlation of axial chirality with centrodissymmetry. Example 2. (*R*)-(−)-2,2-dimethyl-3,4-hexadienal (6) with (*R*)-(+)-3-butyn-2-ol (3)

Example 3

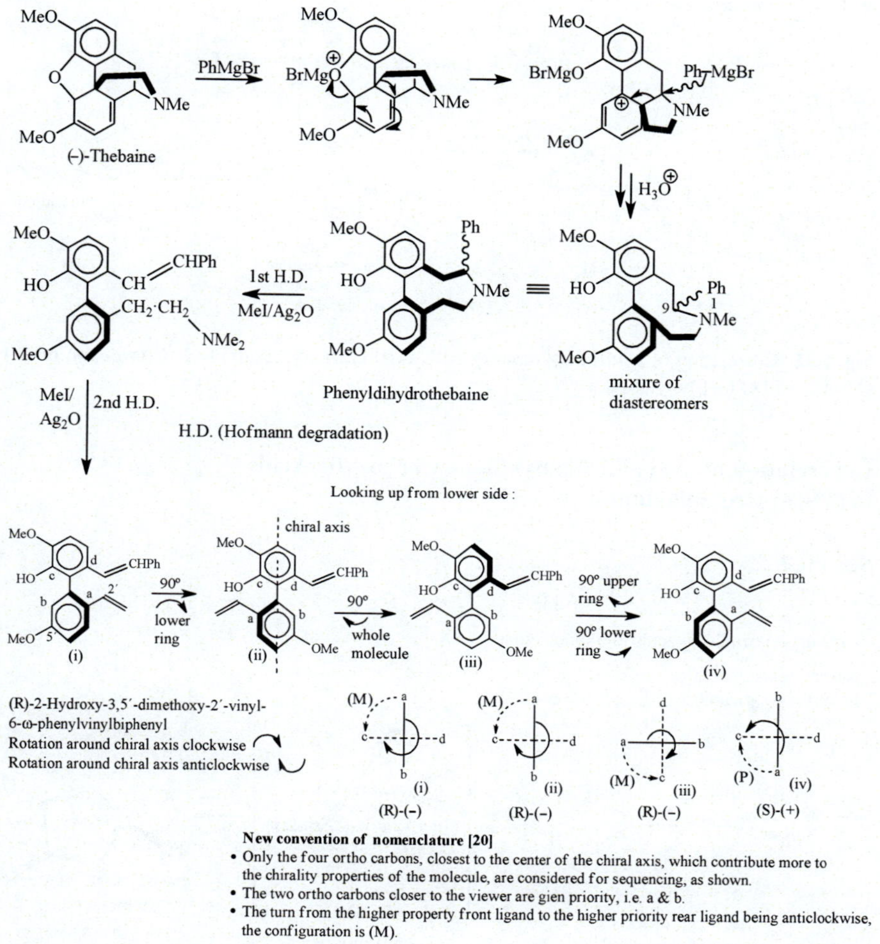

Fig. 6.21 Correlation of axial chirality with centrodissymmetry. Example 3. (*R*)-2-hydroxy-3,5'-dimethoxy-2'-vinyl-6-*E*-ω-phenylvinyl-biphenyl with (−)-thebaine

twice by Hofmann degradations, produces *R*-2-hydroxy-3,5'-dimethoxy-2'-vinyl-6-ω-phenylvinyl-biphenyl, the absolute configuration of which can be determined by application of C.I.P rules. For (*R,S*) specification of axially chiral molecules, see Sect. 10.6.5, [20] and [21] (Fig. 6.21).

Correlation of Centrodissymmetry with Axial Chirality. Example 1. Conversion of (*R*)-Glutimic Acid to (−)-Norcamphor

See Fig 6.22.

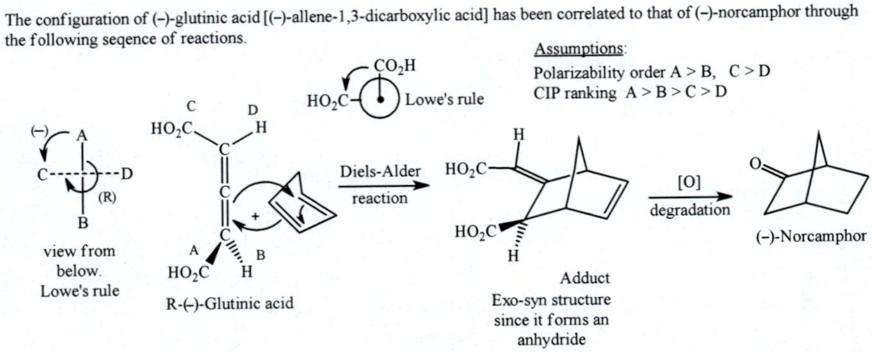

Fig. 6.22 Correlation of centrodissymmetry with axial chirality. Example 1. Conversion of (*R*)-glutinic acid to (−)-norcamphor

Corrrelation of (*S*)-(+)-3-Phenylallene Carboxylic Acids to (*S*)-(+)-Bromolactones

See Fig 6.23.

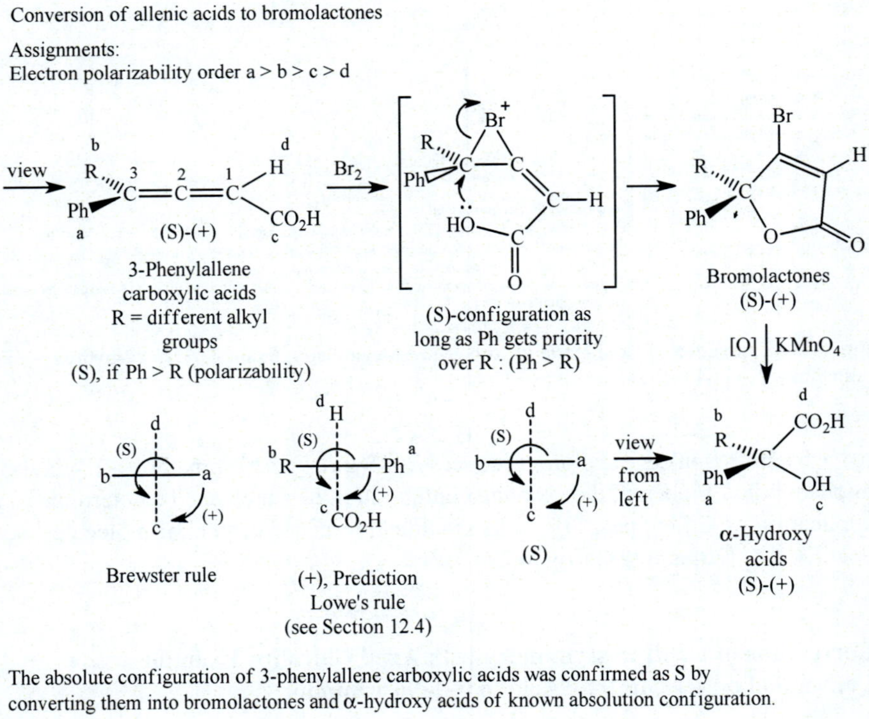

Fig. 6.23 Corrrelation of (*S*)-(+)-3-phenylallene carboxylic acids to (*S*)-(+)-bromolactones

6.3 Baldwin's Rules for Ring Closure

During Professor Baldwin's stay in MIT, Boston enunciated rules of ring closure bearing his name [22, 23]. He classified ring closures by distinguishing "exo" and "endo" cases, as shown in the Fig. 6.24. We are going to discuss the Baldwin's rules based on the summarization of the rules by Eliel and coworkers in their books [24, 25]. The nucleophile at one end of the cyclizing chain attaches itself to the other end of that chain in endo ring closures, whereas in exo ring closures, attachment is at the penultimate atom and the end (terminal) atom remains outside of the newly formed ring.

The reactions are also classified as "tet," "trig," or "dig" depending on whether ring closure involves displacement at a tetrahedral (*sp*$_3$ hybridized) atom or addition

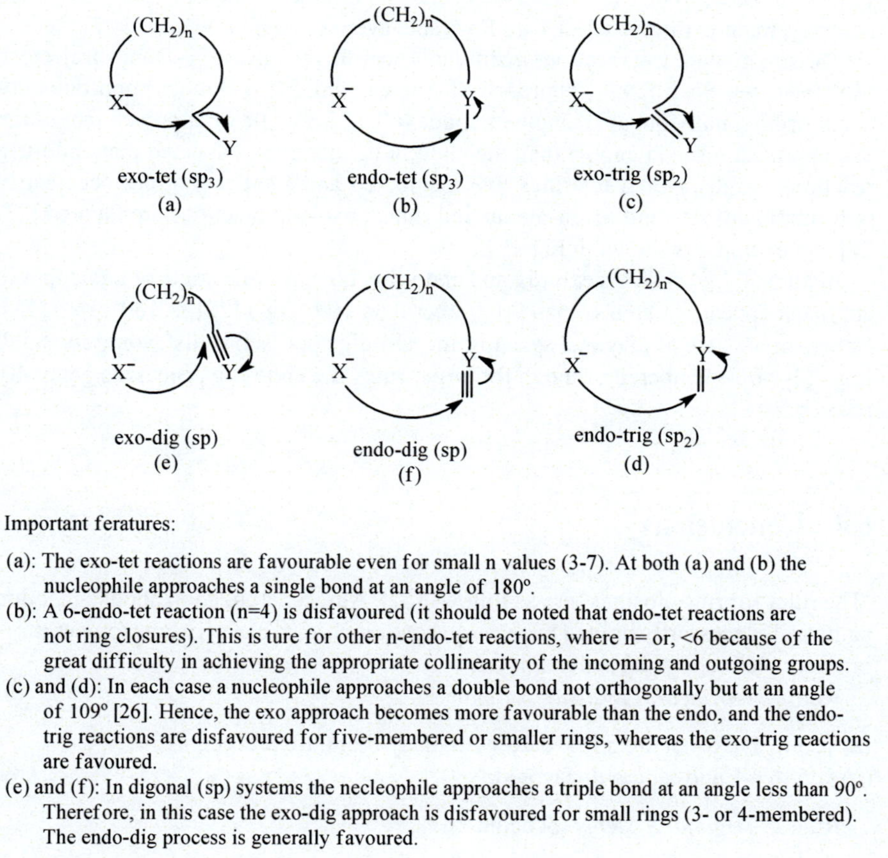

Fig. 6.24 Examples of exo-tet and endo-tet, exo-trig and endo-trig and exo-dig and endo-dig ring closures

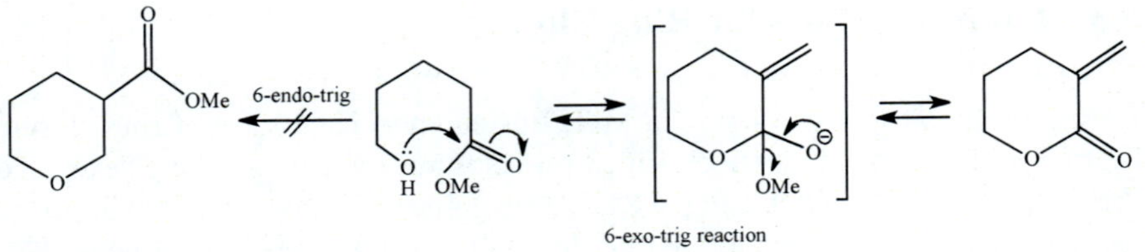

Fig. 6.25 Limitations of Baldwin's rules

to a trigonal (sp^2 hybridized) or diagonal (*sp* hybridized) moiety. The size of the ring to be formed is finally indicated by an appropriate leading "numeral."

6-Endo-tet reaction is disfavored—which would not be a ring closure. Other n-endo-tet reactions where $n < 6$ are unfavorable because of the great difficulty in achieving the appropriate collinearity of the incoming and outgoing groups. On the contrary, the n-exo-tet reactions are favorable even for small n values (3–7).

The situation for trig reactions is similar. According to the Burgi–Dunitz trajectory [26] (also see Sect. 8.4.3), approach of a nucleophile to a double bond does not occur orthogonally but at an angle of about 109°. This angle of approach makes the exo approach more favorable than the endo; it is, therefore, expected that endo-trig reactions are disfavored for rings five membered and smaller (but not necessarily in formation of six- and seven-membered rings); exo-trig reactions are favored [23, 27]. An example is shown in Fig. 6.25.

In digonal (*sp*) systems exo-dig and endo-dig, however, the angle of nucleophilic approach appears to be less than 90°, rather than 109° (sp^2) [22] or 180° (sp^3) [22]. Hence, in the case of diagonal systems, the exo-dig approach is disfavored for small rings (3- or 4-membered), though for larger rings the endo-dig process is generally favored.

6.4 Conclusions

'The rules for ring closures' are as follows [22]. Any generalization about a reaction pathway is necessarily limited by reaction conditions, e.g., at a very high temperature, or under photochemical activation etc.

Rule 1: *Tetrahedral Systems* (Scheme 1).

(a) 3 to 7-Exo-tet are all favored with many literature precedents.
(b) 5 to 6-Endo-tet are disfavored.

Rule 2: *Trigonal Systems* (Scheme 2).

(a) 3 to 7-Exo-trig are all favored processes with many literature precedents.
(b) 3 to 5-Endo-trig are disfavored, 6- to 7-endo-trig are favored.

Rule 3: *Digonal Systems* (Scheme 3).

(a) 3 to 4-Exo-dig are disfavored processes; 5 to 7-exo-dig are favored.
(b) 3 to 7 Endo-dig are favored.

Scheme 1–3 forming rings of different sizes have been delineated by Baldwin [22].

References

1. D.H.R. Barton, Conformation of Steroid Nucleus, Experientia, 316–319 (1950); reprinted in Top. Stereochemistry, 6, 1–4 (1971)
2. S.K. Talapatra, B. Talapatra, *Chemistry of Plant National Products: Stereochemistry, Conformation Synthesis, Biology, and Medicine* ed by K. C. Nicolaou (ed.) vol. 1, Chapters 8, 10, and 11 (Springer-Verlag, Berlin Heidelberg, 2015)
3. M. Anteunis, D. Tavernier, F. Borremans, NMR experiments on Ketals VI. PMR Spectra of Alkyl Substituted 1,3-Deoxanes with rigid or anancomeric structures. Bull. Soc. Chim. Belges, **75**, 396–412 (1966)
4. M.J.O. Anteunis, Some results and limitations in conformational studies at six-membered heterocycles, in *Conformational Analysis, Scope and Present Limitations* ed. by G. Chiordoglu (Academic Press, New York, 1971), p. 32
5. A.J. Gordon, R.A. Ford, *The Chemists' Comanion* (Wiley, New York, 1972)
6. S. Winstein, N.J. Holness, Neighboring carbon and hydrogen. XIX, *t*-Butylcyclohexyl derivatives. Quantitative conformational analysis. J. Am. Chem. Soc. **77**, 5562–5578 (1955)
7. E.L. Eliel, S.H. Wilen, M.P. Doyle, *Basic Organic Stereochemistry* (John Wiley, New York, 2001), pp. 443–445
8. E.L. Eliel, R.S. Ro, Conformational analyses III. Epimerization equilibria of Alkylcyclohexanols. J. Am. Chem. Soc. **79**, 5992–5994 (1957)
9. E.L. Eliel, R.S. Ro, *Chemistry and Industry* (London, 1956), pp. 251–252.
10. E.L. Eliel, C.A. Lukach, Conformational analysis, II. Esterification rates of Cyclohexanols. J. Am. Chem. Soc., 9, 5986–5992 (1957)
11. J.I. Seeman, Effect of conformational change on reactivity in organic chemistry, evaluations, applications and extensions of curtin-hammett/winstein-holness kinetics. Chem. Rev. **83**, 83–134 (1983)
12. J.I. Seeman, The curtin-hammett principle and the winstein-hlness equation, new definition and recent extensions to classical concepts. J. Chem. Educ. **63**, 42–48 (1986)
13. D.Y. Curtin, Stereochemical control of organic reactions. Rec. Chem. Prog. (Kresge-Hooker Sci. Lab.) **15**, 111–128 (1954)
14. D.Y. Curtin, D.B. Kellom, Elimination and replacement reactions of *dl-erythro* and *dl-threo*-2-Deutero-1,2-diphenylethanol and Derivatives. J. Am. Chem. Soc. **75**, 6011–6018 (1953)
15. J.I. Seeman, The curtin-hammett principle and the wistein-holness equation. J. Chem. Educ. **63**, 42–48 (1986)
16. E.L. Eliel, Conformational analysis in mobile systems. J. Chem. Educ. **37**, 126–133 (1960)
17. J.I. Seeman, H.V. Secor, H. Hartung, R. Galzerano, Steric effects in conformationally mobile systems. The Iodomethylation of 1-Methyl-2-arylpyrrolidines Related to Nicotines. J. Am. Chem. Soc. **102**, 7741–7747 (1980)
18. M. Oki, *Methods of Stereochemical Analysis; Application of Dynamic NMR in Organic Chemistry*, Vol. 4, ed. by A.P. Merchannd, (VCH Deerfield Florida, 1984)
19. D. Hafner, S.A. Lesko, G. Binsch, Org. Magn. Reson. **11**, 179 (1978)
20. R.S. Cahn, S.C.K. Ingold, V. Prelog, Specification of molecular chirality, Angew. Chem. Int. Edn. Engl. **5**, 385–415 (1966)
21. R.S. Cahn, C.K. Ingold, V. Prelog, The specification of asymmetric configuration in organic chemistry. Experientia **12**, 1022–1025 (1956)

22. J.E. Baldwin, Rules for ring closure. J.C.S. Chem. Comm. 734–736 (1976)
23. J.E. Baldwin, J. Cutting, W. Dupont, C. Kruse, L. Silberman, R.C. Thomas, 5-Endo-trigonal reactions: a disfavoured ring closure. J.C.S. Chem. Comm. 736–738 (1976)
24. E.L. Eliel, S.H. Wilen, L.N. Mander, *Stereochemisty of Organic Compounds.* (John Wiley, New York, 1994), pp 684–685
25. E.L. Eliel, S.H. Wilen, M.P. Doyle, *Basic Organic Stereochemistry* (Wiley, New York, 2001), pp. 434–436

Chapter 7
Conformation of Saturated Six-Membered Ring Compounds

Check for updates

Abstract This chapter depicts the chair and boat conformations of cyclohexane, their symmetries and enthalpies or potential energies, the energy profile of ring inversion of cyclohexane, conformational energies of monosubstituted cyclohexane and 1,1-disubstituted cyclohexanes; conformational energies, symmetries and optical activity of the nongeminal dimethylcyclohexanes; and effect of dipole moment, H-Bonding and bulky substituents in conformation in some typical disubstituted cyclohexane derivatives.

7.1 Conformational Aspects of Cyclohexane

7.1.1 Geometry of Cyclohexane Chair. Bond Lengths. Bond Angles. Torsion Angles

Baeyer considered that cyclohexane is a strained planar molecule. In 1890, Sachse first pointed out that cyclohexane might be nonplanar puckered chair or boat shaped and unstrained. The physical and chemical properties of axially and equatorially substituted cyclohexane derivatives (especially of steroids) were first pointed out by D. H. R. Barton in his classical epoch-making paper [1] in 1950, which should be read by any interested reader.

As evident from X-ray and electron diffraction experiments by Hassel in 1943, cyclohexane exists almost exclusively in its chair conformation. The electron diffraction experiments reveal the C–C and C–H bond lengths, bond angles, the intraannular (C–C–C–C), and the external H–C–C–H torsion angles, which are shown in (**1A**) and (**2A**) (Fig. 7.1). The C–C–C– bond angle (111.4°) is more than the regular tetrahedral angle (109°) and little less than the C–C–C bond angle (112.4°) in *n*-propane. The (C–C–C–C) torsion angle (55°) deviates from the optimum of 60° in n-propane, because of the constraint in the ring. Such a decrease in the torsion angle is resisted due to an increase in torsional strain involved. A compromise is reached to minimize the total strain when the bond angle slightly increases, and the torsion angle becomes little smaller than that of an open chain molecule like n-propane. Thus, cyclohexane is not an entirely strain-free molecule; some angle strain and torsional strain remain.

S. K. Talapatra and B. Talapatra, *Basic Concepts in Organic Stereochemistry*,
https://doi.org/10.1007/978-3-030-95990-6_7

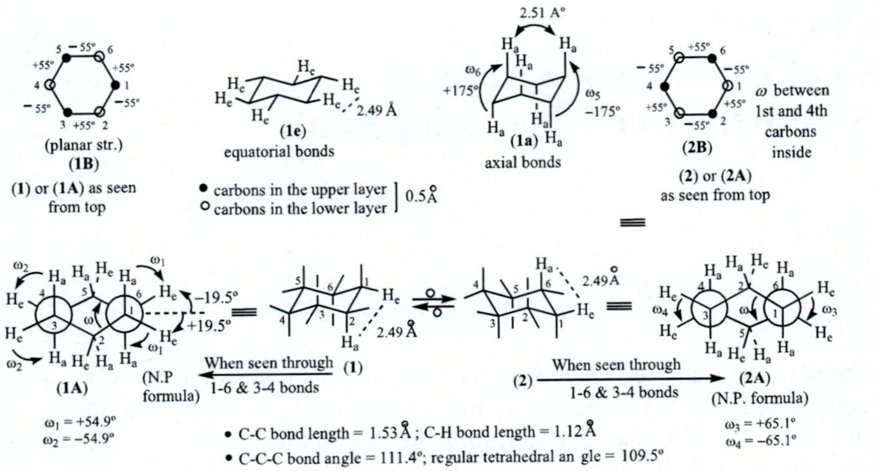

- C-C-C bond angle in propane = 112.4°
- Torsion angle, ω, of C-C-C-C, or ω_1 or ω_2 H_a-C-C-H_e = +54.9° or −54.9°. Torsion angle ω_3 or ω_4, H_e-C-C-H_e = +65.1° or −65.1°
- For determining the sign of ω from chair orboat form (without drawing the N.P formulas) see **Section 2.15.1.2.**
- The distance between hydrogen atoms at positions (i) 1a & 2e = 1e & 2a = 1e &2e = 2.49 Å (diverse directions) (ii) 1a & 3a = 2.51 Å (parallel directions). The distance between substituent(s) in case (i) increases (more in 1e, 2e), but in case (ii) remains unaltered if the substituents are same, but increases slightly when the substituents are different.

Fig. 7.1 Salient features of the geometry of cyclohexane chair and its flipped conformer

7.1.2 Equatorial and Axial Bonds

Cyclohexane has two geometrically different sets of H atoms, six are approximately parallel with the vertical C_3 axis, alternatively up and down, as shown in structure **(1a)** (Fig. 7.1) called axial (a) hydrogen atoms; the remaining six H atoms are distributed around the periphery of the ring making alternately +19.5° and −19.5° angles with the equatorial (horizontal) plane of the molecule and are called equatorial (e) as shown in the structure **(1e)** (Fig. 7.1) (see references [1, 2]

The decrease in the intraannular torsion angle to 54.9° brings with it a decrease in the external H–C–C–H torsion angle of *cis* (*ae or ea*) located H atoms from usual 60° (in ethane) to 54.9° [structure (**1A**)] and that of *trans* (*aa*) located H atoms from usual 180° to 174.9°, and an increase in the corresponding *trans* (*ee*) located H atoms from 60° to 65.1°. In ^{1}H NMR spectrum, the vicinal coupling constants for such *cis* and *trans* protons vary according to Karplus equation {see [3]}.

Since the distances between different pairs of adjacent hydrogens and between 1,3-diaxial hydrogens are 2.49 and 2.51 Å, respectively (Fig. 7.1), which are more than twice the van der Waals radius (1.20 Å) of hydrogen, there is no nonbonded interaction in the chair form of cyclohexane.

7.1.3 *Symmetry of Cyclohexane Conformations*

1. ***Chair*** **(1)**. The principal axis of symmetry is the C_3 axis passing vertically through the center of the chair; this is also an S_6 axis. Additionally, the chair conformation possesses three C_2 axes bisecting pairs of opposite sides (str. **3** and **4**) (Fig. 7.2), a center of symmetry, and also three vertical σ planes intersecting at the C_3 axis and passing through the diagonal (opposite) carbon atoms. Each σ_v plane also bisects the angle between two C_2 axes; hence, the σ_v planes are also known as σ_d. Thus, the cyclohexane chair belongs to the point group $\mathbf{D_{3d}}$ (Fig. 7.2) and cyclohexane molecule is achiral.
2. ***The flexible forms–boat and twist–boat. Symmetry***. The two extremes of the flexible forms of cyclohexane are boat **(5)** and twist–boat **(8)** (Fig. 7.2). These are interconverted by *pseudorotations*, which are low-energy processes, and involve only changes in torsional strain and other nonbonded interactions, but do not involve bond angle variation. This can be verified by molecular model (Dreiding or Fischer) study. In the true boat **(5)**, the bowsprit and flagpole hydrogens are only about 1.8 Å apart, whereas the sum of the van der Waals radii of two hydrogen atoms is about 2.4 Å. Thus, there is a significant *bowsprit–flagpole* (*bs–fp*) interaction.

To minimize the *bs–fp* interaction, the *bs* and *fp* H's are pulled a little apart resulting in the twist–boat in which the eclipsing of the adjacent H's at C2 and C3 and at C5 and C6 are also somewhat alleviated. The important stereochemical features including the symmetry point group of the boat and twist–boat forms are delineated in Fig. 7.2.

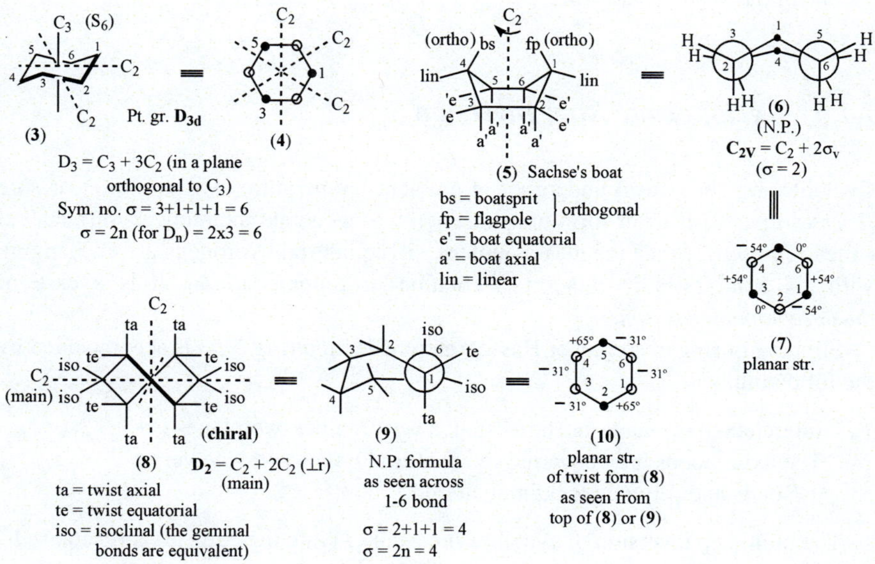

Fig. 7.2 Salient features of the geometry of cyclohexane boat and twist form

The symmetry point groups of boat and skew–boat forms of cyclohexane are $\mathbf{C_{2v}}$ [$C_2 + 2\sigma_v$ orthogonal)] and $\mathbf{D_2}$ [$C_2 + 2C_2$ (orthogonal)], respectively (Fig. 7.2).

7.1.4 *Enthalpy* (H) *or Potential Energy* (E) *Difference*

Here b-ec stands for butane–eclipsing interactions and b–g stands for butane–gauche interactions (see Fig. 7.2).

H_{boat} = two *b-ec* (1,2,3,4 carbons + 4,5,6,1) interactions + 4 *b*-g (The other four successive 4 carbons) = 2 × 4.4 to 6.1) + (4 × 0.9) = (8.8 to 12.2) + 3.6 = 12.4 to 15.8 kcal/mol

H_{chair} = 6 b–g interactions (6 combinations of consecutive 4 carbons) = 0.9 × 6 = 5.4 kcal/mol

$H_{boat}–H_{chair} = \Delta H$ (or ΔE) = 12.4–15.8) – 5.4 = 7.0 to 10.4 kcal/mol

Since potential energy of skew–boat is 1.6 kcal/mol less than that of boat, the energy difference between skew–boat and chair form = 7.9 to 10.4—1.6 = 5.4 to 8.8 kcal/mol.

The enthalpy difference between flexible (skew–boat) form and chair form is estimated to be 5–6 kcal/mol based on heat of combustions and equilibrium determinations.

So the free energy difference $\Delta G = \Delta H - \mathrm{T}\Delta S$ (entropy of flexible form is 5 e.u. greater than that of chair form at 298° K) = 5560 − 5 × 298 = 4000 cals = 4 kcal/mol.

Now from the expression $\Delta G = 4000$ cals $= RT\ln K$, one can calculate the equilibrium constant $K = 1000$ (approx.) which means that only one molecule in a thousand will be in the skew–boat form.

7.1.5 *Cyclohexane Ring Inversion*

Cyclohexane chair form undergoes at ambient temperature *ring inversion* or *ring reversal* to another chair form of same energy (degenerate interconversion). Hassel called it *flipping*. Since the diastereomeric six equatorial hydrogens are exchanging with six axial ones in this transformation or *topomerization*, it is a case of *diastereotopomerization*.

Flipping or ring inversion or Hassel interconversion (Fig. 7.3) is accompanied by the following:

1. Interchange of planes occur (distance between the two planes is ~0.5 Å).
2. The axial bonds are converted to equatorial bonds and vice versa.
3. α-Bonds and β-bonds remain unaltered.

The multistep inversion of cyclohexane chair (**1**) into the equienergetic chair (**1′**) through sequentially the transition state (**TS1**), skew–boat forms, and the transition state (**TS1′**) [enantiomeric with (**TS1**), is depicted in Fig. 7.4. The interconversion

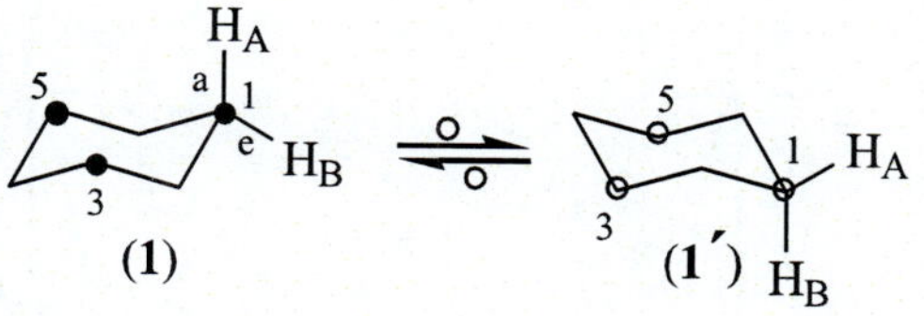

Fig. 7.3 Ring inversion of cyclohexane

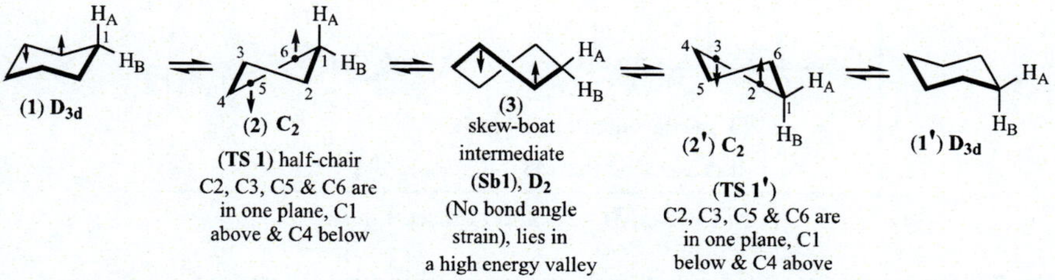

- Arrows show the direction of suitable twisting involving extensive bond angle deformation with increased torsional strain
- The enantiomeric (**TS1**) and (**TS1′**) have a C_2 axis (point group $\mathbf{C_2}$), but no σ-plane.
- (**Sb1**) may be reversibly converted to two other skew-boat forms, one identical and the other enantiomeric, by pseudorotation without change of bond angles, but changing only torsion angles.

Fig. 7.4 Conformational itinerary involved in cyclohexane ring inversion (C_2 pathway)

of the skew–boat forms takes place through the intermediate Sachse's boat form, as evident from manipulation of molecular models. A pathway involving chair → envelop/half chair-like TS → intermediate boat form → inverted envelop/half chair TS → inverted chair is also possible. This is called σ-pathway since both the equivalent transition states and the intermediate boat form retain a symmetry plane of the ground state chair form. *But force field calculations make the existence of the boat form of cyclohexane as an intermediate in the cyclohexane inversion itinerary highly unlikely* [4], and hence the σ-pathway is ruled out. The skew–boat (or twist) form is obtained from the so-called *boat form* in the potential energy diagram by slight deformation to alleviate the *fp* ↔ *bs* interaction and is less stable than the chair conformer by 4.7–6.2 kcal/mol according to various indirect experiments and also by force field calculations. The calculations also suggest that the true boat form is of about 1–1.5 kcal/mol higher energy than that of the twist form, and that the (**TS1**) or (**TS1′**) is of 10.7–11.5 kcal/mol higher energy than the chair form. The boat form is apparently at the energy maximum in the interconversion of the twist (or skew–boat) conformers of which two are mirror images. This pathway (Fig. 7.4) is called C_2 pathway, since the C_2 axis of the ground state chair form is retained along this pathway.

The energy profile diagram of ring inversion of cyclohexane following the C_2 pathway is depicted in Fig. 7.5.

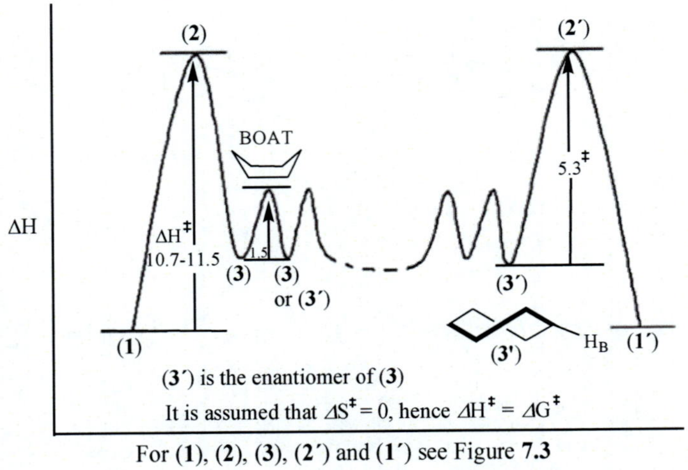

Fig. 7.5 Energy profile ($\Delta H^{\neq}$ in kcal/mol) of ring inversion of cyclohexane

7.1.6 Stable Boat or Skew–Boat Conformers

The chair form of cyclohexane is usually the most stable conformer. However, under certain conditions of structures the boat or the skew–boat forms can be stabilized in preference to the chair form. In case of certain molecules, as shown in Fig. 7.6, the configurational requirement does not allow the chair form of a particular ring. This is also true in cases of *trans-cisoid-trans*-perhydrophenanthrene (in Fig. 9.11), *trans-transoid-trans*-perhydroanthracene (in Fig. 9.15), the D/E rings in multiflorenol and friedelin, and the E ring in bauerenol and eapacannol (see Fig. 10.33 of [3]), also see eupacannone in Fig. 8.7.

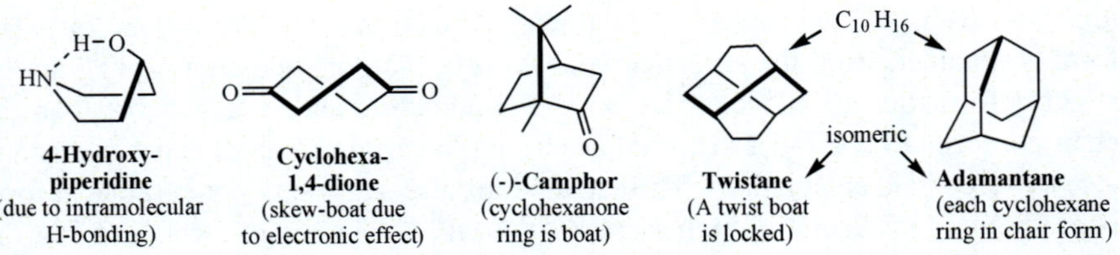

Fig. 7.6 Some compounds having fixed boat or skew–boat conformation of the cyclohexane ring

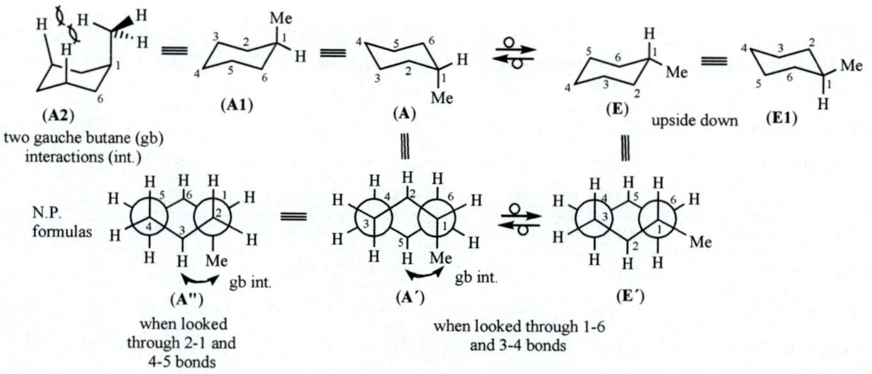

Note: In (**A′**) or (**A′′**) (like A2) the distance between axial methyl protons and synaxial protons is ~ 1.83 A (~ van der Waals radii of two H's)

Fig. 7.7 Conformers of methylcyclohexane

7.2 Monosubstituted Cyclohexanes. Conformational Energy

Monosubstituted cyclohexanes with the more stable equatorial substituent (E) flip into its nonequivalent diastereomeric less stable chair conformations with the substituent axial, as shown for methylcyclohexane in Fig. 7.7. Corresponding Newman projection formulas are also shown. The energy barrier of ring inversion in substituted cyclohexanes remains practically unaffected. The energy profile diagram is similar to that of cyclohexane ring inversion (Fig. 7.5); of course, the two chair conformers have different enthalpies and the rates of forward and backward interconversions are different. For example, in case of methylcyclohexane the equatorial conformer (E) will have two gauche–butane interactions (~1.8 kcal/mol) less enthalpy than the corresponding axial conformer (A), as shown in Fig. 7.7, both in chair and Newman projection (N.P.) formulas.

7.3 Conformational Energy

The energy (enthalpy or potential energy or free energy) of a given conformational isomer over and above that of the conformational isomer of minimum energy is called conformational energy.

As already mentioned before, the two diastereomeric chair forms (**A**) and (**E**) (Fig. 7.7) are of unequal free energy and hence their populations are different. Their equilibrium constant K is given by the equation:

$$\Delta G^{\circ} = -RT\ln K \ldots \text{(equation } \mathbf{a}\text{), where } K = [(\mathbf{E})]/[(\mathbf{A})].$$

Table 7.1 Conformational free energies ($-\Delta G°$) in kcal/mol (approx)[a]

Substituent $-\Delta G°$ (~*t*, °C)	Substituent $-\Delta G°$ (~*t*, °C)	Substituent $-\Delta G°$ (~*t*, °C)	Substituent $-\Delta G°$ (~*t*, °C)
D 0.006 (25)	CN 0.2 (−79)	OH[b] 0.60 (25)	Me 1.74 (27)
T 0.011 (−88)	NH_2[c] 1.23 (−80)	OH[d] 1.04 (−83)	Et 1.79 (27)
F 0.25 (−86)	NH_2[e] 1.7 (20)	OMe 0.63 (−93)	$CH(Me)_2$ 2.21 (27)
Cl 0.53 (−80)	NO_2 1.1 (−80)	$OC(Me)_3$ 0.75 (36)	$C(Me)_3$ 4.8 (−120)
Br 0.48 (25)	COOMe 1.2 (25)	OPh 0.65 (−93)	Ph 2.8 (−100)
I 0.47 (−78)	COOH 1.4 (25)	OCOMe 0.68 (25)	CH_2Ph 1.68 (−71)

[a]More detailed tabulations have been compiled in the textbook by Eliel, Wilen, and Doyle [5]
[b]in cyclohexane
[c]in toluene-d_6
[d]in CS_2
[e]in $MeOCH_2CH_2CH_2OH$

The difference of free energy between equatorial (**E**) and axial (**A**) conformers, $\Delta G°$, is negative (the more stable conformer to be written on the right side of the equilibrium). *Thus, $-\Delta G°$, which is positive, is known as the conformational free energy of the substituent* (also known as *A*-value). Usually steric grounds (as explained in Fig. 7.7 in case of methylcyclohexane), in some cases electronic factors may dictate the more stable conformer. In the Newman projection (**E′**) for the equatorial conformer no extra gauche–butane interaction is present, whereas in Newman projections (**A′**) and (**A″**) of the axial conformer one extra *gb* is seen in each; the chair form (**A2**) shown both *gb* interactions. Thus, the enthalpy difference, ΔH, between the two diastereomeric conformers = 0.8 × 2 = 1.6 kcal/mol (approx.). If $\Delta S = 0$, $\Delta G = \Delta H$ (from the equation $\Delta G = \Delta H - TAS$) (equation **b**). This value corresponds to over 90% population of the e conformer (**E**) (but NMR studies indicate it to be 95%).

The conformational free energies ($-\Delta G°$ values) of a number of common substituents are listed in Table 7.1.

Table 7.1 reveals the following interesting facts.

1. In the halogen series the effective bulk of F being least, $-\Delta G°$ value is least, $-\Delta G°$ values for Cl, Br, and I are almost equal due to the increase in their bond lengths resulting in increase in the distance from the *syn*axial hydrogens. Moreover, larger atomic volume causes the increased polarizability of electrons and the ease of deformation to reduce the energy of the axial conformer. In general, the elements of the first two rows in the Periodic Table with relatively short C–X bonds and low polarizability show larger $-\Delta G°$ values than the heavier elements with longer bonds to carbon and higher polarizability.
2. Groups like >NH, –NHMe, OH, etc. which may form H-bonds have different $-\Delta G°$ values in protic and aprotic solvents (for OH, see reference [6]).
3. Substituents at the second atom (in italics), for example, *O*Me, *O*Et, *O*Ts, *O*COMe do not significantly affect the effective bulk. The $\Delta G°$ values of Me,

Et, *i*Pr do not differ as expected, but it increases sharply for *t*-Bu group having greatest effective bulk. The *t*-butyl substituted cyclohexanes have almost exclusively equatorial conformation and have been called ***anancomeric***, meaning *fixed in one conformation* (derived from Gk *unankein* meaning to fix by some fate or law).

4. The $\Delta G°$ values are approximately additive, and they may be used for di- and polysubstituted cyclohexanes.
5. The additivity of $\Delta G°$ values does not hold good in geminally substituted cyclohexanes [7].

7.4 1,1-Disubstituted Cyclohexanes

1,1-Disubstituted cyclohexanes having a plane of symmetry (**C**$_S$) passing through C1 and carrying achiral substituents do not show any configurational isomerism even if the substituents are different. They exist in two interconvertible conformers when the substituents are different. The energy barrier is usually of the similar order as that of cyclohexane. For example, in 1-methylcyclohexanol (**1**) the ratio of the two diastereomeric conformers (**1a**) and (**1e**) (Fig. 7.8) should correspond to the difference in conformational free energies of the two substituents (1.74 − 1.04 = 0.70 kcal/mol). But actually (**1**) exists as a 70.30 mixture of (**1a**) and (**1e**) conformers in Me_2SO at 35 °C, corresponding to a free energy difference of 0.48 kcal/mol. Here the bulkier group Me in (**1a**) is predominating but to a much less amount due to some leveling effect.

In case of 1-methyl-1-phenylcyclohexane (**2**) (Fig. 7.8), contrary to expectation, the conformer (**2a**) with axial phenyl and equatorial methyl is preferred over the other conformer (**2e**) with equatorial phenyl and axial methyl by 0.32 kcal/mol in spite of the fact that the difference of their conformational free energy (2.8 − 1.74 = 1.06 kcal/mol) would lead to the opposite conclusion. This anomaly may be explained as follows (Fig. 7.8). In the conformer (**2a**), the phenyl ring is so oriented that the ortho H's are away from axial H3 and axial H5 and so the *synaxial* interactions of the phenyl with these H's become minimal. However, there may be some interactions between the ortho H's and the adjacent e-H2 and e-H6. On the contrary, the phenyl

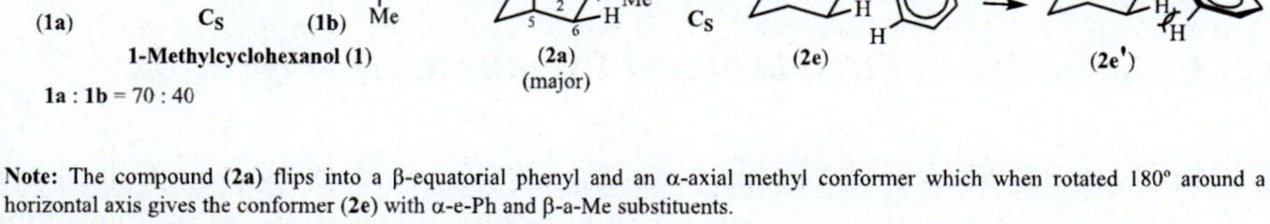

Note: The compound (**2a**) flips into a β-equatorial phenyl and an α-axial methyl conformer which when rotated 180° around a horizontal axis gives the conformer (**2e**) with α-e-Ph and β-a-Me substituents.

Fig. 7.8 Conformations of two geminally substituted cyclohexanes group in the flipped conformer (**2e**) (see **Note** in the Figure) would interact strongly with the Me H's. Hence, the phenyl group must rotate by 90° (*cw*) about its pivotal bond to give the conformer (**2e′**), in which also strong interactions between the ortho H's and the adjacent *e*–H's exist. Thus, the additivity of ΔG° values does not hold good in geminally substituted cyclohexanes [5].

7.5 Nongeminal Disubstituted Cyclohexanes

Disubstituted cyclohexanes, other than the germinal ones (Sect. 7.4), exist in three sets of positional isomers, viz. 1,2-, 1,3-, and 1,4-isomers. Each set has a *cis*–*trans* pair of diastereomers. Each diastereomer can flip into interconvertible chair conformers. The relative enthalpy of each conformer depends upon the steric interactions present. Depending upon its symmetry property, a particular isomer may exhibit enantiomerism. These features in case of all dimethylcyclohexanes are illustrated in Fig. 7.9. Calculation of relative free energy and entropy for each isomer is also shown as **Notes** in the figure. Some important stereochemical features of dimethylcyclohexanes having *different substituents* are stated below.

cis-1,2-Isomers. The *cis*-1,2-isomer having *different substituents* is resolvable since the flipped conformers are different entities and not enantiomers. The bulkier group exists predominantly in the equatorial orientation.

1,3-Isomers. Both *cis*- and *trans*-isomers with *different substituents* are resolvable, since each of them has C_1 pt gr. and each flips into a different chiral variety. The *cis*-isomer exists in a preferred *ee* conformation and the *trans*-isomer in an *ea* conformation with the bulkier group predominantly in the *e* orientation.

Though the 1,3-*cis*-isomer exist almost 100% in the diequatorial (*ee*) conformation, if necessary it can adopt the diaxial (*aa*) conformation by ring inversion to bring the substituents within reacting distance, *e.g.*, formation of an anhydride from *cis*-1,3-cyclohexanedicarboxylic acid or formation of intramolecular H-bond in *cis*-1,3-cyclohexanediol.

1,4-Isomers. The *cis*- and *trans*-1,4-isomers are always meso because they possess a vertical σ-plane passing through C1 and C4, even if the substituents are different. The *cis*-1,4-isomer exists preferentially in the conformation having the bulkier group in the *e* conformation, whereas the *trans*-1,4-isomer exists almost exclusively in the *ee* conformation.

7.5.1 Some Typical Disubstituted Cyclohexanes (Fig. 7.10)

1. In *trans*-1,2-dihalocyclohexanes, the diaxial (*aa*) conformer (which does not exist in dimethyl series) is substantially populated, sometimes as the major conformer (more than the *ee* conformer (**2**)). The percentage of *aa* conformer (**1**) having no dipole repulsion between the halogen atoms increases in the series:

Planar Structure (Optical activity)	Conformation (symmetry pt. gr.)	Inter-actions	Relative enthalpy ΔH kcal/mole	Kcal/mole **ΔH exptl. (ΔG exptl.)**	Flips into	Regarding chirality
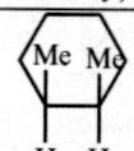 cis-1,2 σ-plane (meso)	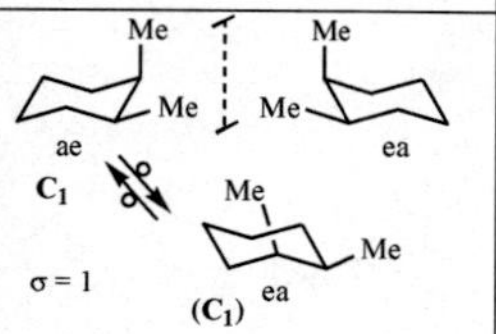	2 (axial Me) + (1e ↔ 2a Me's) = 3 gb.	2.7 } ΔH (calc) 1.8	1.87 (1.66)	A non-superposable mirror image (energy barrier much less than 10 kcal/mole	Unresolvable (±)-pair
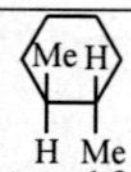 trans-1,2 C_2 pt.gr. (active)	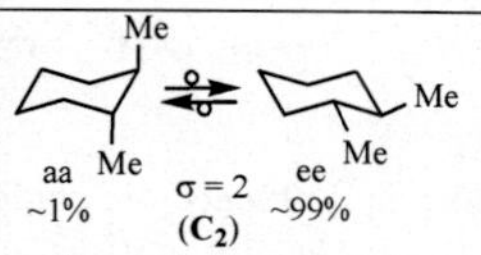 (The C_2 passes through the midpoints of 1-2 & 4-5 bonds)	ee 1 gb aa 4 gb	0.9 3.6		another chiral variety of the same enantiomer (lacks an σ-plane or i)	Resolvable (±)-pair
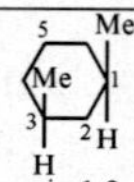 cis-1,3 (σ-plane) (meso)	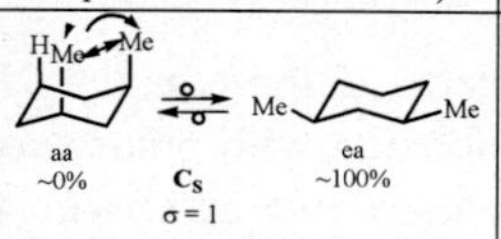 Each variety has a vertical σ-plane passing through C2 and C5 and orthogonal to the puckard plane	aa syn axial MeMe, + 2 gb	5.4 0 } −1.8	−1.96 (−1.56)*	another achiral variety	True meso
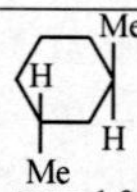 trans-1,3 C_2 (active)	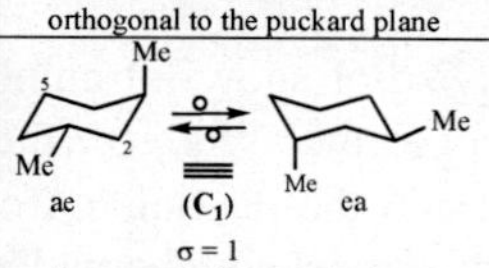 (possess only C_1 symmetry)	2 gb for one a-Me	1.8		a superposable (homotopic form)	Resolvable (±)-pair
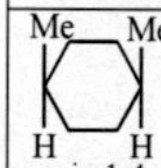 cis-1,4 (meso) σ-plane through C1 & C4	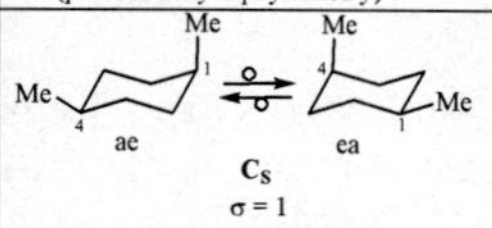 (A vertical σ-plane passess through C1 and C4)	2 gb for one a-Me	1.8 } 1.8	1.90 (1.55)*	a superposable (homotopic form)	True meso
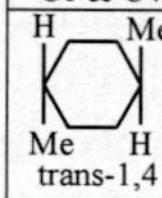 trans-1,4 (meso) σ-plane through C1 & C4	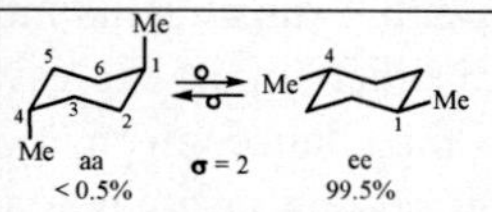 (C_{2h} for each): C_2 axis biscet 2-3 and 5-6 bonds; σ_h orthogonal to C_2, passing through C1 and C4)	No gb for ee form 4 gb for aa form	0 3.6		another achiral variety	True meso

Notes: • Symmetry consideration of the planar structures gives the correct inference regarding resolvability.
- G = Free energy; H = Enthalpy (or Potential Energy E); S = Entropy
- ΔG (Relative free energy) = $G_{cis} - G_{trans}$; $\Delta H = H_{cis} - H_{trans}$; $\Delta S = S_{cis} - S_{trans}$
- ΔG = ΔH − TΔS (ΔG*exptl. is calculated from experimental ΔH and ΔS data)
- Total relative entropy = S_{calcd}

S_{calcd} (for trans-1,2) = −Rln σ (σ, symmetry no. is 2) + Rln (entropy of mixing) (due to dl) + S (due to ee, aa equilibrium)
= (−1.38) + 1.38 + 0.11

S_{calcd} (for cis-1,2) = 0 + 1.38 + 0 = 1.38

*Entropy of mixing = $-R(x_1 \ln x_1 + x_2 \ln x_2)$ where x_1 & x_2 are mole fractions of each component in the mixture
= $-R(\frac{1}{2}\ln\frac{1}{2} + \frac{1}{2}\ln\frac{1}{2}) = R\ln 2 = 1.38$ cal/deg. mole

Fig. 7.9 Conformations energies, symmetries, and optical activity of the nongeminal dimethylcyclohexanes

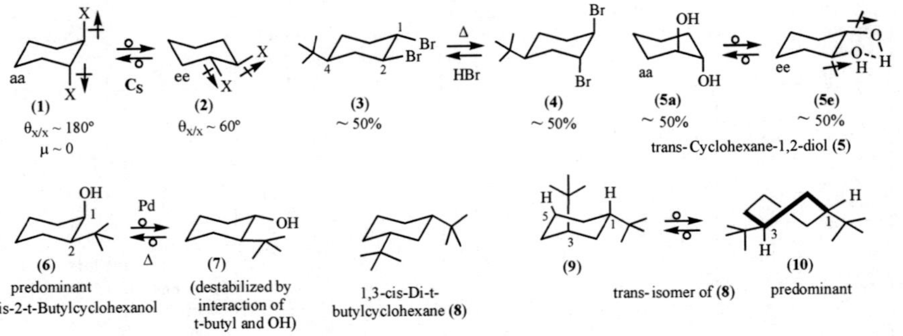

Fig. 7.10 Effect of dipole moment, H-bonding, and bulky substituents on conformation in cyclohexane derivatives

Cl < Br < I, since the increased bond length decreases the synaxial interactions. The *ee* conformer (**2**) is also stabilized by solvation with polar solvents. The effects of dipole–dipole repulsion in (**3**) and synaxial interactions in (**4**) result in the 50% of each isomer in the equilibrium mixture of these two diastereomers of 4-t-butyl-1,2-dibromocyclohexane.

2. The *cis*- and *trans*-isomers of cyclohexane-1,2-diol show intramolecular H-bonding (little stronger in *cis*). The *ee* conformer (**5e**) of the *trans*-2-halocyclohexanol is stabilized by H-bonding, and the diaxial conformer (**5a**) is free from dipole–dipole repulsion resulting in almost equal population of the two conformers.
3. Equilibration of *cis*-2-*t*-butylcyclohexanol (*anancomeric* system) with Raney Ni or with Pd upon heating gives a predominant amount of the *cis*-isomer (**6**) over the *trans*-isomer (**7**). The interaction between a methyl group of the t-Bu and OH destabilizes the *trans*-isomer. The *cis*-isomer of 1,3-di-*t*-bulylcyclohexane (**8**) exists in the chair conformation with both bulky substituents in equatorial orientation. However, the corresponding *trans* isomer almost exclusively exists in the skew–boat conformation (**10**), since in the chair conformation (**9**) one *t*-butyl group have to occupy axial orientation involving severe interaction with H1 and H5.

References

1. D.H.R. Barton, Conformation of steroid nucleus. Experientia, 316–319 (1950); reprinted in Top. Stereochem. **6**, 1–4 (1971)
2. D.H.R. Barton, O. Hassel, Conformational analysis—the fundamental contributions of D.H.R. Barton and O. Hassel. Top. Stereochem. **6**, 1–17 (1971)
3. S.K. Talapatra, B. Talapatra, *Chemistry of Plant Natural Products: Stereochemistry, Conformation, Synthesis, Biology, and Medicine*, vol. 1 (Springer-Verlag, Berlin Heidelberg, 2015), pp. 294–295

4. G.M. Kellie, F.G. Riddell, Non-chair conformation of six-membered rings. Top. Stereochem. **8**, 225–269 (1974)
5. E.L. Eliel, S.H. Wilen, M.P. Doyle, *Basic Organic Stereochemistry* (John Wiley, New York, 2001), pp. 688, pertinent pages 443–445.
6. E.L. Eliel, S.H. Schroeter, Conformational analysis. IX. Equilibrations with Raney Nickel. The conformational energy of hydroxyl group as a function of solvent. J. Am. Chem. Soc. **87**, 5031–5038 (1965)
7. N.L. Allinger, S.J. Angyal, G.M. Morrison, *Conformational Analysis* (Wiley, New York, 1965)

Chapter 8
Cyclohexanone

Abstract This chapter concerns the Conformational Features of Cyclohexanone, its Geometry, Stability, and Torsion Angles; 2-Alkyl, 3-Alkyl, and 4-Alkyl Ketone Effects with examples; Stereochemical Aspects of Addition of Nucleophiles to Cyclohexanones—PDC (PSC) and SAC (SSC); Torsional Strain—Role of C-2 and C-6 Axial Hydrogens, Participating Orbitals in Cieplak Hypothesis, Hydride Reduction of 5-Substituted 2-Adamantanones, during Metal Hydride Reduction Mehta's Electronic Induction by the 2,3-endo,endo Substituents of 7-Norbornanones supporting Cieplak Hypothesis; Stereoselective Birch Reduction of Saturated cyclohexanones by Dissolving Metals—its Mechanism by Pradhan; Johnson and Malhotra's Rationalization of Stereochemistry of Exocyclic Enolate Protonation by Avoiding Allylic$^{(1,3)}$ Strain; Conformation of Cyclohexene, Allylic$^{(1,2)}$ Strain, Conformational Preference of Substituted Cyclohexenes, Synthesis of Solenopsin A—Application of A$^{(1,2)}$ Strain Concept; and Pseudoallylic 1,2 Strain in Enamines.

Cyclohexanone exists almost exclusively (99% at 25 °C) in the chair form and only a small amount (~ 1%) in skew–boat forms (Fig. 8.1). The conformational features of cyclohexanone are shown briefly in the Figure.

8.1 Torsion Angles, Stability

Torsion angles between pairs of adjacent C–C bonds (Fig. 8.1) show flattening of the ring at the region of the carbonyl group. The chair conformation has only a vertical σ-plane passing through C1 and C4 and belongs to the point group $\mathbf{C_S}$. The skew–boat forms belong to chiral point groups $\mathbf{C_1}$ and $\mathbf{C_2}$. Due to flattening at the site of the CO group, the *e*-Hs at C2 and C6 are partially eclipsed with carbonyl oxygen ($\theta = 4.3°$), whereas the corresponding a-H's lean slightly outwards. Cyclohexanone is *slightly destabilized* relative to cyclohexane due to the angle strain and torsional strain. The equilibrium of cyclohexanone cyanohydrin (sp^2 sp^3 equilibrium) lies more toward cyanohydrins side than that of di-*n*-octyl ketone indicating the lower thermodynamic stability of cyclohexanone. Again, cyclohexanone is reduced with

S. K. Talapatra and B. Talapatra, *Basic Concepts in Organic Stereochemistry*,
https://doi.org/10.1007/978-3-030-95990-6_8

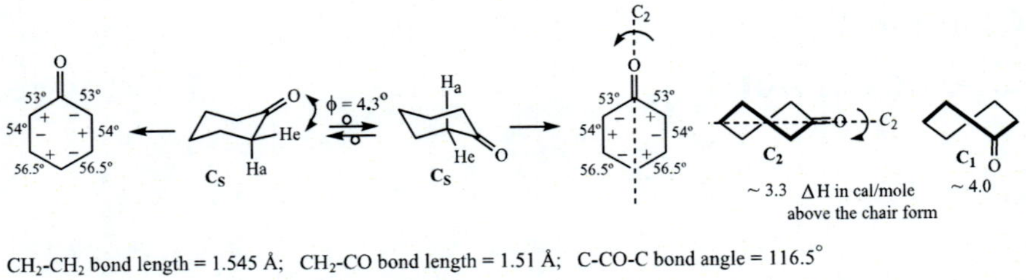

Fig. 8.1 Conformations, geometry, and torsion angles of cyclohexanone

$NaBH_4$ at a rate 355 times as fast as di-*n*-hexylketone manifesting the lower kinetic stability of cyclohexanone. The twist–boat forms with $\mathbf{C_2}$ and $\mathbf{C_1}$ point groups are having enthalpies (calcd.) 3.3 and 4.0 kcal/mol (approx.), respectively, above that of the chair form (Fig. 8.1).

8.2 Ring Inversion

The free energy of activation $\Delta G^{\neq}$ (exptl.) [1] for the ring inversion in cyclohexanone at $-$ 170° is ~ 5.0 kcal/mol, much lower than in cyclohexane. This is because of lower torsional barrier around an sp^3–sp^2 C–C bond than an sp^3–sp^3 one. Pseudorotation among the flexible forms is more facile in cyclohexanone than in cyclohexane, and the transition states for $\mathbf{C_2}$ and $\mathbf{C_1}$ skew–boats are not equivalent.

8.3 Alkylketone Effects

The alkylketone effect of the substituent $R = (-\Delta G_R)$ (in alkylcyclohexane)—$(-\Delta G'_R)$ (in 2, 3, or 4-alkylcyclohexanone) = the decrease in the conformational free energy of a substituent *R* in the alkylcyclohexanone with respect to the conformational free energy of the alkyl substituent in the alkyl cyclohexane.

8.3.1 2-Alkylketone Effect (Fig. 8.2)

The equatorial (*e*) substituents at C2 and C6 are nearly eclipsed with respect to CO oxygen and may destabilize the *e* conformer due to additional steric repulsion—thus decreasing the conformational free energy (ΔG between axial (*a*) and *e*) in comparison to that in alkylcyclohexane. Thus, in free energy, the term 2-alkylketone effect was first proposed by Walker in 1955 and Klyne in 1956. The 2-methyl ketone effect was estimated to be 1 kcal/mol, reducing the normal difference (ΔG_{Me} ~

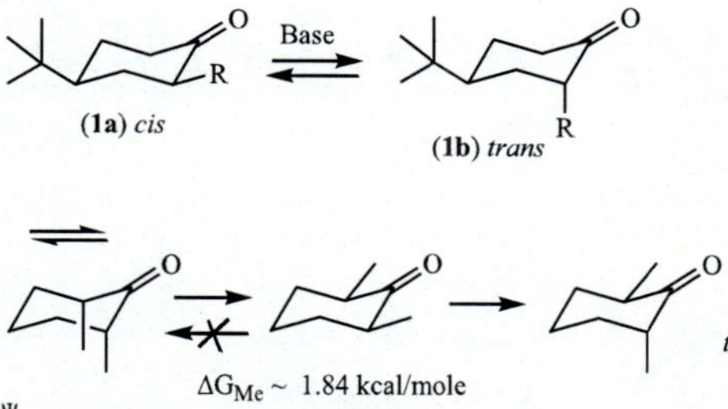

$^{\Psi}$–ΔG$_R$ values are taken from **Table 7.1**

R	kcal/mole		
	$-\Delta G_R{}^{\Psi}$	$-\Delta G'_R$	2-Alkyl-Ketone effect
Me	1.74	1.84	-0.10
Et	1.79	1.10	0.69
iPr	2.21	0.60	1.61

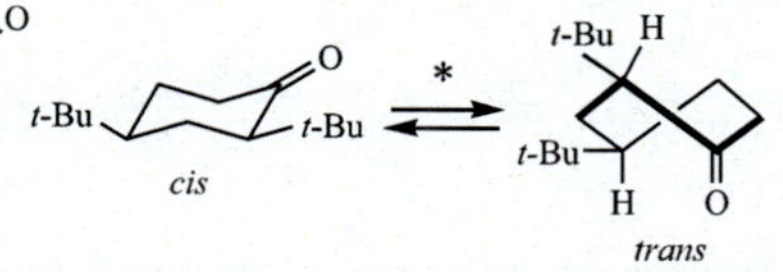

*cis-2,4-Di-t-butylcyclohexanone undergoes base catalyzed equilibration to give trans-2,4-di-t-butylcyclohexanone in skew-boat form (and not in chair form, and hence 2-t-butylketone effect cannot be measured).

Fig. 8.2 Measurement of 2-alkylketone effects

1.7 kcal/mol) by this amount making the $\Delta G'_{\mathrm{Me}}$ (in 2-methyl cyclohexanone) as 0.7 kcal/mol.

But Allinger, in 1961, demonstrated that 2-methylketone effect is absent. The reason for the absence of 2-methylketone effect is that the *e*-Me is too far (due to greater C–CH_3 bond length) from CO oxygen to have any van der Waals repulsion (steric interaction) between them; on the other hand, eclipsing of C–Me and C=O is electronically slightly favorable (cf. preferred conformation of propanal is having C=O and C–Me eclipsed). Thus, the 2-methylketone effect appears to be slightly negative, – 0.10 kcal/mol. 2-Alkylketone effect can be estimated from the equilibrium data of 2-alkyl-4-*t*-butylcyclohexanones or 2,6-dialkylcyclohexanones (Fig. 8.2). A few alkylketone effects thus obtained are given in the Figure.

8.3.2 3-Alkylketone Effect

In 3-alkylcyclohexanone the axial conformer has only one alkyl-hydrogen synaxial interaction (gauche) instead of two such interactions in the axial conformer of alkyl cyclohexane. Thus, there is a decrease in $-\Delta G°$ value [of the alkyl (R) group], which is known as 3-alkylketone effect. In case of 3-methylcyclohexanone (R = Me), this effect is equivalent to one gauche–butane interaction (0.89 kcal or 3.75 kJ/mol). However, in view of the van der Waals repulsion between axial Me and CO oxygen in the axial conformer this value is reduced to a calculated value of 0.6 kcal/mol. Experimental study of the equilibrium between *cis*- and *trans*-isomers of 3,5-dimethylcyclohexanone (**a** in Fig. 8.3) suggests a value of 0.37 kcal/mol (1.73–1.36 kcal/mol) (taking $\Delta S = O$) for the decrease (3-methylketone effect) in $-\Delta G_{\mathrm{Me}}$. The equilibrium data of 2,5-dimethylcyclohexanone (between the *trans*-(**1**) and *cis*-(**2**) isomers) (**b** in Fig. 8.3) gives a value of 0.5 kcal/mol (Eliel 1965) which is explained in the Figure.

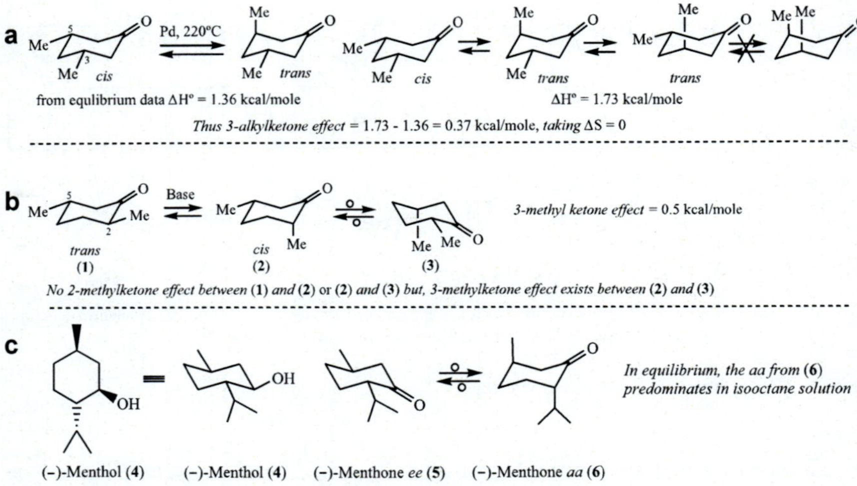

Fig. 8.3 3-Alkylketone effect. Few examples

In (−)-menthone [(**5**) in Fig. 8.3] 2-isopropylketone effect (1.8 kcal/mol) and 3-methylketone effect (0.5 kcal/mol) cooperate with each other in the *aa* conformer (**6**) with respect to the *ee* conformer (**5**)—so much so that in a solvent of low polarity, e.g., isooctane, the diaxial conformer (**6**) predominates, as evident from its CD spectrum. In the polar solvent, the *ee* form (**5**) is also stabilized by dipole–dipole interaction with the solvent.

8.3.3 4-Alkylketone Effect

4-Alkylketone effect depends upon the relief of strain due to synaxial interactions between an axial 4-alkyl group and the axial H or substituent at C2 and C4, when C1 is converted from the tetrahedral (>CHOH) to the trigonal (>C = 0) configuration, since this conversion leads to an outward motion (leaning) of the axial H's or substituents at C2 and C4.

This effect has been invoked to explain (Fig. 8.4).

(a) The greater rate of CrO_3 oxidation of 4,4-dimethylcyclohexanol as compared to that of cyclohexanol (cf. triterpene 3β-ol vs. steroidal 3β-ol).
(b) The larger dissociation constant (K) of 4,4-dimethylcyclohexanone cyanohydrin as compared to that of cyclohexanone cyanohydrin.

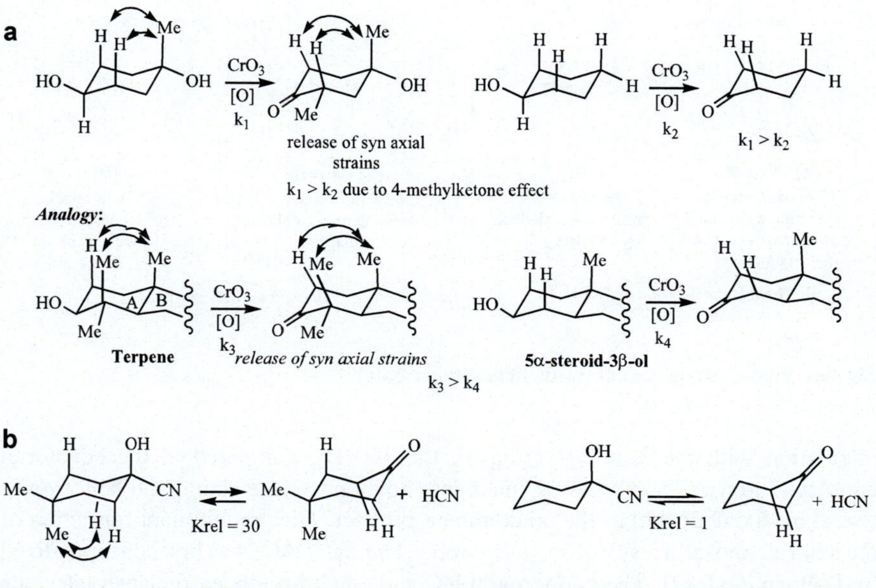

Fig. 8.4 Invocation of 4-alkylketone effect to explain some oxidation rates and relative equilibrium constants

8.4 Addition of Nucleophiles to Cyclohexanones. Stereochemical Aspects

The nucleophile can approach the carbonyl carbon of a cyclohexanone derivative from the axial side producing the equatorial alcohol and from the equatorial side producing the axial alcohol.

8.4.1 PDC (PSC) and SAC (SSC)

Dauben et al. in 1956 [2] first introduced the concept that in case of unhindered cyclohexanones having no substituent at C3 or C5 position, axial approach of the nucleophile does not encounter any steric hindrance and hence is preferred, producing the more stable equatorial alcohol in major amount (~90%) (Fig. 8.5), as a result of *product development control* (PDC). This is illustrated with an example of the sodium borohydride reduction of an anancomeric cyclohexanone like 4-*t*-butylcyclohexanone [3] through the transition state (**A**) in Fig. 8.5, which energetically resembles the product *trans*-alcohol. If the attack takes place from the equatorial side, it involves the TS (**B**) of higher energy due to synaxial interactions with the developing axial OH and hence is formed in minor amount (~10%).

On the other hand for hindered anancomeric cyclohexanones like *cis*-3-methyl-4-*t*-butylcyclohexanone, the axial approach becomes more hindered due to synaxial

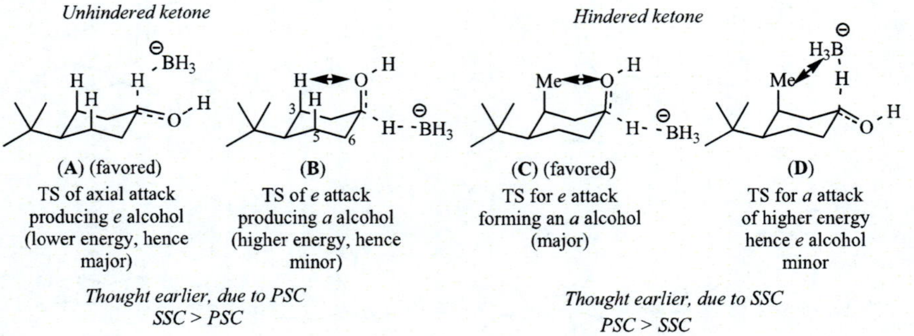

Fig. 8.5 Product strain control versus steric strain control

interaction with the 3-methyl group in the TS (**D**), compared to the equatorial approach through TS (**C**) having much less nonbonded interaction and hence gives rise to the axial alcohol as the predominant product. The predominant formation of the axial alcohol as a result of *steric approach control* (SAC) was first conceptualized by Dauben et al. [2]. The two terms, PDC and SAC, have been replaced later one by Brown and Deck [4] by *product stability control* (PSC) and *steric strain control* (SSC) to give more emphasis on the transition state rather than on events prior to that the term PSC refers to a product-like or a late TS along the reaction coordinate, while the term SSC refers to a reactant-like or an early transition state.

8.4.2 Observations Against PSC

The role of SSC on the stereochemical course of addition of nucleophiles to cyclohexanones has never been in doubt, but the role of PSC has always been controversial. Some observations against the validity of PSC are as follows:

(a) Usually, a higher proportion of an equatorial alcohol is obtained in lithium aluminum hydride reduction than that corresponding to thermodynamic equilibrium, e.g., 90% versus 80% in the reduction of 4-*t*-butylcyclohexanone.
(b) The rate constants of a number of sterically hindered ketones, relative to that of 4-*t*-butylcyclohexanone, support the concept of SSC, but show little effect of PSC [5].
(c) Reduction of ketones with hydrides is highly exothermic; hence, according to Hammond's postulate the transition states are expected to be reactant-like (cf. SSC) and not product-like.
(d) The deuterium isotope effect (k_H/k_D) in borohydride/borodeuteride reduction of cyclohexanones with varying degrees of steric hindrance [6] is almost constant in spite of widely varying ratios of the equatorial and axial alcohols

indicating that the extent of B–H bond breaking at the TS is similar for cyclohexanones having different degrees of steric hindrance—a fact inconsistent with the PSC concept.

While teaching it occurred to the present authors long ago that PSC involves product strain which itself cannot have any role in the TS and activation energy in any reaction of kinetic control, and hence, the specific contribution of PSC is unexpected, as is substantiated by the above observations. The same is true for reduction of acyclic carbonyl groups regulated only by steric factors such as Cram's rule and Prelog's rule (Chap. 5).

8.4.3 *Torsional Strain. Role of C2 and C6 Axial Hydrogens. Burgi–Dunitz Trajectory*

In 1965, it was suggested by Richer [7] that the axial Hs at C2 and C6 offer steric resistance to the equatorial approach of a nucleophile (torsional strain) which competes with that offered by the axial H's at C3 and C5 to axial approach (Fig. 8.6). The relative strength of the resistances would, however, depend upon the bulk of the hydride reagent and the exact position of the TS in the reaction coordinate. If C…Nu is shorter than 1.6 Å, the equatorial approach would be more hindered in case of reduction with hydride reagent.

The most satisfactory and widely accepted alternative to PSC is the concept of *torsional strain* introduced in 1968 by Chérest and Felkin [8]. Two competing factors, namely (1) the steric interaction of the incoming nucleophile with the 3,5-diaxial groups (SSC) in the *axial attack* and (2) the *torsional strain* which arises between the semiformed bond of the nucleophile and the two axial CH bonds at C2 and C6 during the *equatorial approach* (Fig. 8.6) arise, as already stated.

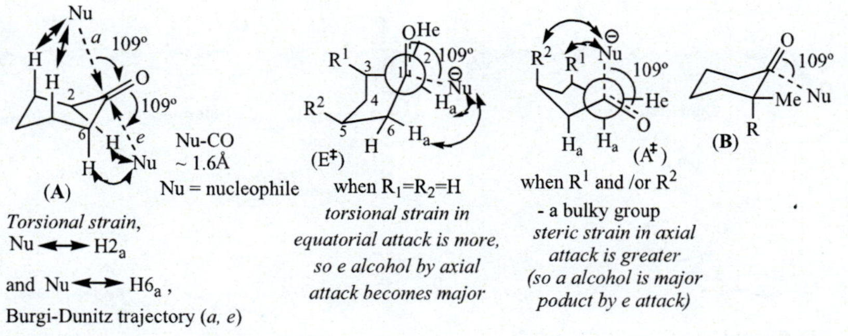

Note: When GE‡ > GA‡ Æ e Alcohol is major. When GA‡ > GE‡ Æ a Alcohol is major

Fig. 8.6 Felkin transition states for addition of a nucleophile to a cyclohexanone, torsional strain versus steric strain. Formation of equatorial alcohol versus axial alcohol

In the absence of any steric hindrance (factor 1) (in case of unhindered ketones), the effect of the torsional strain prevails leading to a preferential axial attack forming the equatorial alcohol. On the contrary, in the presence of one bulky axial substituent at C3 or C5 (or at both), the steric strain control (SSC) (2) prevails reversing the steric course and forming the axial alcohol predominantly. The bulk of the hydride reagent will increase the SSC, more so if C3 and C5 substituents are present and will definitely increase the proportion of axial alcohol. In Fig. 8.6, appropriate transition states are shown for both cases.

This torsional strain effect is not evident in the reduction of 2-methylcyclohexanone (**B**, R=H) and 2,2-dimethylcyclohexanone (**B**, R=Me) indicating that torsional strain is relatively insensitive to the bulk of the group involved (cf. rotational barrier of propane is barely larger than that of ethane, Sect. 1.3.4).

The application of Bürgi–Dunitz trajectory [9–11] proposed in 1973 and 1974, based on molecular orbital interactions (Sect. 5.4.1) which states that a nucleophile attacks the carbonyl carbon at an angle of 109° with the carbonyl plane, makes the torsional strain stronger by making it almost like eclipsing strain (Fig. 8.6). Moreover, the torsional strain theory also successfully explains the stereochemistry of the reduction of acyclic ketones (see Sect. 5.4.1, Felkin-Anh Open Chain Model). Both steric strain and torsional strain can be minimized for acyclic substrates.

Several examples of sodium borohydride/lithium aluminum hydride reduction of hindered cyclohexanones are given in Fig. 8.7. The more hindered is the ketone the more equatorial attack leads to the formation of greater percentage of the axial alcohol, and the less is the rate of the reaction.

Hydride reductions of cyclohexanones.

The percentages of *a:e* alcohols are given in parenthesis.

Studies on π-facial selectivity in the addition of nucleophiles to double bonds, especially the carbonyl-based systems have generated more models and hypotheses (~10) (not necessarily mutually exclusive) than any other subject in the field of

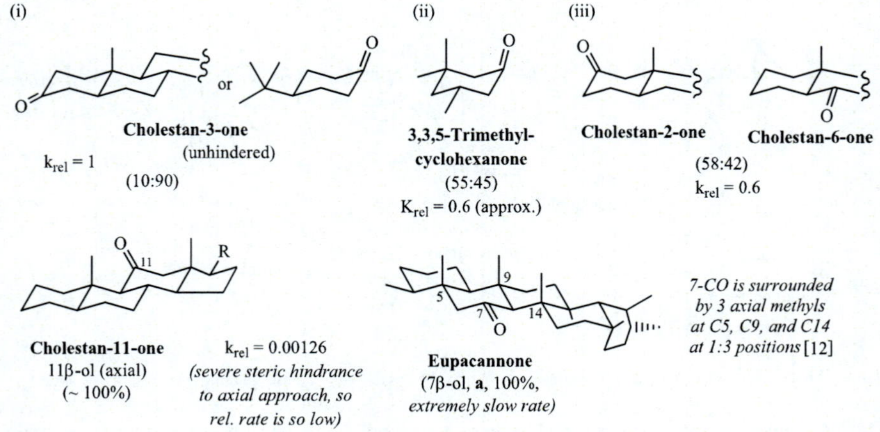

Fig. 8.7 Increased amount of axial alcohol by increased SSC in hindered cyclohexanones

stereoselective synthesis [13], some of which, e.g., Cram, Prelog, Felkin–Anh, have been discussed earlier.

8.5 Cieplak Hypothesis

Cieplak in 1981 suggested [13] that the stereochemistry of nucleophilic addition to cyclohexanones is determined by a combination of steric factor, which as usual favors the equatorial approach, and the stereoelectronic factor which involves electron donation from the cyclohexanone C–C and C–H σ bonds into the vacant σ* orbital of the incipient bond formation between the nucleophile and the carbonyl carbon (Fig. 8.8). Since axial C–H bonds (at C2 and C6) next to the carbonyl group are better electron donors than the 2–3 and 5–6 σ bonds, the transition state with axially oriented incipient bond (axial approach) is preferred over the one with the equatorially oriented incipient bond (Fig. 8.8). Thus, the hypothetical model proposed by Cieplak is based on the concept that the carbonyl group undergoes extensive pyramidialization, and that the outcome is primarily a consequence of the aforesaid interactions between the occupied vicinal σ-orbitals by electron donation into the vacant σ*-orbital leading to the stabilization of the transition state.

Cheung et al. [14] in 1986 reported that in the nucleophilic addition of $NaBH_4$ to the carbonyl groups of 5-substituted 2-adamantanones, depletion of electron density on the C1–C9 and C3–C4 bonds by halogen or phenyl groups possessing *para* electron-withdrawing substituents favored *syn* approach forming (*E*) (*anti*) alcohols. On the contrary, the electron-donating groups like *p*-aminophenyl and *p*-hydroxyphenyl analogs favored the *anti*-approach forming the (**Z**) (*syn*) alcohols as major products (Fig. 8.9).

The above observations [14] strongly supported Cieplak view of electronic effects in such asymmetric induction through preferential interaction with newly developing σ* orbital with electron-rich *anti*-periplanar bonds and vice versa.

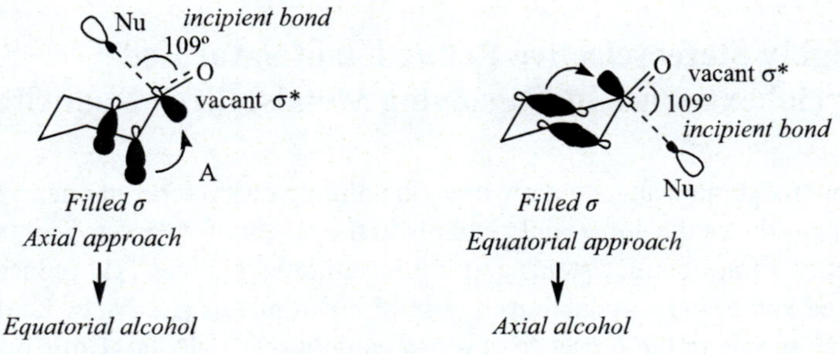

Fig. 8.8 Participating orbitals in the Cieplak model

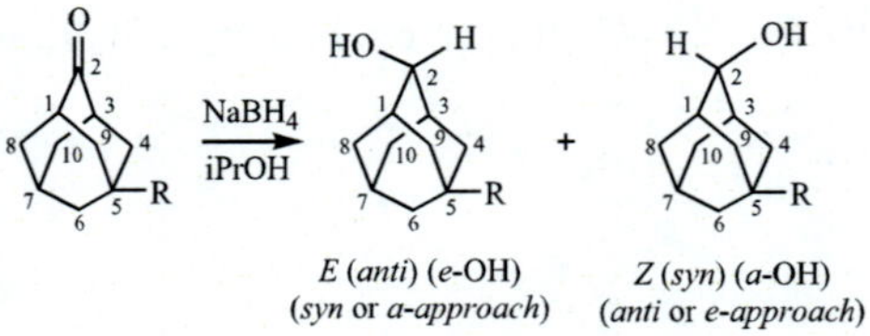

	R	E/Z
EWG	Cl p-Cl-Ph Ph	59:41 60:40 58:42
EDG	p-HO-Ph p-H$_2$N-Ph	44:56 34:66

Fig. 8.9 Hydride reduction of 5-substituted 2-adamantanones

Cieplak et al. in 1989 [15] observed that the electronegative substitution of the substrate by CF_3 at C3 of cyclohexanone or in the reagent, e.g., sodium tri-isopropoxyborohydride reverses the π-facial diastereoselection by increasing the relative proportion of the axial attack leading to the favored formation of the equatorial alcohol. The same type of favored axial attack has been found in a number of widely different reactions including alkyllithium addition and also in cases of peracid epoxidations of methylenecyclohexenes. The findings appear inconsistent with the predictions of Felkin, Klein, Ashby, and Anh models of stereochemistry of reactions in cyclohexane-based systems, but are consistent with Cieplak model [14].

In 1990, Mehta and Khan [16] demonstrated that electronic induction by the 2,3-*endo, endo* substituents reverses the π-facial selection (resulting in the reversal of the *E*:*Z* ratio) in the nucleophilic addition of hydrides and methyllithium to 7-norbornanone, as shown in a tabular form in Fig. 8.10. The predominant approach of nucleophiles to the *syn* face in **1a** and to the *anti*-face in **1e** is fully consistent with the prediction based on Cieplak's hyperconjugative model, according to which delocalization of σ electrons in the electron-rich antiperiplanar σ bond into the incipient σ* orbital lowers the transition state energy (cf. Fig. 8.10). The unexpected *anti*-face selectivity in the cases of **1b** and **1c** (with 2- and 3-substituents supposed to be electron-withdrawing groups) (see the Table) may be attributed to *through space donation in a perpendicular conformation* as shown for **1c** in (**G**) in Fig. 8.10. For an authoritative background review and a comprehensive list of references, see [17].

8.6 Highly Stereoselective Reduction of Saturated Cyclohexanones by Dissolving Metals. Birch Reduction

Reduction of saturated anancomeric or rigid unhindered cyclohexanones by $LiAlH_4$ or $NaBH_4$ produces the equatorial alcohols to the extent of 90% (cf. Sect. 8.4).

Reduction of aromatic systems and α,β-unsaturated ketones (1:4 reduction) and of saturated ketones (1:2 reduction) by use of alkali metals (Li, Na or K) dissolved in liquid ammonia in the presence of a proton donor-like alcohol (little amount) is known as **Birch reduction**, which is of versatile synthetic utility [18, 19]. Saturated

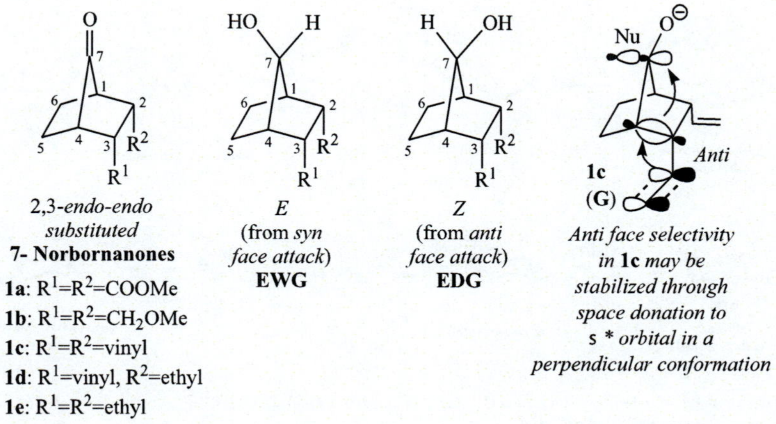

Product Ratios in the Metal Hydride Reductions and Methyllithium Addition

E:Z distribution

Substrate	$NaBH_4$	$LiAlH_4$	$Li(t\text{-}BuO)_3AlH$	MeLi
1a	84:16	87:13	77:23	>90:<10
1b	40:60			34:66
1c	36:64	35:65	34:66	27:23
1d	25:75			
1e	20:80	21:79	29:71	17:83

Fig. 8.10 Metal hydride reductions of and methyllithium additions to 2,3-*endo*, *endo* substituted 7-norbornanones. *E*:*Z* ratios of the products support Cieplak hypothesis

anancomeric or rigid cyclohexanones are reduced to equatorial cyclohexanols (~99% diastereoselectivity) by Birch reduction.

The mechanism of such reduction of 4-*t*-butylcyclohexanone (**A**) proposed by Pradhan [20] is delineated stepwise in Fig. 2.115, based on *Fukui effect* [21], rationalized by Fukui's Frontier Molecular Orbital (FMO) approach. The stereoelectronic effect provides a *kinetic preference of the axial attack of* an electron of the metal (M) on the carbonyl carbon, by a single electron transfer (SET) mechanism leading through a stabilized transition state (TS), to a resonance stabilized ketyl radical; this effect has been termed as Fukui effect. In the Birch reduction 4-t-butylcyclohexanone (**A**) carbonyl carbon, the resonance-stabilized ketyl radical (**B**) with a singly occupied molecular orbital (SOMO) is formed by axial approach of an electron transfer from the metal (SET mechanism). In steps 2–4, the species (**C**), (**D**), and (**E**) are formed successively and the stabilizations of species (**B**) to (**E**) have been indicated in the Figure. The species (**E**) equilibrates to the more stable species (**F**), and

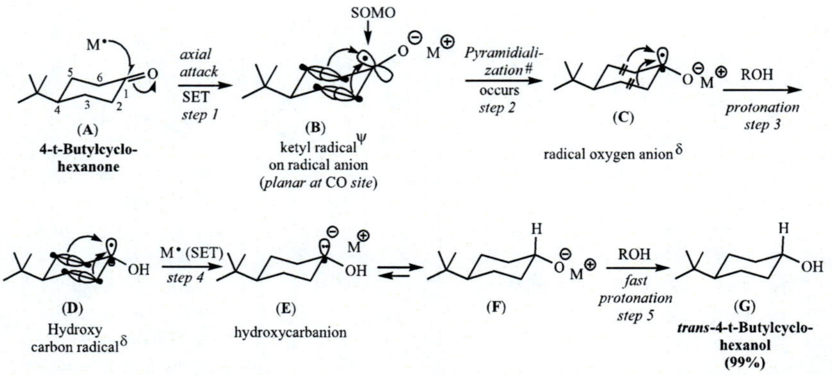

ψ70% unpaired electron density is located at C1; and 2 -3 and 5-6 σ-bonds (HOMO) are favorably disposed to interact with singly occupied π*-orbital (SOMO)

#SOMO is extended in the axial direction

δOM in (**C**) or OH group in (**D**) is oriented exclusively in the equatorial direction and is antiperiplanar with interacting 2-3 and 5-6 σ-bonds

Fig. 8.11 Stepwise mechanism of Birch reduction of an anancomeric saturated cyclohexanone

either species undergoes protonation very fast to form the final product *trans*-4-t-butylcyclohexanol (**G**) (99%), giving no chance of the carbanion (**E**) to undergo inversion at C1 (Fig. 8.11).

8.7 Alkylidene Cyclohexanes. Allylic$^{(1,3)}$ Strain

The C=O group of cyclohexanone is replaced by $C{=}CR^1R^2$ to form alkylidenecyclohexanes. The same geometry as in cyclohexanone (Fig. 8.1) is assumed for alkylidenecyclohexanes (Fig. 8.12). Thus, in a cyclohexane ring containing an exocyclic double bond and an allylic substituent, the steric interference between the allylic equatorial substituent at C_γ position and the ethylenic *syn* (*Z*) substituent (at C_α position) (Fig. 8.12) is given the trivial designation allylic 1,3- or $A^{(1,3)}$ strain for the sake of semantic simplicity since the groups involved are at the 1 and 3 positions of the allylic system. This is one of the two general stereochemical theorems introduced by Johnson and Malhotra [22, 23] in 1965; the other one, $A^{(1,2)}$ strain will be discussed in Sect. 8.8.

In the equilibrium between the equatorial (**E**) and the axial (**A**) conformers, R and R^1 in (**E**) are nearly eclipsed (the spatial arrangement of $R{-}C_\gamma{-}C_\beta{=}C_\alpha{-}R^1$ is almost planar); the dihedral angle θ between $R^1{-}C_\alpha$ and $C_\gamma{-}R$ is ~4°. *The allylic 1,3 strain (*$A^{(1,3)}$ *strain) between two substituents/atoms is almost same as the synaxial interaction between them.* Thus, *in the equatorial conformer* (**E**), *even when* R *and* R^1 are *only moderate in size (medium to large)*, the axial conformer (**A**) predominates.

Some examples illustrating $A^{(1,3)}$ strain follow.

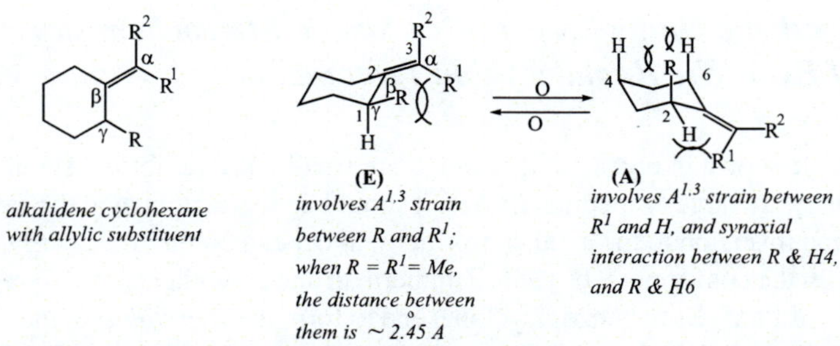

When $R = R^1$ = Me the distance between them = 2.45 Å; enthalpy difference between **(E)** and **(A)** conformers = $A^{1,3}$ Me/Me interaction – [2(1,3 diaxial Me/H interactions + $A^{(1,3)}$ Me/H interaction.] ≈ 1(1,3-diaxial Me/Me interaction) –[2(1,3 dixial Me/H interaction. +$A^{(1,3)}$Me/H interaction] ≈ 3.7 – 3(.9) = 3.7–2.7 = 1 kcal/mole.

The axial conformer **(A)** should, therefore, exist to the extent of ~ 85% and the equatorial conformer **(E)** ~ 15% in the ground state (as obtained from the equation $\Delta G = -RT\ln K$, taking $\Delta S = 0$)

Fig. 8.12 Alkylidenecyclohexane. $A^{1,3}$ strain

8.7.1 Conformational Preference

Of the two geometrical isomers of *2-methylcyclohexylideneacetic acid*, the *Z*-diastereomer exists predominantly in the Me axial (**1a**) conformer, but the *E* diastereomer exists predominantly in the Me equatorial (2**e**) conformer, as evident from ^{1}H NMR study. These facts can be explained in terms of the interactions including $A^{(1,3)}$ strain present in the molecules (Fig. 8.13). Likewise, it may be predicted that in case of 3-*methylcyclohexylidenepropionic acid* (underlined H to be replaced by Me in all the structures of Fig. 8.13), both Z- and *E*-diastereomers exist predominantly in the axial conformer (due to severe $A^{(1,3)}$ Me/Me interaction present in the equatorial conformer of the *E* isomer).

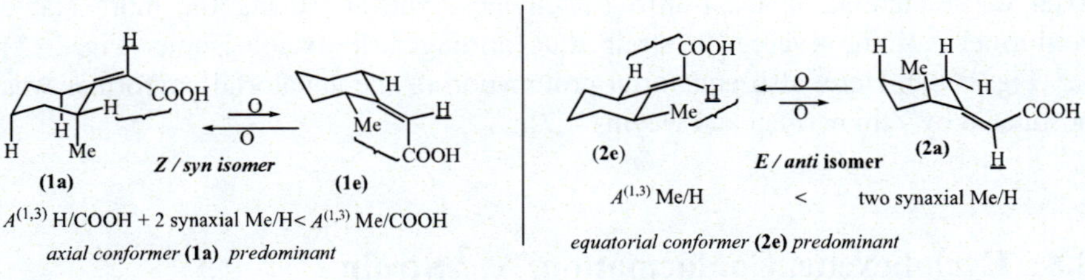

Fig. 8.13 Predominant conformers of the diastereomers of 2-methylcyclohexylidene acetic acid

8.7.2 *Synthetic Utility [24] of $A^{(1,3)}$ Strain. Stereochemistry of Exocyclic Enolate Anion Protonation*

In the example shown in Fig. 8.12, if R and R′ are small, the equilibrium should lie to the left; if they are large or medium in size, it should lie to the right, and the extent of conformation inversion to the axial conformer will depend on the free energy difference between the conformers. In 1955, Zimmerman and coworkers [25, 26] reported the addition of PhMgBr to benzoylcyclohexene to form the *less stable isomer*, *cis*-1-benzoyl-2-phenylcyclohexane (**4**). It was suggested that 1,4 addition of PhMgBr to benzoylcyclohexene takes place by its equatorial approach at C2, avoiding the steric hindrance in axial approach due to synaxial H4 and H6, to form the bromomagnesium enolate anion (**3**). They thought that C-protonation of the conformer (**3e**) with an acid takes place from the less hindered equatorial direction to give (**4**), since steric interference arising from axial H's at C3 and C5 seriously hindered axial protonation. However, according to $A^{(1,3)}$ strain, (**3**) should exist mainly as the conformer (**3a**) in the ground state, and the protonation at C1 should take place from the axial side in order to have continuous overlap with the π-orbital (stereoelectronic requirement, **SER**).

Johnson and Malhotra's rationalization [24] has been delineated in Fig. 8.14, invoking the avoidance of $A^{(1,3)}$-strain and axial protonation of the exocyclic enolates to form the less stable *cis*-isomer (**4a**). The easy bromination of the *cis*-isomer and resistance to bromination of the *trans*-isomer under the same condition is also explained by invoking $A^{(1,3)}$ strain during the formation of its enol.

8.7.3 *Another Example of the Use of $A^{(1,3)}$ Strain Concept*

Another Example of the Use of $A^{(1,3)}$ Strain Concept will be evident from the axial C-protonation of 1-*aci*-nitro-2-phenyhlcyclohexane using the more stable conformer avoiding severe $A^{(1,3)}$ strain, thus forming the less stable isomer (Fig. 8.15) (cf. Fig. 8.14). Here, also equatorial protonation of the equatorial conformer was postulated by Zimmerman and Nevins [27].

8.8 Cyclohexene. Conformation. $A^{1,2}$ Strain

8.8.1 *Conformation of Cyclohexene. Torsion Angles*

Cyclohexene with two sp^2 carbons possesses a half-chair conformation (**A**) or (**A′**) (Fig. 8.16) confirmed by X-ray crystallographic data of cyclohexene derivatives

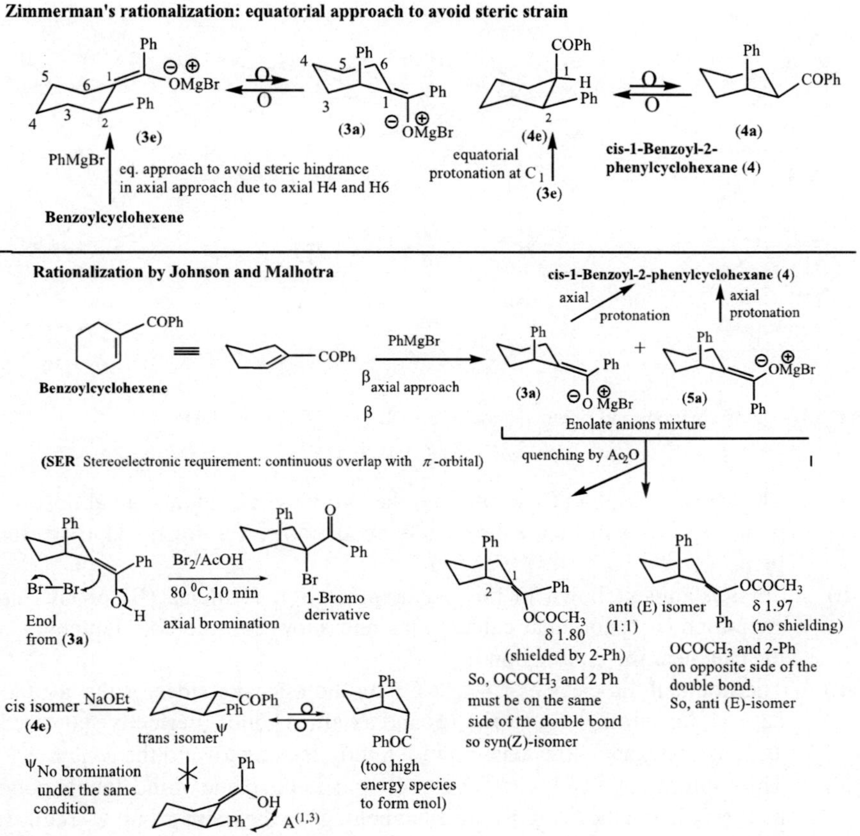

Fig. 8.14 Stereochemistry of protonation of exocyclic enolate anion (avoidance of $A^{(1,3)}$ strain and SER to be satisfied)

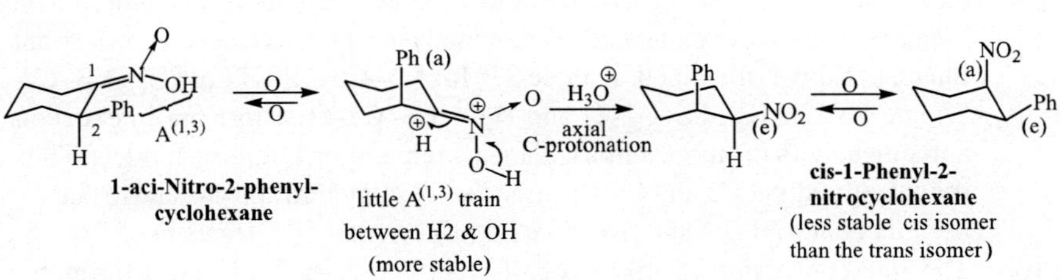

Fig. 8.15 Axial C-protonation of 1-*aci*-nitro-2-phenylcyclohexane. Use of $A^{(1,3)}$ concept

and also by electron diffraction studies of cyclohexene vapor. The geometry of cyclohexene possesses the following characteristic features:

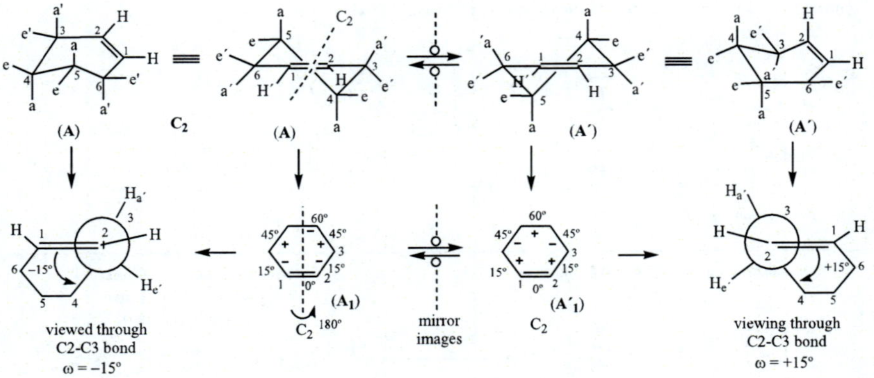

Fig. 8.16 Conformations and ring inversion of cyclohexene. Torsion angles

(i) The C1, C2, C3, and C4 atoms and the two vinylic H atoms are almost in a plane. The C4 and C5 are alternately up and down (or down and up) of this plane.

(ii) Torsion angles shown in the planar projection formulas (**B**) or (**B′**) are supported by theoretical calculations and show considerable flattening of the ring near the double bond.

(iii) The homoallylic carbons C4 and C5 on the opposite sides of the π-plane have almost normal equatorial (e) and axial (a) bonds (perfectly staggered) as in cyclohexane—the axial bonds slightly leaning toward the center.

(iv) The conformation (**A**) or (**A′**) has a C_2 axis in the plane of the double bond, bisecting it and the C4–C5 bond, and belongs to the point group $\mathbf{C_2}$ (chiral).

(v) The conformation (**A**) is convertible to its enantiomer (**A′**) by ring inversion and vice versa, and so they form an inseparable (±)-pair. They are shown in a different perspective by structures ($\mathbf{A_1}$) and ($\mathbf{A_1'}$), respectively.

(vi) At the allylic carbons C3 and C6, the axial and equatorial character of the bonds is considerably changed. Vector analysis of cyclohexene by Corey and Sneen [28] gives the dihedral angle 37° for $H_{e'}$–C6–C1–H (and $H_{e'}$–C3–C2–H) and 83° for $H_{a'}$–C6–C1–H and H_a + C3–C2–H. These are in excellent agreement with the more empirical measurements on Dreiding models. Thus, the substituents at C3 and C6 are imperfectly staggered and are said to occupy pseudoequatorial (e′) and pseudoaxial (a′) orientations, respectively.

(vii) The interconversion of the two half-chair forms probably goes through a TS with boat-like conformation. The energy barrier ($\Delta G^{\neq}$) determined by ^{1}H NMR from the coalescence of 1,2,3,3,6,6-d$_6$-cyclohexene at −164 °C is 5.3 kcal/mol.

(viii) Johnson and Malhotra has proposed allylic 1,2-strain ($A^{(1,2)}$-strain) in 1,6-disubstituted cyclohexene system, the discussion on which follows:

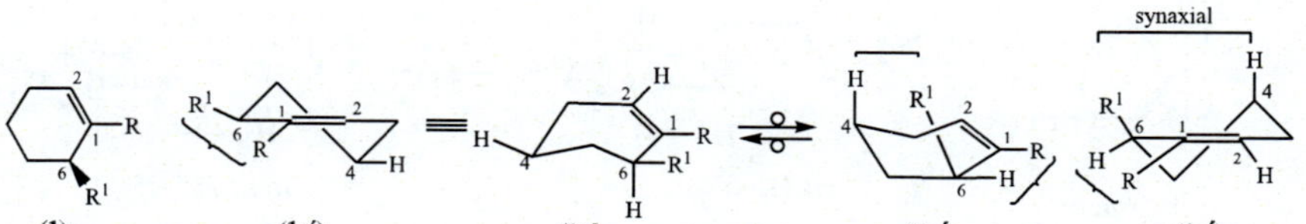

Notes: The position and intensity of the pseudoallylic t-proton at C6 in NMR gives an estimate of the relative abundance of (**1a′**) and (**1e′**) conformers. The splitting patterns also help the assignment. A$^{(1,2)}$ Interference being weaker than synaxial interaction, will exert a controlling effect only in absence of synaxial interaction

Fig. 8.17 Ring inverted conformers of 1,6-disubstituted cyclohexene

8.8.2 Allylic 1,2-Strain (A$^{(1,2)}$-Strain)

The strain or instability in 1,6-disubstituted 1,2-cyclohexenes (**1**) due to unfavorable interaction between ethylenic substituent and allylic pseudoequatorial substituent of the adjacent C6 is given the trivial designation A$^{(1,2)}$ strain [22, 23] (Fig. 8.17). The dihedral angles between R^1 and R in the equatorial (**1e**) and axial (**1a**) conformers (Fig. 8.17) are 37° and 83°, respectively.

A$^{(1,2)}$ *strain is not a strong effect* and is manifested only when the groups are moderately large or bulky. However, in 6-methylcyclohexene, the equatorial conformer is more stable, since it does not possess any A$^{(1,2)}$ strain or synaxial strain. A$^{(1,2)}$ strain between two substituents is much weaker than the synaxial interaction between them.

A$^{(1,2)}$ strain operates best when R conjugates with the double bond (e.g., phenyl group). The *ortho*-H of the phenyl group becomes planar with the double bond and comes still much closer to give stronger A$^{(1,2)}$ strain.

Some applications of A$^{(1,2)}$ strain are discussed in the sequel.

Conformational Preference

The outstanding examples of A^{1,2} strain in simple 6-disubstituted 1-phenyl (and methyl) cyclohexenes (Fig. 8.18) have been provided by Garbisch [29, 30]. He determined the preferred conformations for C6 substituents in a number of such compounds (Table 8.1) from the widths at half height of the C6 proton magnetic resonance bands [29, 30]. Synaxial interaction between same groups or groups of similar bulk are stronger than the A^{1,2} strain and between them and the former exerts a controlling effect. However, in the absence of synaxial interactions, the presence of A^{1,2}-strain makes the particular conformer less predominant. The preferred conformer in each case can be rationalized comparing all the interferences involved in the C6-pseudoaxial (a′) and C6-pseudoequatorial (e′) conformers. The preferred conformations of C_6 substituents of 1-phenylcyclohexenes are delineated in Table 8.1.

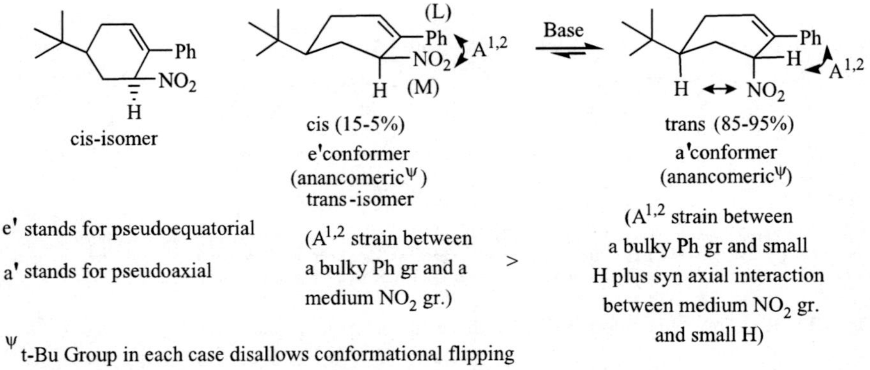

Fig. 8.18 Isomerizational preference in base-catalyzed equilibration of *cis*- and *trans*-4-t-butyl-6-nitro-1-phenylcyclohexenes

Table 8.1 Preferred conformations of the C6 substituents of 1-Phenyl (or Methyl)-6-substituted or 4,6-substituted cyclohexenes

Preferred conformation of 6R′	1,6 or *trans*-1,4,6-substituted cyclohexenes	
Pseudoaxial	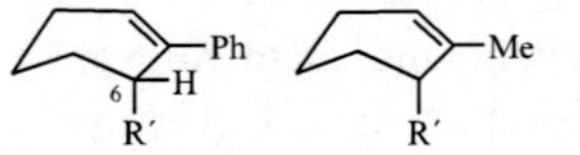 R′ = *t*-Bu, Ph, $Me_2C(OH)$ or NO_2; R′ = NO_2 or Br	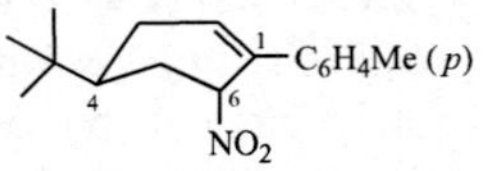 only conformer
Pseudoequatorial	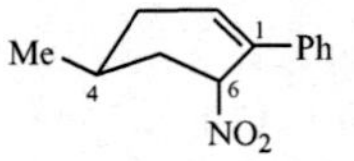 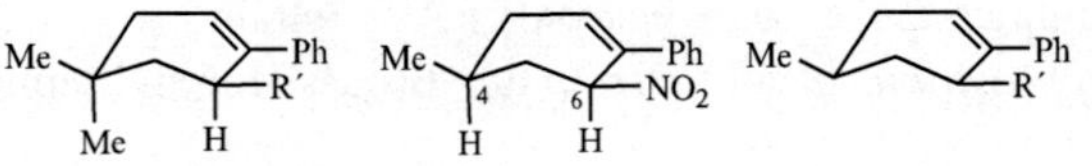 R′ = Me_2C (OH), COMe or, NO_2; *cis*-4-Me-6-NO_2- **Note**. Syn axial R′ ↔ Me interaction in the a′ conf. overrides $A^{1,2}$ interaction between NO_2 and Ph or R′ and Ph groups in the e′-conformer, leading to the preferred e′ conformer	
No conformational preference (NCP)	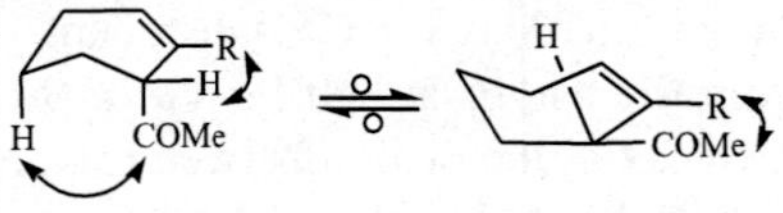 synaxial; R = Me or Ph	

Isomerizational Preference

One of the best examples is the base-catalyzed equilibration of *cis*- and *trans*-4-*t*-butyl-6-nitro-1-phenylcyclohexenes leading to an equilibrium composition of 85–95% of the *trans*-isomer (pseudoaxial nitro group) (Fig. 8.18).

Pseudoallylic 1,2-Strain in Enamines

The value of $A^{1,2}$ strain becomes apparent only from its generality. If we replace the ethylenic substituent of 1-alkyl- or 1-aryl-6-substituted cyclohexene by a nitrogen or an oxygen atom, we get an enamine (**2**) or an enol ether. Thus, cyclohexanone can be converted to 1-pyrrolidinyl-6-methylcyclohexene (**2**), which may be regarded as a pseudoallylic system (Fig. 8.19). This reaction is known as Stork reaction [31, 32]. Now, the pseudoallylic 1,2-strain, Me ↔ pyrrolidinyl α-H in (**2e′**), is greater than the sum of the pseudoallylic 1,2-strain, e′H ↔ pyrrolidinyl α-H, and the synaxial interaction between 6-Me and 4-H in (**2a′**). Thus, the pseudoaxial conformer (**2a′**) becomes predominant in the ring inversion of (**2**) (Fig. 8.19). The enamine upon hydrolysis forms pure 2-methylcyclohexanone without any further methylation at C2 or C6—this being the best method of its formation. The absence of any vinylic proton in the NMR establishes the nonformation of (**3**), which would involve severe interaction between the olefinic methyl and a planar pyrrolidinyl α-H, as shown in (3′) in Fig. 8.20.

Experimental verification of the pseudoaxial conformation of the 6-Me group in (**2**) is obtained by complete conversion of the more stable *cis*-isomer (**1**) of 2-methyl-4-*t*-butylcyclohexanone to the less stable *trans*-isomer (**2**) [33], which can be explained in terms of the nonformation of the enamine (**3′**) from the *cis*-isomer due to

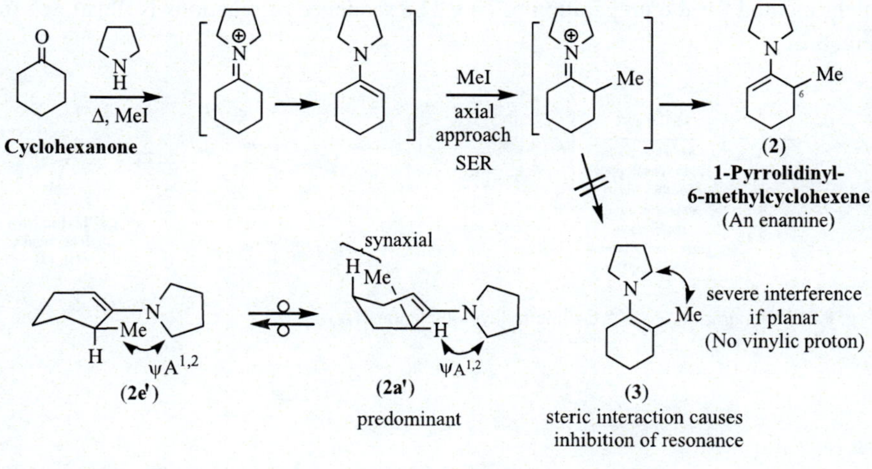

Fig. 8.19 Pseudoallylic 1,2-strain in a 6-methylenamine derivative

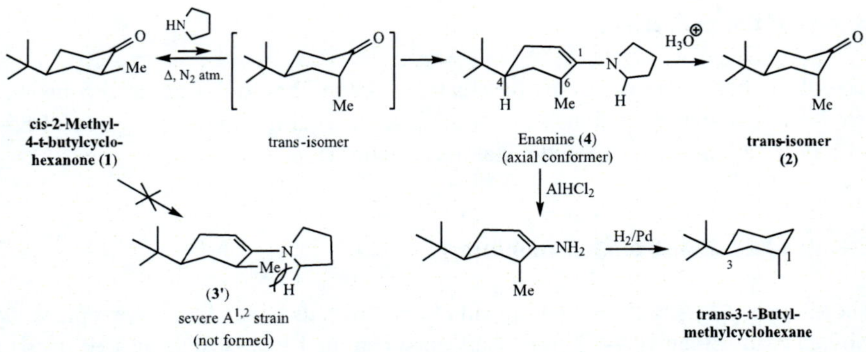

Note. All compounds including the starting material are racemates, but, as usual, only one enantiomer is written in each case.

Fig. 8.20 Complete conversion of the *cis*-2-methyl-4-t-butylcyclohexanone into the corresponding less stable *trans*-isomer

severe $A^{1,2}$ strain (Fig. 8.20). The base (pyrrolidine) catalyzed equilibration produces the intermediate *trans*-isomer (~5%), which with axial Me is converted to the stable enamine (**4**) involving much weak $A^{1,2}$ strain between He and α-H of pyrrolidine, and the equilibrium is shifted till the whole amount of *cis*-isomer is transformed to (**4**). The latter upon hydrolysis gives the *trans*-isomer (**2**) quantitatively. The last two steps establish that the acid hydrolysis did not cause any change in orientation (here axial) of the Me group. Thus, the enamine (**4**) must have the Me group axial.

Another proof of pseudoaxial orientation of 6-substituent of an enamine:

The 6-methylenamine derivative enantiomer of (**2a′**) of Fig. 8.19 may undergo further methylation (or alkylation), but with difficulty, and the resulting Schiff base (**A²**) (Fig. 8.21) produces upon acid hydrolysis *cis*-2,6-dimethylcyclohexanone (**A**) (without any other accompanying polymethylated cyclohexanone). This is a convenient method for the preparation of (**A**). The sequence of reactions is illustrated in Fig. 8.21.

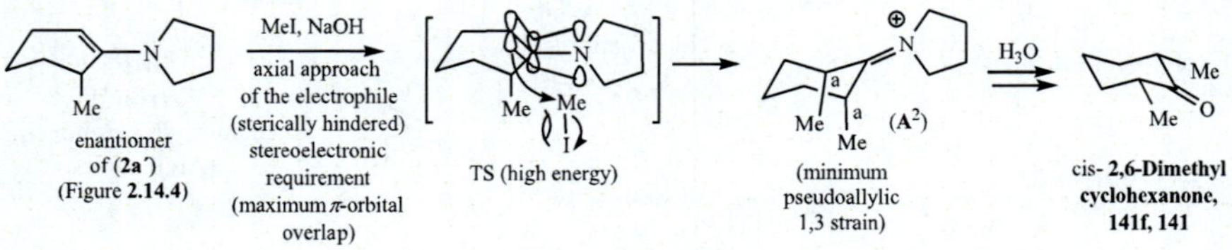

Fig. 8.21 Preparation of *cis*-2,6-dimethylcyclohexanone (**A**)

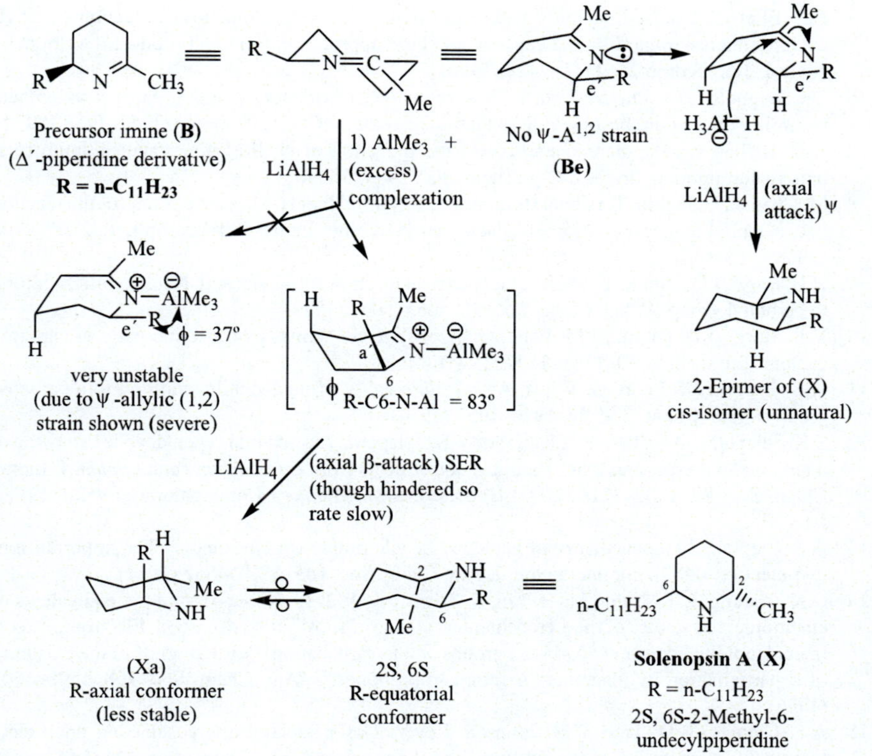

Fig. 8.22 Synthesis of solenopsin A by application of $A^{(1,2)}$ concept

Synthesis of Solenopsin A. Application of $A^{1,2}$ Strain Concept [34]

The concept of $A^{1,2}$ strain has been elegantly applied for the synthesis of solenopsin A(**X**), starting from the precursor imine (**B**), as delineated in Fig. 8.22.

References

1. F.A.L. Anet, G.N. Chmurny, J. Krane, Ring inversion in cyclohexanone. J. Am. Chem. Soc. **95**, 4423–4424 (1973)
2. W.G. Dauben, G.J. Fonken, D.S. Noyce, The stereochemistry of hydride reductions. J. Am. Chem. Soc. **78**, 2579–2582 (1956)
3. D.C. Wigfield, Stereochemistry and mechanism of ketone reductions by hydride reagents. Tetrahedron **35**, 449–462 (1979)
4. H.C. Brown, H.R. Deck, Selective reductions. VIII, The stereochemistry of reduction of cyclic and bicyclic ketones by the alkoxy substituted lithium aluminium hydrides. J. Am. Chem. Soc. **87**, 5620–5625 (1965)

5. E.L. Eliel, Y. Senda, Reduction with metal hydrides—XIX competitive reduction of alkylcyclohexanones: regarding the cencepts of steric approach control and product development control. Tetrahedron **26**, 2411–2428 (1970)
6. D.C.Wigfield, D.J. Phelps, Transition state analysis: evidence against product development control in the sodium borohydride reduction of ketones. Chem. Commun. 1152–1153 (1970)
7. J.-C. Richer, On the stereochemistry of the reduction of cyclic ketones with lithium tri-*t*-butoxyaluminium hydride. J. Org. Chem. **30**, 324–325 (1965)
8. M. Chérest, H. Felkin, Torsional strain involving partial bonds. The steric course of the reaction between allyl magnesium bromide and 4-t-butylcyclohexanone. Tetrahedron Lett. 2205–2208 (1968)
9. H.B. Bürgi, J.D. Dunitz, E. Shefter, Geometrical reaction coordinates. II. Nucleophilic, addition to carbonyl group. J. Am. Chem. Soc. **95**, 5065–5067 (1973)
10. H.B. Bürgi, J.D. Dunitz, J.M. Lehn, G. Wipff, Stereochemistry of reaction paths at carbonyl centers. Tetrahedron **30**, 1563–1572 (1974)
11. H.B. Bürgi, J.M. Lehn, G. Wipff, An *ab initio* study of nucleophilic addition to a carbonyl group. J. Am. Chem. Soc. **96**, 1956–1957 (1974)
12. S.K. Talapatra, D.S. Bhar, R. Chakraborty, B. Talapatra, Eupacannol, a pentacyclic triterpene of a new skeletal type, and other chemical constituents from *Eupatorium cannabinum.* J. Indian Chem. Soc. **80**, 1209–1216 (2003) (Professor S.M. Mukherji CommemorationVolume (Part II))
13. A.S. Cieplak, Stereochemistry of nucleophilic addition to cyclohexanone. The importance of two-electron stabilizing interaction. J. Am. Chem. Soc. **103**, 4540–4552 (1981)
14. C.K. Cheung, L.T. Tseng, M.-H. Lin, S. Srivastava, W.J. Le Noble, Geometric equivalents of enantiomers in studies of the stereochemical course of substitution at carbon. Electronic effects in nucleophilic addition to carbonyl groups and to carbocations. Virtual proof of the existence of σ participation by unstrained carbon-carbon bonds. J. Am. Chem. Soc. **108**, 1598–1605 (1986)
15. A.S. Cieplak, B.D. Trait, C.R. Johnson, Reversal of π-facial diastereoselection upon electronegative substitution of the substrate and the reagent. J. Am. Chem. Soc. **111**, 8447–8462 (1989)
16. G. Mehta, F.A. Khan, Electronic control of π-facial selectivities in nucleophilic addition to 7-norbornanones. J. Am. Chem. Soc. **112**, 6140–6142 (1990)
17. G. Mehta, J. Chandrasekhar, Electronic control of facial selection. Chem. Rev. **99**, 1438–1467 (and references cited)
18. A.J. Birch, G.S.R. Subba Rao, Reductions by metal-ammonia solutions and related reagents, in *Advances in Organic Chemistry: Methods and Results*, ed. by E.C. Taylor, vol. 8 (Wiley, New York, 1972), pp. 1–65
19. A.J. Birch, The birch reduction in organic synthesis. Pure Appl. Chem. **68**, 553–556 (1996) (and references cited)
20. S.K. Pradhan, Mechanism and stereochemistry of alkali metal reductions of cyclic saturated and unsaturated ketones in protic solvents. Tetrahedron 6351–6388 (1986) (Tetrahedron Report No. R212)
21. K. Fukui, *Theory of Orientation and Stereoselection* (Springer, Berlin, 1975)
22. F. Johnson, S.K. Malhotra, Steric interference in allylic and pseudo-allylic systems. 1. Two stereochemical theorems. J. Am. Chem. Soc. **87**, 5492–5493 (1965)
23. F. Johnson, Allylic strain in six-membered rings. Chem. Rev. **68**, 375–413 (1968)
24. F. Johnson, S.K. Malhotra, Steric interference in allylic and pseudo-allylic systems. II. Stereochemistry of exocyclic enolate anion protonation. J. Am. Chem. Soc. **87**, 5493–5494 (1965) (Zimmerman & coworkers' work interpreted)
25. H.E. Zimmerman, The stereochemistry of ketonization reaction of enol. J. Org. Chem. **20**, 549–557 (1955)
26. H.E. Zimmerman, T.W. Cutshall, The stereochemistry of ketonization. X. Enols from α-haloacids X. J. Am. Chem. Soc. **81**, 4305–4308 (1959) (and earlier papers)

27. H.E. Zimmerman, T.E. Nevins, The stereochemistry of ketonization. V. aci-Nitro tautomerism. J. Am. Chem. Soc. **79**, 6559–6561 (1957)
28. E.J. Corey, R.A. Sneen, Calculation of molecular geometry by vector analysis. Application to six-membered alicyclic rings. J. Am. Chem. Soc. **77**, 2505–2509 (1955)
29. E.W. Garbisch Jr., Conformations. 1. Synthesis, proton magnetic resonance spectra, and ultraviolet spectra of substituted 1-phenylcyclohexenes. J. Org. Chem. **27**, 4243–4249 (1962)
30. E.W. Garbisch Jr., Conformations. II. Proton magnetic resonance. Determination of C-6 substituent conformations in 6-substituted 1-phenylcyclohexenes. J. Org. Chem. **27**, 4249–4253 (1962)
31. G. Stork, R. Terrell, J. Szmuszkovicz, A new synthesis of 2-alkyl and 2-acyl ketones. J. Am. Chem. Soc. **76**, 2029–2030 (1954)
32. G. Stork, A. Brizzolara, H. Landesman, J. Szmuszkovicz, R. Terrell, The enamine alkylation and acylation of carbonyl compounds. J. Am. Chem. Soc. **85**, 207–222 (1963) (and references cited)
33. F. Johnson, A. Whithead, The stereochemistry of 2-substituted cyclohexanone enamines and the corresponding Schiff's bases. Tetrahedron Lett. 3825–3829 (1964)
34. K. Marnoka, T. Miyazaki, M. Ando, Y. Matsumura, S. Sakane, K. Hattori, H. Yamamoto, Organoaluminium—promoted Beckmann rearrangement of oxime sulfonates. J. Am. Chem. Soc. **105**, 2831–2843, vide page 2836 (1983)

Chapter 9
Fused Ring Systems

Abstract This chapter portrays the conformations and symmetry point groups of *trans*-decalin and *cis*-decalin, Talapatra's rule for torsion angle sign determination across the bonds of decalins, *cis*-decalin: steroid and nonsteroid conformations, ring inversion in *cis*-decalin, some coalescence temperatures, entropy and enthalpy difference in decalins, Auwers–Skita Rule, 9-methyldecalins and 9,10-dimethyldecalins conformations of *trans*-1-decalone and *cis*-1-decalone, conformational analysis of *trans*-2-*trans*-decalols and *trans*-2-*cis*-2-decalols; conformations. *Cis*-2-decalol *p*-nitrobenzoate diastereomers, octalins—their conformers. Perhydrophenanthrene and perhydroanthracene diastereomers, perhydrophenanthrones and all perhydrodiphenic acids—their conformations, relative stability, optical activity, resolvability, stereochemistry, and configurational assignments.

9.1 Decalins

Of the fused carbocyclic systems with 6-membered rings like 6–3, 6–4, 6–5, 6–6, and 6–7 systems, the most important fused system is the 6–6 system, [4.4.0]bicyclodecane, trivially named as *decalin* (=decahydronaphthalene).

This system is very useful for consideration of conformation and reactivity, more quantitatively than other bicyclic systems. The decalin system occurs abundantly in natural products, like diterpenes, triterpenes, steroids, and alkaloids.

9.1.1 Brief History

The existence of *cis*- and *trans*-decalins with two puckered six-membered rings was first predicted by Mohr in 1918. He also predicted chair/chair conformation (1) (Fig. 9.1) for *trans*-decalin and boat–boat conformation (2) (Fig. 9.2) for *cis*-decalin. On the contrary, two planar six-membered rings, as postulated by Baeyer, can only be fused as *cis* and not as *trans*.

S. K. Talapatra and B. Talapatra, *Basic Concepts in Organic Stereochemistry*,
https://doi.org/10.1007/978-3-030-95990-6_9

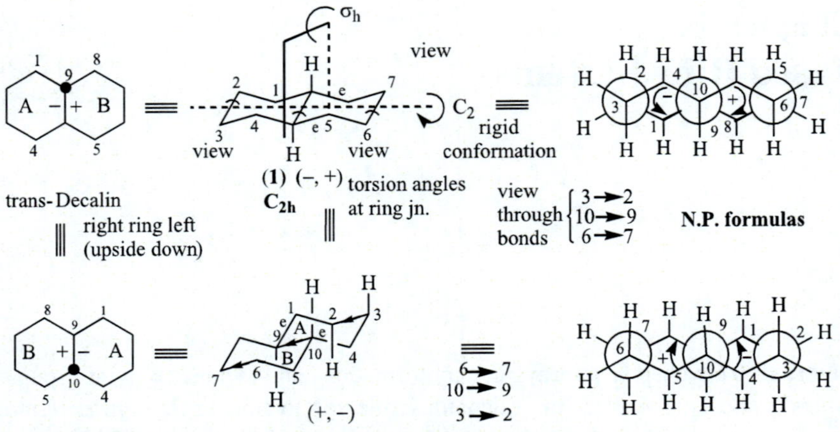

Upon monosubstitution at any position C_2 disappears, and σ also disappears, if the substitution is in a position other than C-9 and C 10, and the substituted product is optically active.

Fig. 9.1 Conformation and symmetry point group of *trans*-decalin

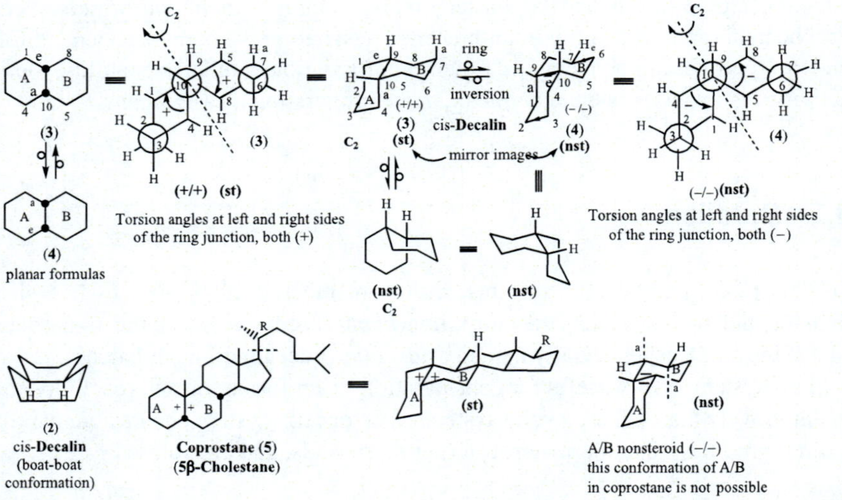

Note. Monosubstitution product of cis-decelin at any position other than C9 and C10 gives a diastereomer upon inversion and hence is optically active.

Fig. 9.2 Steroid (st) (+/+) and nonsteroid (nst) (–/–) conformations of *cis*-decalin

- In 1925, Mohr's prediction was confirmed by isolation of *cis*- and *trans*-decalins by Hückel, and *trans*-isomer was found to be more stable.
- In 1943, Hassel recognized that the chair form of cyclohexane is at least 5.6 kcal/mol more stable than its boat/skew–boat form, and for the first time predicted the chair–chair conformation for both *trans*-decalin (**1**) (Fig. 9.1) and *cis*-decalin (**3**) (Fig. 9.2).

- This was also substantiated by electron diffraction data in 1946.

The conformations of the two diastereomeric forms, *trans*-decalin and *cis*-decalin, are discussed in the sequel.

9.1.2 Trans-*Decalin. Conformation. Torsion Angles. Symmetry*

In *trans*-decalin the ring junction H's at C9 and C10 are on the opposite sides. *The sign of any torsion angle involving three consecutive bonds is written against the central bond.* Torsion angles around all C–C bonds in both *trans*-decalin and *cis*-decalin are ~ 55–56°, the same as in cyclohexane. In *trans*-decalin, the two rings (chair) are fused through *e,e* bonds and cannot undergo ring inversion. From the Newman projection formula of *trans*-decalin, obtained by looking through the bonds mentioned in Fig. 9.1, it is quite clear that the torsion angles (ϕ or ω) in rings ***A*** and ***B*** on the left and right sides of the common central bond are (−) and (+), respectively (see also Sects. 1.2.1 and 7.1.3). The signs are reversed if the ***B*** ring is on the left side (Fig. 9.1).

Talapatra's Rule for Torsion Angle Sign Determination Across the Bonds of Decalins

The sign of torsion angle (ϕ) can be determined directly from the chair–chair conformation of the *trans*-decalin or *cis*-decalin using a rule [1], stated below. One does not have to draw the Newman projection formula across the specified bond for this purpose, as is shown on the right sides of the Figs. 9.1 and 9.2. According to the frequent suggestions of many readers of the reference [1], henceforth this rule is referred to as "Talapatra's rule".

Talapatra's rule of determining the sign of the torsion angle (ω or ϕ): While moving our eyes in a *clockwise direction from the top of a ring* (in a slanting manner to see the chair or boat or skew–boat conformation, *the β-axial H or substituent* in that ring *is preceded by* (+) sign of the torsion angle ω or ϕ, and *is followed by* (−) sign. Any molecular model or drawing the Newman projection viewing across a bond (as shown in Fig. 9.10) (cf. also Figs. 9.1 and 9.2) would not be necessary to find the sign of the torsion angle at that bond.

The reader will be convinced about the validity of Talapatra's rule by examining with a molecular model.

Corollary: While looking in the above way, *the α-axial H or substituent* in that ring is preceded by (−) sign and is followed by (+) sign of the torsion angle ω or ϕ.

Symmetry: The chair–chair conformation of *trans*-decalin (Fig. 9.1) has a center of symmetry at the midpoint of C9–C10 bond and is achiral. Additionally, it has a C_2 axis passing through the midpoints of C2–C3, C9–C10, and C7–C6 bonds, and a σ_h plane passing through the ring junction, and perpendicular to the C_2 axis. Thus, its point group is $\mathbf{C_{2h}}$.

9.1.3 Cis-*Decalin. Conformations. Torsion Angles. Symmetry (Fig. 9.2)*

In the *cis*-decalin the ring junction H's at C9 and C10 are on the same side of the puckered plane of the molecule, and the two rings are fused through *ea* or *ae* bonds. Thus, ring inversion takes place when the ring fusion bonds *ea* [to a ring, say(***A***)] (***3***) are converted to ae (***4***). From the Newman projection formulas of *cis*-decalin (Fig. 9.2), it is clear that torsion angle ϕ at the left and right sides of the common central C9–C10 bond in rings ***A*** and ***B*** are (+, +) for the left-side formula (***3***) and (−, −) for the right-side formula (***4***), which are undergoing rapid interconversion. Such a concerted change in torsion angles of a *cis* fused junction is quite feasible. One can easily see this interconversion in its Dreiding or Fischer model.

The signs of the torsion angle ϕ at the ring junction in the rings ***A*** and ***B*** can be determined directly from either chair–chair conformations (***3***) (+/+) and (***4***) (−/−), by application of Talapatra's rule (Sect. 9.1.2.1).

The conformation (**3**) of *cis*-decalin is called *steroid* conformation since it resembles the A/B rings of 5β-steroids like coprostane (***5***), whereas the conformation (***4***) which cannot exist in 5β-steroids is called *nonsteroid.*

Symmetry: *cis*-Decalin chair–chair conformation (***3***) or (***4***) is chiral since either has no reflection symmetry (no σ plane or center of symmetry). Each conformer possesses a C_2-axis which passes through the midpoint of C9–C10 bond to which it is orthogonal, and it bisects the dihedral angle between 9-H and 10-H bonds (ϕ 9-H/10-H). So the point group is $\mathbf{C_2}$ (chiral). But it forms an *unresolvable* (±) *pair* because of the rapid interconversion of the two flipped forms: *steroid and nonsteroid (which are enantiomers), at even low temperatures,* the energy barrier between them being much lower than 15 kcal mol^{-1}. Although it is appreciably higher than that in cyclohexane, the barrier to ring inversion $\Delta G^{\neq}$ in *cis*-decalin is *12.3–12.6* kcal mol^{-1} (51.5–52.7 kJ mol^{-1}).

The inversion path involves the following transformations: chair–chair (CC) → chair–twist (CT) → twist–twist (TT) → alternate TT → CT → CC; the alternate TT is reconverted to the CT and CC by a reversal (mirror imaged) of the path by which it was formed (cf. Sect. 7.1.5).

	~ Coalescence temp.	Energy barrier (kcal $mol^{-1)}$
cis-Decalin	≥ −30°C	12.3 – 12.5
Cyclohexane[#]	– 66.7°C	10.1
cis-Hydrindane	– 121°C	6.5

[#]The rate of chair inversion is 105 sec^{-1}.

Fig. 9.3 Some coalescence temperatures

9.1.4 Ring Inversion in Cis*-Decalin*

In rigid *trans*-decalin the equatorial and axial protons are distinguishable which appear in ^{1}H NMR as two broadbands due to spin–spin coupling.

cis-Decalin and its derivatives undergo ring inversion (cf. cyclohexane)—already discussed. This can be studied by ^{1}H NMR. The equatorial and axial protons are averaged out due to ring inversion and appear as a narrow band at ambient temperature. However, at temperatures lower than the coalescence temperature, they can be split up into two broadbands. The *coalescence temperature* gives a value of 12.76 kcal mol^{-1} at – 18° in CS_2 for the free energy barrier of ring inversion. 2,2-Difluoro-*cis*-decalin exhibits a barrier of 12.26 kcal mol^{-1} at – 30 °C in ^{19}F NMR. The energy barrier of *cis-decalin* is thus considerably higher than that of cyclohexane and *cis*-1,2-dimethylcyclohexane ($\Delta G^{\neq} = 10$ kcal/mol). Ring inversion data of many cyclic compounds are available in a monograph by Oki [2]. Some coalescence temperatures and energy barriers are given in Fig. 9.3.

9.1.5 Entropy Difference in Decalins

Symmetry number of *cis*- and *trans*-decalin is 2. The difference in entropy of *cis*-decalin results from the fact that the *cis*-isomer is a (±)-pair (although non-resolvable). The total entropy difference is shown in Fig. 9.4.

Thus, ΔS experimental value (0.55 cal K^{-1} mol^{-1} ≡ 2.3 J K^{-1} mol^{-1}) is much less than the calculated value 1.38 cal K^{-1} mol^{-1} ≈ 5.8 J K^{-1} mol^{-1}.

Isomer	Symmetry no. σ	Entropy cal. / deg. mole^{-1}				
		– R ln σ	R ln 2 due to ± forms	Total S	Calcd. ΔS $S_{cis} - S_{trans}$	Exptl. ΔS
cis-Decalin	2	– 1.38	1.38	0	+ 1.38	0.55
trans-Decalin	2	– 1.38	–	– 1.38		

Fig. 9.4 Entropy difference in decalins

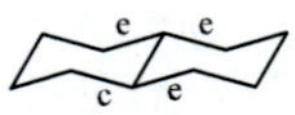

trans-**Decalin**

No additional gauche butane unit (over those in two cyclohexane rings)

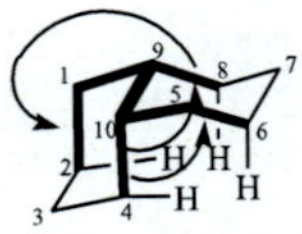

cis-**Decalin**

additional gauche butane units exist, namely,
4-10-5-6
4-10-9-8
8-9-1-2
} shown by thick bonds

$\therefore \Delta H_{calc.} = H_{cis} - H_{trans} = 3$ gauche butane interactions

$= 3 \times 0.8$ kcal mol^{-1} = 2.4 kcal mol^{-1} (= 10.05 k J mol^{-1})

Good agreement with $\Delta H_{exptl.}$ (2.10 to 2.72 kcalmol^{-1}) (≡ 8.8 to 11.4 k J mol^{-1}) determined by temperature dependence of cis ⇆ trans equilibrium and also from heat of combustion data.

According to **Auwers-Skita rule or**conformational rule, cis-decalin having higher enthalpy has higher b.p., density, refractive index than those for trans-decalin.

cis-Decalin : b.p. 193°C.743 mm; d_4^{20} 0.895; n_D^{20} 1.4811

trans-Decalin 184°C/747 mm; 0.870; 1.4697

Fig. 9.5 Enthalpy and physical constants difference of *cis*- and *trans*-decalins

This is contrary to the popular belief that *cis*-decalin is more flexible than *trans*-decalin; in that case, ΔS would have been more rather than less. Here ring inversion ability possessed by *cis*-decalin should not be confused with the flexibility of the system as in case of *trans*-decalin.

9.1.6 Enthalpy and Physical Constants. Auwers-Skita Rule

Difference in enthalpies of *cis*-decalin and *trans*-decalin can be counted in terms of gauche interactions arising out of carbon atoms of two different rings. In *trans*-decalin, each ring is fused through *e* bonds, and hence no additional gauche unit is introduced. The calculated and experimental results of ΔH (enthalpy difference) are presented in Fig. 9.5.

9.1.7 Free Energy Difference in Decalins

The free energy difference has been determined experimentally from the equation $\Delta G = \Delta H - T\Delta S$ as 2.39 kcal mol^{-1}, which is in good agreement with the equilibrium data for 1-decalones [*trans* ⇆ *cis* (5–10%) from $\Delta G = -\mathrm{RT} \ln K$]. The comparison is, however, not strictly valid because of 2- and 3-alkylketone effects (cf., Sect. 8.3) in 1-decalone.

9.1.8 *Effect of Introduction of Angular Methyl Group/s*

The enthalpy difference between *cis*-decalin and *trans*-decalin after introduction of 9-methyl group and 9,10-dimethyl groups are delineated in **a** and **b** of Fig. 9.6, respectively. The additional gauche–butane interactions between an axial methyl group and synaxial hydrogen atoms are shown in each case. The axial hydrogen atoms at the concave face of *cis*-decalin sustaining three additional gauche–butane interactions are also shown (cf. Fig. 9.5).

trans-Decalins unsubstituted at positions other than bridgeheads are always *achiral*, having a σ-plane vertical to the common 9–10 bond. *cis*-Decalins substituted only at C9 or C10 or at both carbons although chiral exist as unresolvable

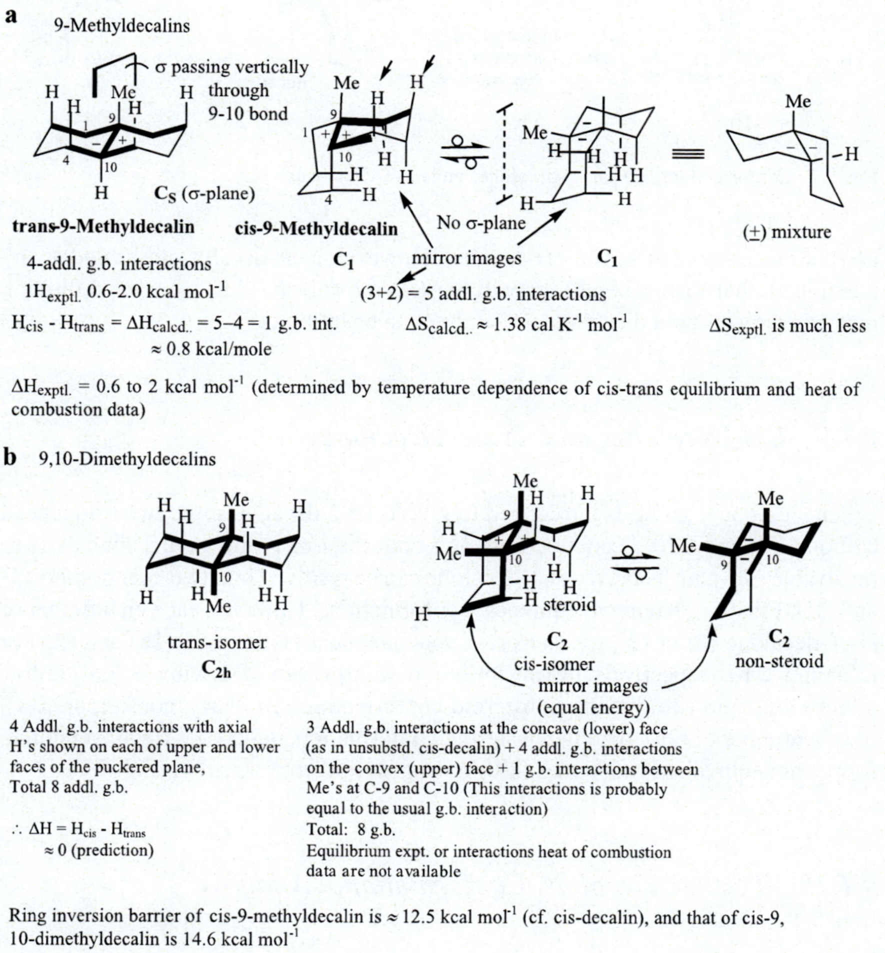

Fig. 9.6 Conformations and enthalpy differences of 9-methyldecalins and 9,10-dimethyldecalins

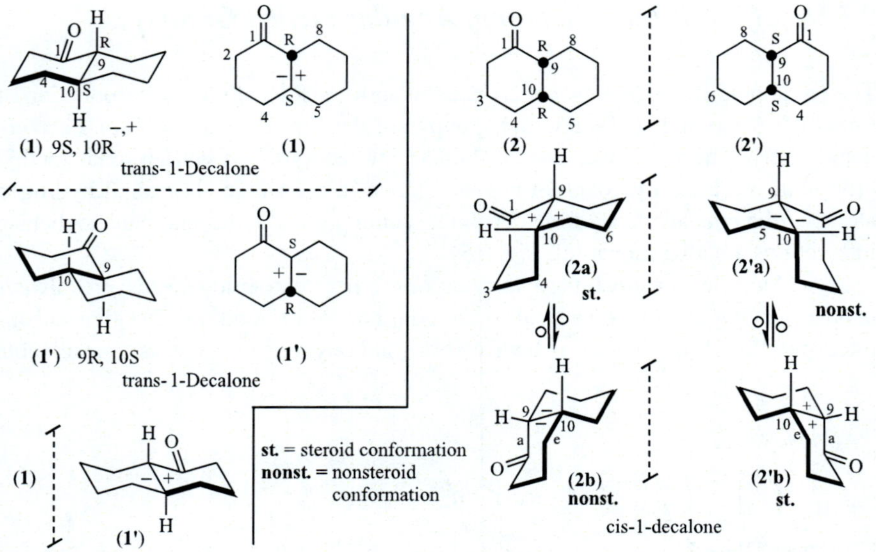

Fig. 9.7 Conformations of *trans*-1-decalones and *cis*-1-decalones

(±)- or racemic mixtures (cf. *cis*-decalin). However, a *cis*-decalin substituted at any position/s other than bridgeheads will exist as a resolvable (±)-pair, since flipping now converts it into a diastereomer (and not an enantiomer).

9.1.9 Cis-*Decalones and* Trans-*Decalones*

When a carbonyl group is introduced to give 1- or 2-decalones, the two bridgehead carbons, C9 and C10, become chiral, and both the *trans*- and *cis*-decalins exist as resolvable (±)-pair. Thus, *trans*-1-decalone can be resolved into two enantiomers (**1**) and (**1′**) (Fig. 9.7), having unequivocal conformations. However, each enantiomer of *cis*-1-decalone (**2**) or (**2′**) exists as two nonequivalent conformers (**2a**) and (**2b**) or (**2′a**) and (**2′b**), respectively, by ring inversion. Conformer (**2a**) having steroid conformation flips into (**2b**) having nonsteroid conformation. Similarly, conformer (**2′a**), the enantiomer of (**2a**), having nonsteroid conformation, undergoes ring inversion to form a nonequivalent conformer (**2′b**) having steroid conformation (Fig. 9.7).

9.1.10 Trans-*2-Decalols. Conformational Analysis*

Trans-2-Decalol exists as two diastereomers, viz. *trans*-2-*trans*-decalol (**1**) and *trans*-2-*cis*-decalol (**2**) (Fig. 9.8); they have the hydroxyl group *trans* and *cis*, respectively,

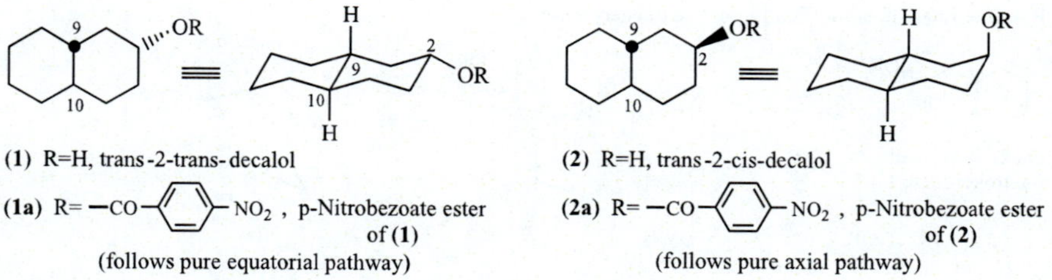

Fig. 9.8 Conformational analysis of *trans*-2-decalol diastereomers

to the 9-H atom. Since *trans*-decalin possesses unique rigid conformation (ring inversion being not possible), the OH groups in the diastereomers (**1**) and (**2**) (Fig. 9.8) must be equatorial and axial, respectively. Thus, the conformational analysis here is easy and gives definite conclusions. Compound (**1**) is converted to its *p*-nitrobenzoate ester (**1a**) at a faster rate than compound (**2**) under the same condition, *because of the steric hindrance* in the latter case. Likewise, the saponification of the ester (**1a**) follows pure equatorial pathway, whereas the reactions of (**2**) or (**2a**) follow pure axial pathway involving steric hindrance.

9.1.11 Cis-2-Decalols. Conformational Analysis

Let us now consider the relative reactivity of the *p*-nitrobenzoates of isomeric *cis*-2-dacalols [3]. The configurations of these epimeric decalols were established by Dauben and Hoerger [4]. Because of the flexibility (due to ring inversion) of *cis*-decalols, conformational analysis does not give definite conclusions regarding stereochemistry of the molecules. The relative rates of saponification of the diastereomers are delineated in Fig. 9.9 in terms of the equatorial **1e** and axial **1a** conformers arising due to ring inversion of **1** and in terms of the equatorial **2e** and axial **2a** conformers arising out of the ring inversion of **2** and by the application of Winstein–Holness equation (i) (Fig. 9.9); here k_e and k_a denote the specific rate constants of the equatorial and axial conformers, and Ne and Na denote the mole fractions of the equatorial and axial conformers, respectively.

9.2 Octalins (Octahydronaphthalenes)—Their Conformations

There are *six* octalins depending upon the position of the double bond and also upon the *trans* and *cis* junctions, namely $\Delta^{9,10}$-octalin (1) and $\Delta^{1,9}$-octalin (2) (Fig. 9.10), *trans*-$\Delta^{1,2}$-octalin (6) and *trans*-$\Delta^{2,3}$-octalin (Fig. 9.11), and *cis*-$\Delta^{1,2}$-octalin (7) and *cis*-$\Delta^{2,3}$-octalin (8) (Fig. 9.12). The conformations of these octalins have been delineated by Nasipuri [5].

Relative rates of saponification of the diastereomers

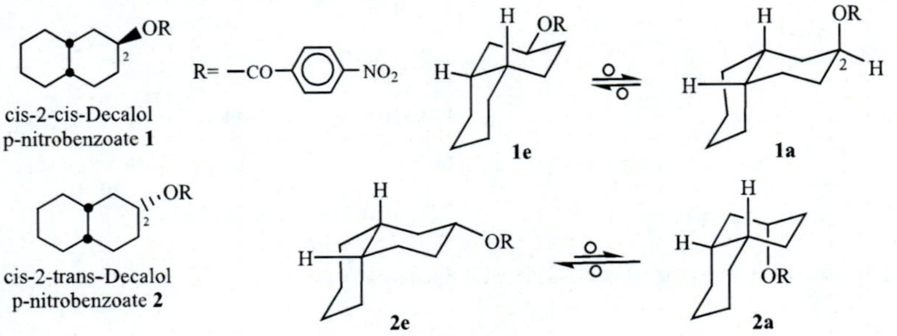

Winstein-Holness equation: overall rate $k = k_e N_e + k_a N_a$

In case of compound **1**

$k_1 = k_{1e}N_{1e}$ + $k_{1a}\, N_{1a}$

↑ ↑ (both greater, e pathway) | ↑ ↑ (both smaller, but not negligible; a pathway)

∴ The ester **1** undergoes saponification by both e and a pathways

$G_{1e} \approx G_{2e} < G_{1a} << G_{2a}$

$\therefore N_{1e} > N_{1a};\ N_{2e} >> N_{2a}$

Again, $\Delta G^{\neq}_{1e} < \Delta G^{\neq}_{1a}$ $\therefore k_{1e} > k_{1a}$

and $\Delta G^{\neq}_{1e} < \Delta G^{\neq}_{2a}$

(cases of steric hindrance)

$\therefore k_{2e} >> k_{2a}$

In case of compound **2**

$k_2 = k_{2e}\, N_{2e}$ + $k_{2a}\, N_{2a} \approx k_{2e}\, k_{2a}$ (approx.)

↑ ↑ (both much greater) ↑ ↑ (both negligibly smaller, and the product is even more negligible)

Note: The equatorial pathway for the saponification of an ester is faster than the axial pathway (case of steric hindrance).

∴ Compound 2 undergo saponification by equatorial pathway.

∴ k_2 should be greater than k_1.

Experimentally it has been found [94] that the ester **2** is saponified 1.5 times more rapidly than that of **1**.

Fig. 9.9 Conformational analysis of *cis*-2-decalol *p*-nitrobenzoate diastereomers

$\Delta^{9,10}$-Octalin (1) has the double bond at the junction with the two cyclohexene rings in half-chair conformation (1a) (Fig. 9.10).

$\Delta^{1,9}$-Octalin having a chiral center at C-10 exists as a (±)-pair and belongs to the point group $\mathbf{C_1}$. Ring A is fused with ring B through a fixed 9-8 bond and a flexible *pseudoequatorial* 10-5 bond as in (2a) (almost flat) or a *pseudoaxial* 10-5 bond as in (2e) (Fig. 9.10). Their Newman projection formulas viewed along C-9 and C-10 show torsion angles at the junction of (2a) of opposite signs (+ and −), almost like *trans*-decalin and is, therefore, called a *quasi-trans* form. The same torsion angles of the other conformer (2e), as shown by its Newman projection (2d), are of the same sign (+, +). The conformer (2e) is thus folded like *cis*-decalin (see Sect. 9.1) and is called a *quasi-cis* form. The torsion angle signs may be directly determined from the respective conformations by application of Talapatra's rule (see Sect. 9.1.2.1).

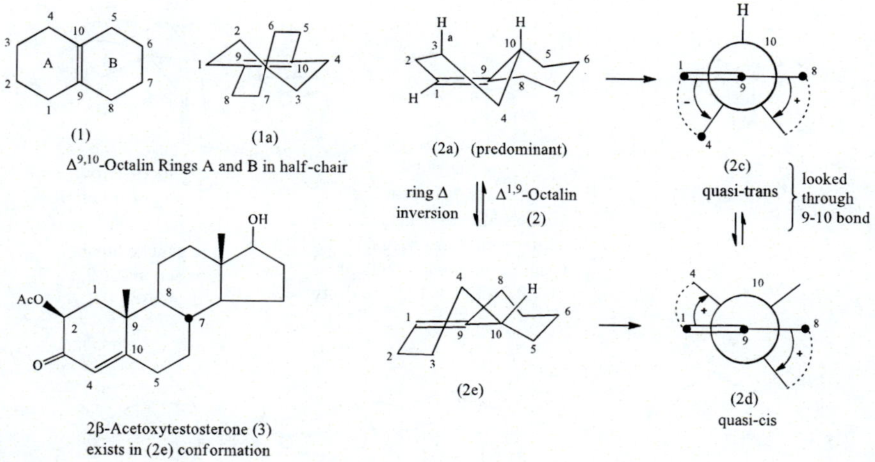

Fig. 9.10 Conformations of $\Delta^{9,10}$-octalin (1), $\Delta^{1,9}$-octatin: *quasi-trans* and *quasi-cis* fusions. An example. 2β-acetoxytestosterone (3)

Fig. 9.11 Torsion angle signs of *trans*-decalin (4), *trans*-$\Delta^{1,2}$-octalin (5), and *trans*-$\Delta^{2,3}$-octalin (6)

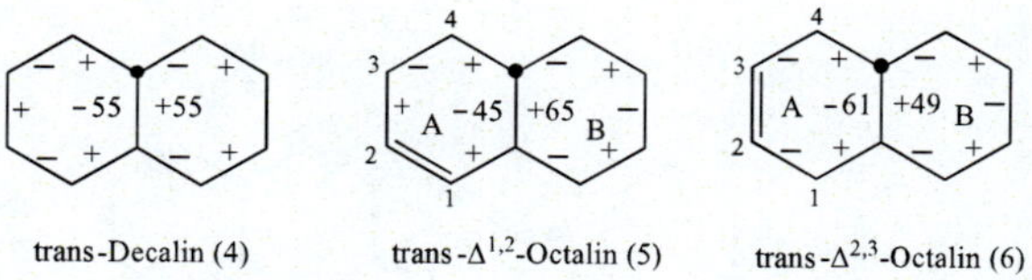

In absence of any additional steric situation, the conformer (2a) (*quasi-trans*) is predominant. But in case of 2β-acetoxytestosterone (3) in order to avoid the *syn-axial* interaction with its 9-*axial* Me group, the 2-acetoxy group assumes equatorial orientation in *quasi-equatorial* form like (2e). The *quasi-trans* and the *quasi-cis* may also be called as *transoid* and *cisoid*, respectively.

9.2.1 *Six Octalins. $\Delta^{9,10}$-Octalin. $\Delta^{1,9}$-Octalin and Its Conformations and Torsion Angles. 2Δ Acetoxytestosterone (Fig. 9.10)*

9.2.2 Trans-*$\Delta^{1,2}$-Octalin and* Trans-*$\Delta^{2,3}$-Octalin. Torsion Angles at the Ring Junction and Their Relative Stability (Fig. 9.11)*

Trans-$\Delta^{1,2}$-Octalin (5) and *trans*-$\Delta^{2,3}$-octalin (6) have the torsion angle signs as shown in Fig. 9.11.

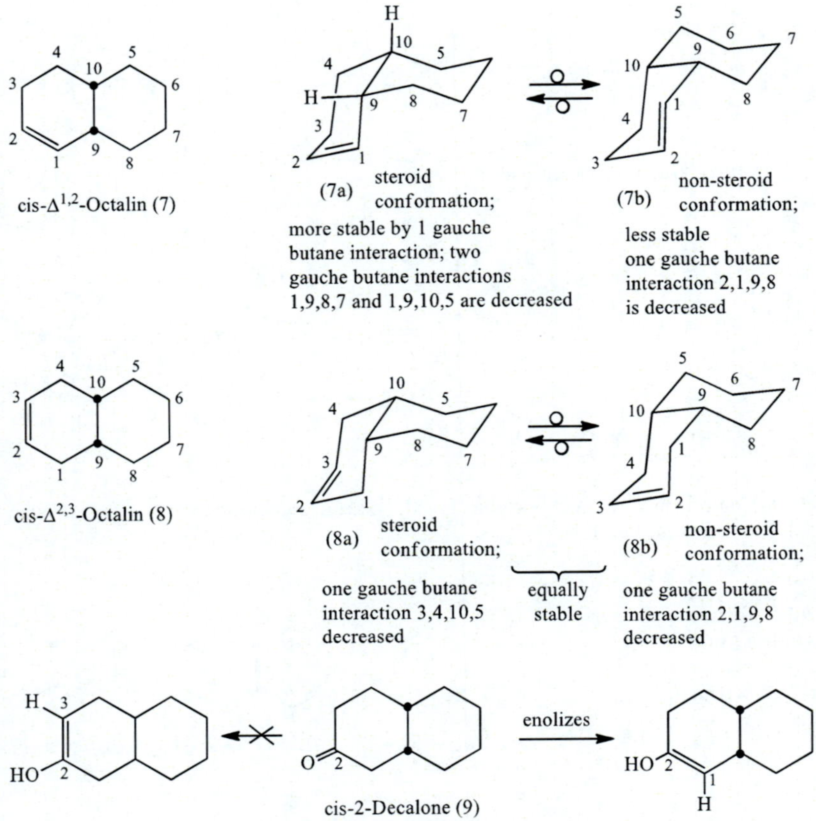

Fig. 9.12 Conformations of *cis*-$\Delta^{1,2}$-octalin (7) and *cis*-$\Delta^{2,3}$-octaline; preferential enolization of *cis*-2-decalone (9)

Compared to *trans*-decalin (4) (cf. Sect. 9.1.2), the torsion angle in the saturated ring (A) of *trans*-$\Delta^{1,2}$-octalin (5) has undergone an opening of 10° (from + 55° to + 65°), while the same in the saturated ring (B) of *trans*-$\Delta^{2,3}$-octalin has been reduced by 6° (from 55° to 49°). This means that the double bond at the 2,3 position is more favourable than at the 1,2 position in *trans*-decalin. The calculated enthalpy difference between (5) and (6) is approximately 3.0 kJ mol^{-1}, which is close to the experimental value obtained by their equilibration. This may explain the tendency of *trans*-2-decalone to enolize towards C-3 than C-1.

9.2.3 Cis-$\Delta^{1,2}$-Octalin and Cis-$\Delta^{2,3}$-Octalin. Their Conformations and Relative Stability (Fig. 9.12)

From a consideration of the torsion angles at the junction, the relative stability of *cis*-$\Delta^{1,2}$-octalin (7) and *cis*-$\Delta^{2,3}$-octalin (8) cannot be evaluated (Fig. 9.12). In case of (7), there is equilibration between the steroid conformer (7a) and the non-steroid conformer (7b); in the former two *gauche* butane interactions are decreased, while in the latter one *gauche* butane interaction is decreased (Fig. 9.12) (cf. Sect. 9.1.6). So, the conformer (7a) becomes more stable than its conformer (7b).

While, in case of *cis*-$\Delta^{2,3}$-octalin both the steroid and non-steroid conformers involve the decrease of one *gauche* butane interaction and hence are equally stable.

Hence, *cis*-$\Delta^{1,2}$-octalin (7) is more stable than *cis*-$\Delta^{2,3}$-octalin, a result opposite to that for the *trans*-decalin). This may explain the tendency of *cis*-2-decalone (9) to enolize to give a 2,3-double bond rather than a 1,2-double bond.

9.3 Perhydrophenanthrenes (PHP's). Stability. Point Groups. Optical Activity

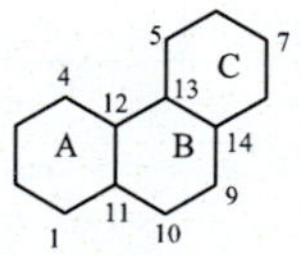

PHP

Three cyclohexane rings when fused successively to an angular arrangement give rise to perhydrophenanthrene (PHP). It constitutes the ABC rings of terpenoids and steroids. It contains two equivalent pairs of chiral centers—the four bridgehead carbons, C11 and C12 and C13 and C14. Thus, it constitutes an ***ABBA*** system and can have four pairs of enantiomers and two meso forms. (cf. CH_2OH–$(CHOH)_4$–CH_2OH); total *10 stereoisomers* and *six diastereomers* (Fig. 3.16).

In the nomenclature of the diastereomers, the prefixes *cis* and *trans*, here abbreviated as ***c*** and ***t***, define the stereochemistry of the ring junctions A/B or B/C, whereas the other terms *cisoid* and *transoid*, abbreviated as ***ci*** and ***tr***, define the steric relationship of the nearest bridgehead atoms (which also denotes the orientation of the terminal rings with respect to each other).

The six perhydrophenanthrene diastereomers (**1**), (**2**), (**3**), (**4**), (**5**), and (**6**) are included in Fig. 9.13 in order of increasing enthalpy, i.e., decreasing stability. This basic ring system is of wide occurrence in nature, and the relative stabilities of the diastereomers are of interest. In this figure, the calculated relative enthalpies ($\Delta\Delta H^0$) based on nonbonded interactions (**C**) [6], and on molecular mechanics [force field

Planar structure (A) (Point group) (chirality)	Conformation (B) e equatorial a axial	No. of gauche butane interactions (C)	Relative ΔΔ H^0 calcd. kcal mol^{-1} [89] (D)	Relative ΔΔ H^0 calcd. Molecular mechanics [90] (E)	ΔΔG^0 exptl. [91] (F)	σ plane in planar projection formula† (G)
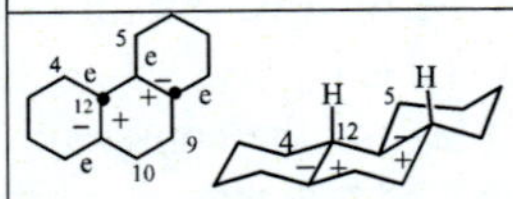 (1) t-tr-t (chiral) (C_1)	four e ring fusion bonds (rfb) 11α, 12β, 13α, 14β R S S R	1 (C4-C12-C13-C5)	0	0	0	Absent so resolvable
		Note: A dot at a ring juncture indicates a forward or upward hydrogen, whereas the absence of a dot indicates a backword or downward hydrogen. In case of (±) pairs only one enantiomer with H12 β- is shown.				
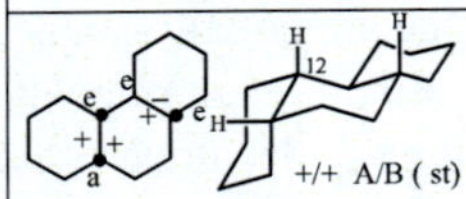 (2) c-tr-t (chiral) (C_1)	+/+ A/B (st) three e rfb S S S R	1+3 (for cis-decalin)	2.4	2.44	2.25	Absent, so resolvable
		†the chirality or otherwise can be judged by looking for a vertical σ-plane passing through the 9-10 and 12-13 bonds in the planar structure				
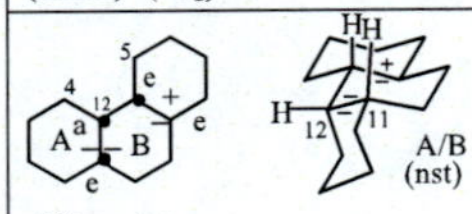 (3) c-ci-t	A/B (nst) three e rfb (chiral) (C_1)	1+3 (for cis-decalin)	2.4	2.57	2.66	Absent so resolvable

(A)	(B)	(C)	(D)	(E)	(F)	(G)
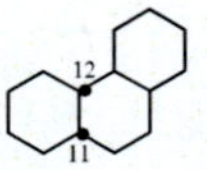 (4) c-tr-c (C_1) chiral two e rfb)	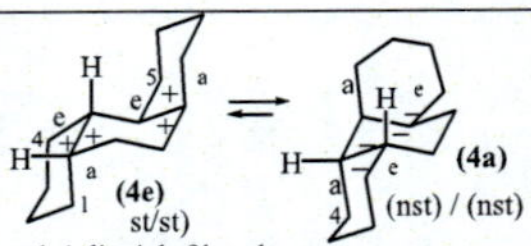 (4e) st/st) 1,4,diaxial rf bonds. A/B B/C +/+ +/+ ⇌ (4a) (nst) / (nst) 1,2-diaxial rf bonds −/−, −/−	1+6(two cis-decalins) ⇅ 6(two cis-decalins)	4.8 **(4e)** (4,5 di eq) ⇅ 4.0 **(4a)**	4.01	4.6	Absent, so resolvable; flips into a diastereomer

Fig. 9.13 Perhydrophenanthrene (PHP) diastereomers: conformations, torsion angles, energies, resolvabilities

calculations **(D)**] [7], and the experimental relative free energies ($\Delta\Delta G^0$) **(E)** [8] are expressed in kcal/mol (approx.)

Conformations of the enantiomers, (marked with superscript 1) of the chiral perhydrophenanthrene diastereomers (**1**), (**2**), (**3**), and (**4**) are shown in Fig. 9.14.

It should be noted that the torsion angles of the central ring at 11–12 and 13–14 bonds at the ring junctions are the same in (**1**) to (**4**) and (**6**), as expected for chair conformation. But in the isomer ***t-ci-t*** (**5**), these signs are opposite which suggest a non-chair conformation for the central ring. The same conclusion is also drawn from the orientation (*a*, *e*) of the ring fusion bonds. If the ring B were a chair, the

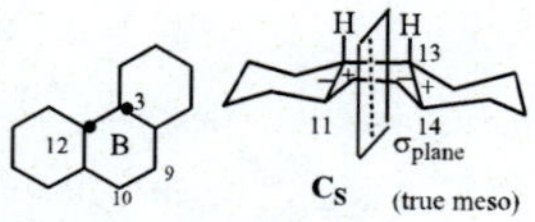 (5) t-ci-t **Achiral** (4be rfb) In ring B the sign at 11-12 bond is + " " 13-14 " is −, (should have been + in chair form) Impossible in chair conformation; so B ring is boat.	B-ring true boat [89]	7.2	7.03	8.98	Present in planar as well as conformational structure. So true meso
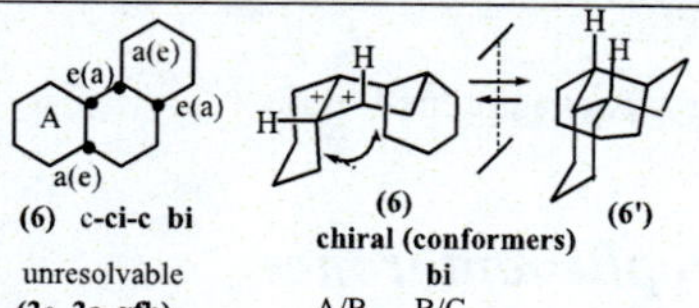 (6) c-ci-c bi unresolvable (2e, 2a, rfb) (6) chiral (conformers) bi (6') A/B B/C +/+ +/+ st st A/B B/C −/− −/− nst nst	1+6+ syn axial $CH_2 \leftrightarrow CH_2$ ¥ interaction 	4.8+ >3.6¥ = >8.4 total (5.2-1.6 = 3.6) for Me ↔ Me	9.01	7.43	present in planar str, so unresolvable;. actually unresolvable (±) pair formed by flipping.(cf. cis-decalin). So not true meso

Notes : Approx. agreements between H^0(calcd.) and G^0(exptl.) energies are satisfactor y in view of the differences, though small, in entropy of mixing and entropy of symmetry between isomers.

- For the highly stra ined isomers (**5**)and (**6**) the calculated. energies are th e reverse of the experimental.
- ¥ This is between 1, 3 synaxial ring fusion bonds; being rigi d it would be more than synaxial Me□ Me interaction.

Fig. 9.13 (continued)

Planar str. Conformation	Planar str. Conformation
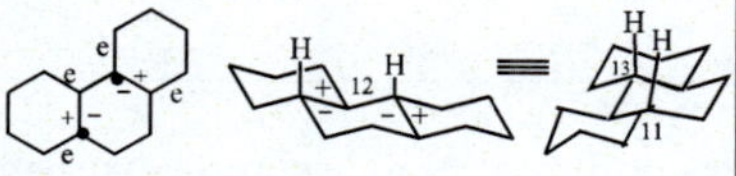 (1^1) t-tr-t	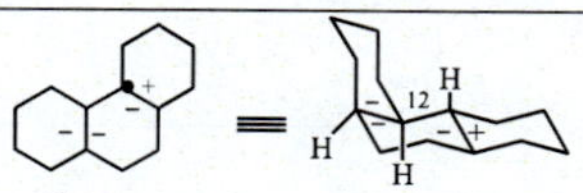 (2^1) c-tr-t A/B, nst
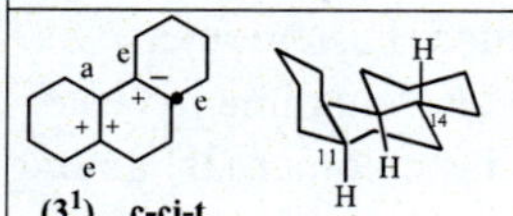 (3^1) c-ci-t A/B **st.**	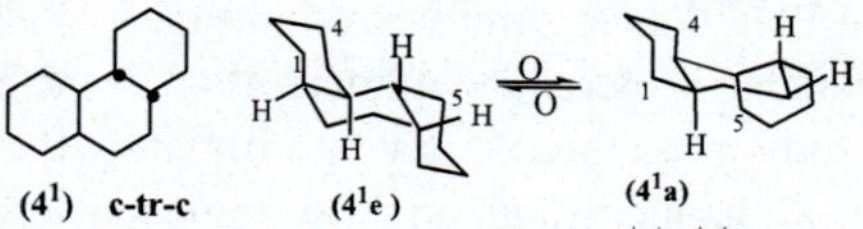 (4^1) c-tr-c (4^1e) −/−, −/− m. images A/B, B/C **nst, nst** (4^1a) ++, ++ A/B, B/C **st, st**

Fig. 9.14 Conformations of the enantiomers of the active perhydrophenanthrene diastereomers of Fig. 2.136

ring fusion bonds would be *e, e, a, a* with respect to B, and it is known that the *a, a* fusion in the *trans*-decalin moiety is sterically not possible. Because of the rigidity of the ***t-ci-t*** (**5**) diastereomer, the central ring assumes a *true boat* form (instead of a flexible boat).

The predicted order of stability of the PHP's has been confirmed by equilibration of the 9-ketones in the cases shown in Fig. 9.15.

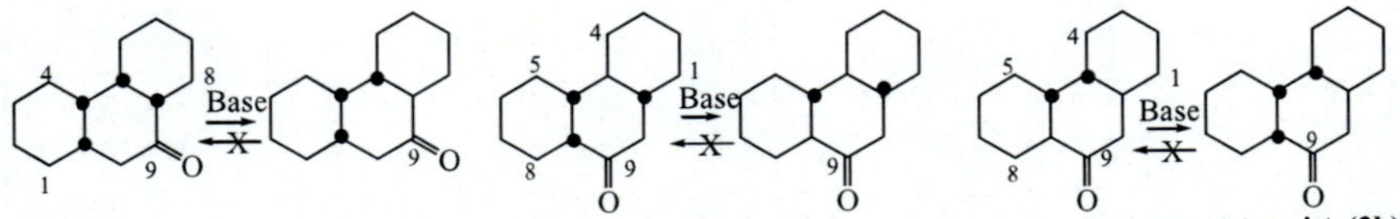

(±) c-ci-c (6') (7.43) | (±) c-ci-t (3a) (2.66) | (±) c-tr-t (2a) (2.25) | (±) t-tr-t (1a) (0) | (±) t-ci-t (5a) (8.98) | (±) c-ci-t (3b)

Approx. rel. enthalpy (**H**) in kcal/mole (within parenthesis)

Fig. 9.15 Confirmation of the predicted stability of perhydrophenanthrene-9-ones

9.3.1 *Stereochemistry of Some Perhydrophenanthrones and All Perhydrodiphenic Acids (PHDPAs)*

An elegant piece of work by Linstead and coworkers [9–15] presents a challenging exercise in stereochemical reasoning.

Several perhydrophenanthrones were correlated with PHDPAs in the following way:

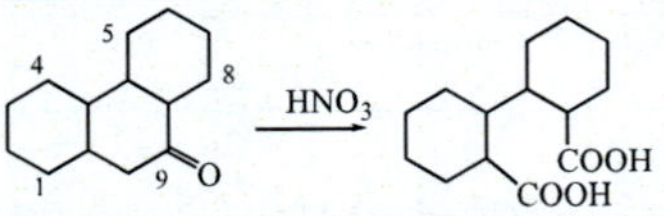

Epimerization in this reaction must be guarded if same configuration is to be assumed

Syntheses of PHDPAs and the process of their configurational assignments from the study of catalytic hydrogenation of diphenic acid giving *cis*-isomers (**A**), (**B**), and (**C**) of PHDP's are summarized in Fig. 9.13. The hydrogenated diastereomers were subjected to preferential epimerization of the monomethyl esters and dimethyl esters with sodium methoxide giving ultimately the most stable diastereomer (**D**) among the *cisoid* isomers, and the most stable diastereomer (**F**) among the *transoid* isomers (Fig. 9.16). The following important points should be noted:

- Formation of *cis*-hexahydrophthalic acid from hexahydrodiphenic acid (**9**) and the conversion of the latter to (**A**) by further hydrogenation indicates that ring A in (**A**) is *cis* 1,2-disubstituted.
- Epimerization takes place at the α-carbon to the ester and not at the backbone carbons. Thus, (**A**), (**B**), and (**D**) belong to the same backbone configuration; this is also valid for (**C**), (**E**), and (**F**).
- Stoermer–Steinbach principle was applied to unresolvable (**A**) and (**D**) as follows: Both of them were converted to the monomethyl ester (dissymmetry introduced), which could be resolved. An active monoester was converted to the inactive diester (plane of symmetry again introduced).
- Thus, **PHDPAs**, (**A**) and (**D**), were established to be meso/unresolvable. Hence, (**A**), (**B**), and (**D**) must belong to cisoid backbone configuration.

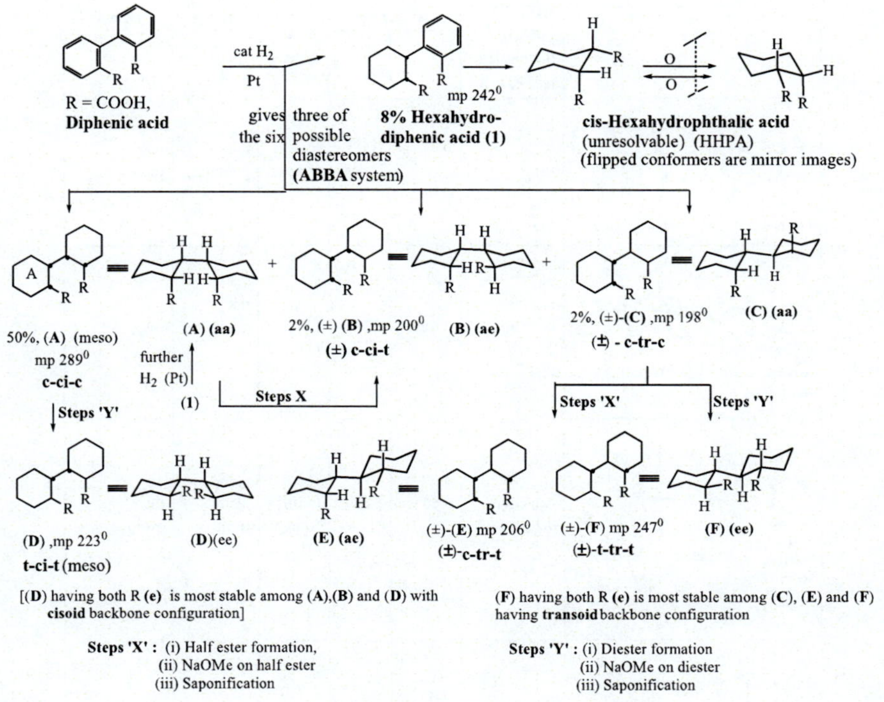

Fig. 9.16 Synthesis and configurational assignments of the perhydrodiphenic acids

- The PHDPAs (**C**) and (**F**) were resolved, so (**C**), (**E**), and (**F**) must belong to the *transoid* backbone configuration.
- Stability order in each series easily determines the relative configuration of the compounds in both *cisoid* and *transoid* series.
- For designation of PHDPAs, the two COOH groups must be on the same side of the backbone C–C bond (cf. the nomenclature of the perhydrophenanthrenes).

9.4 Perhydroanthracenes: Relative Stability. Torsion Angles. Point Group. Optical Activity

In perhydroanthracene (**PHA**) diastereomers (Fig. 9.14), the rings are fused in a linear arrangement. In the nomenclature of the diastereomers, the prefixes *c* and *t* define the stereochemistry of the ring junctions A/B and B/C, whereas the other terms *cisoid* and *transoid*, abbreviated as *ci* and *tr*, define the steric relationship of the nearest bridgehead atoms. Here, all the four chiral centers are equivalent, corresponding to **AAAA** system and there are five diastereomers—two of them are enantiomeric pairs and the other three are *meso*. The diastereomers (A), (B), (C), (D), and (E) in Fig. 9.17 are in order of increasing enthalpy and hence decreasing stability.

Planar str.	Conformation	No. of gauche int (s) (kcal/mole), relative to A[a]	Δ H⁰ kcal/mole Calcd.[b]	Δ H⁰ kcal/mole Exptl.[c]	Sym. pt. gr. (Sym.no σ)	Flipping	Resolvability (opt. active)
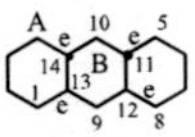 t-ci-t (A) four e rfb	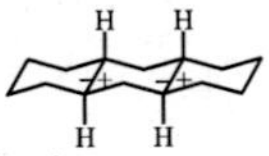 (meso) rfb : ring fusion bonds	0 (0)	0	0	C_{2h} (2)	not possible	no (no) meso
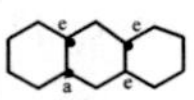 (B) c-ci-t or c-tr-t (three e rfb)	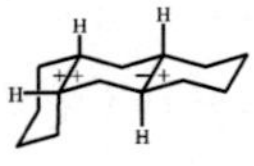 (±)(chiral)	one cis-decalin, 3 (2.7)	2.62	2.76	C_1 (1)	not possible	yes (yes)
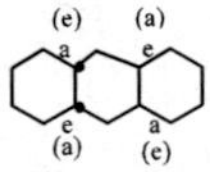 (C) c-tr-c (two e rfb)	(meso) 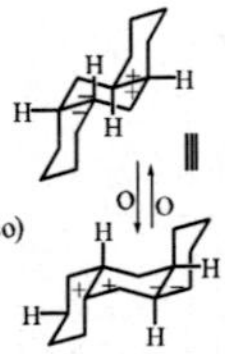	two cis-decalins, 6 (5.4)	5.56	5.58	C_i (both forms) (1)	gives an identical form	no (no) meso
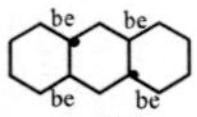 (D) t-tr-t (four boat e rfb)	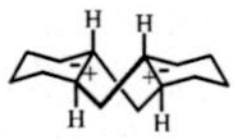 (±) (chiral) D_1 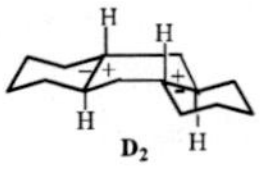 D_2	skew boat (5.6)	5.86	4.15	C_2 (2)	not possible	yes (yes)
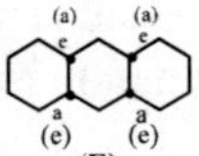 (E) c-ci-c (two e rfb) (meso) 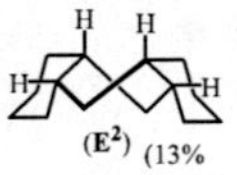 (E^2) (13% at 271 °C)	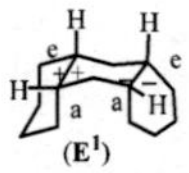 (E^1) 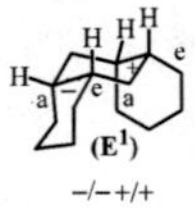 (E^1) −/−+/+	two cis-decalins + syn-diaxial int. (5.4+3.6 = 9.0) (E_2) is of slightly higher energy than (E^1)	8.13	8.74	C_s (1)	gives an identical form	no (no) meso

Notes: [a]ΔH⁰ calculated based on the number of gauche interactions [89]
[b]ΔH⁰ calculated using molecular mechan ics (force field method) [98]
[c]ΔH⁰ determined experimentally from the composition of an equilibrium mixture of the hydrocarbons as a function of temperature [98].
[d]Both **c-ci-c** isomers of PHP (**5**) and PHA (**E**) are the least stable diastereomers of PHP and PHA respectively.

Fig. 9.17 Conformation, stability, and optical activity of perhydroanthracenes

The relative stability of the isomers was originally predicted by Johnson [6]. Their planar structures, conformations, the relative enthalpies (calculated by molecular mechanics and determined experimentally by Allinger and Wuesthoff [16]) and resolvability are outlined in Fig. 9.17. In contrast to the least stable PHP diastereomer, here the ***t-ci-t*** isomer (**A**) is the most stable isomer having no extra gauche interaction among the atoms of different rings. It possesses a plane of symmetry and hence is meso. The ***c-ci-t*** isomer (**B**) may also be called ***c-tr-t*** isomer, if one looks at the bridgehead atoms in the clockwise or anticlockwise direction, respectively. The ***c-tr-c*** isomer (**C**) flips into an identical conformer and possesses a center of symmetry and hence is *meso*. The instability of the isomer (**D**) is due to the twist–boat conformation of the central ring. The ***c-ci-c*** isomer (**E**) possesses 1,3-*syn*diaxial methylene interaction and two *cis*-decalin interactions and is the least stable one. Interestingly, an alternate conformer of the isomer (**E**) with the central ring as a twist–boat ($\mathbf{E^2}$) is of only slightly higher energy than the chair ($\mathbf{E^1}$) and contributes to the extent of 13% at 271 °C. The signs of torsion angles of the two junctions at 11–12 and 13–14 bonds in the central ring are consistent with the chair form in each isomer excepting (**D**). In case of the isomer (**D**), the same sign at the ring junctions 11–12 and 13–14 in the central ring in ($\mathbf{D_2}$) and in ($\mathbf{D_1}$) also indicates its boat or better twist–boat conformation.

References

1. S.K. Talapatra, B. Talapatra, *Chemistry of Plant Natural Products: Stereochemistry, Conformation, Synthesis, Biology, and Medicine* (Foreword by Professor K.C. Nicolaou), vol. 1 (Springer, Berlin, 2015), p. 143
2. M. Oki, *Applications of Dynamic NMR Spectroscopy to Organic Chemistry*, vol. 4 (VCH, Deerfield Beach, Florida, 1985)
3. W.G. Dauben, K.S. Pitzer, Conformational analysis, in *Steric Effects in Organic Chemistry*, ed. by M.S. Newman, Chap. 1 (Wiley, New York, 1956), pp. 1–9 pertinent page 50
4. W.G. Dauben, E. Hoerger, The *cis*-Decahydro-2-naphthoic acid and their relationship to the *cis*-2-Decalols and *cis-2*-Decalylamines. J. Am. Chem. Soc. **73**, 1504–1508 (1951)
5. D. Nasipuri, *Stereochemistry of Organic Compounds (Principles and Applications)*, 2nd edn. (New Age International (P) Limited, 1991), pp. 312–313
6. W.S. Johnson, Energy relationships of fused ring systems. J. Am. Chem. Soc. **75**, 1498–1500 (1953)
7. N.L. Allinger, B.J. Gordon, I.J. Tyminski, M.T. Wuesthoff, Conformational analysis. LXX. Perhydrophenanthrenes. J. Org. Chem. **36**, 739–745 (1971)
8. H. Hőnig, N.L. Allinger, The stereoisomers of perhydrophenanthrene. J. Org. Chem. **50**, 4630–4632 (1985)
9. R.P. Linstead, W.E. Doering, S.B. Davis, P. Levine, R.W. Whetstone, The stereochemistry of catalytic hydrogenation. I. The stereochemistry of the hydrogenation of aromatic rings. J. Am. Chem. Soc. **64**, 1985–1991 (1942)
10. R.P. Linstead, W.E. Doering, The stereochemistry of catalytic hydrogenation. II. The preparation of the six inactive perhydrodiphenic acids. J. Am. Chem. Soc. **64**, 1991–2003 (1942)
11. R.P. Linstead, W.E. Doering, The stereochemistry of catalytic hydrogenation. III. Optically active perhydrodiphenic acids. A proof of the configuration of the backbone. J. Am. Chem. Soc. **64**, 2003–2006 (1942)

12. R.P. Linstead, S.B. Davis, The stereochemistry of catalytic hydrogenation. IV. Hexahydrodiphenic acids. J. Am.Chem. Soc. **64**, 2006–2009 (1942)
13. R.P. Linstead, S.B. Davis, R.R. Whetstone, The stereochemistry of catalytic hydrogenation. V. The assignment of *cis*- and *trans*-configurations. J. Am. Chem. Soc. **64**, 2009–2014 (1942)
14. S.B. Davis, R.P. Linstead, The stereochemistry of catalytic hydrogenation. Part IX. Confirmatory evidence of the configurations of the perhydrodiphenic acids. J. Chem. Soc. 1425–1427 (1950)
15. R.P. Linstead, R.R. Whetstone, The stereochemistry of catalytic hydrogenation. Part X. The hydrogenation of *trans*-as-octahydrophenanthrene derivatives. A new form of the perhydrophenanthrene ring as a model for the inversion of the 6-ketosteroids. J. Chem. Soc. 1428–1432 (1950)
16. N.L. Allinger, M.T. Wuesthoff, Conformational analysis. LXXII. The perhydroanthracenes. An equilibration study. J. Org. Chem. **36**, 2051–2053 (1971)

Chapter 10
Stereoisomerism: Axial Chirality, Planar Chirality, (*R, S*) Notations. Helicity

Abstract This chapter deals with Stereochemistry of Allenes, Chiral Spiranes and Analogs, Chiral Adamantoids, Chiral Catenanes and Their Configurational Nomenclatures; Biphenyl Derivatives and Atropisomerism, Energy Profile Diagram; Orders of Steric Hindrance and of Buttressing Effect, Chiral Biphenyls (R, S or aR, aS), Axially Chiral Molecules—interesting examples; Planar Chirality—(R, S) Specification; Helicity—P, M-designation due to Molecular Overcrowding; and Double P-Helix in Nucleic Acids.

10.1 Introduction

Different aspects of stereoisomerism in organic compounds, having chiral centers and pseudoasymmetric centers acting as stereogenic units, have been discussed in the preceding sections. Other elements of chirality, viz. axis, plane, and helicity, also act as stereogenic units. Appropriately substituted allenes, spiranes, alkylidenecycloalkanes, biaryls, and adamantoids behave as stereogenic units, due to the presence of chiral axis. Likewise, appropriately substituted *trans*-cycloalkenes, cyclophanes, and their analogs display stereoisomerism due to the presence of a chiral plane. A helix is nonsuperposable on its mirror image; it possesses an inherent chirality, a stereogenic unit known as helicity.

The configurational nomenclature of these stereogenic units will be illustrated briefly with examples in this and the following sections.

10.2 Stereochemistry of Allenes. Configurational Nomenclature

Of the three carbons of allene C1 and C3 are sp^2 hybridized and C2 is sp hybridized. The orbital picture of allene (**1**) is shown in Fig. 10.1. The shaded *p* orbitals as well as the unshaded p orbitals overlap with each other separately to form orthogonal π-bonds placing the ligands at C1 in a plane orthogonal to that of the ligands at C3.

S. K. Talapatra and B. Talapatra, *Basic Concepts in Organic Stereochemistry*,
https://doi.org/10.1007/978-3-030-95990-6_10

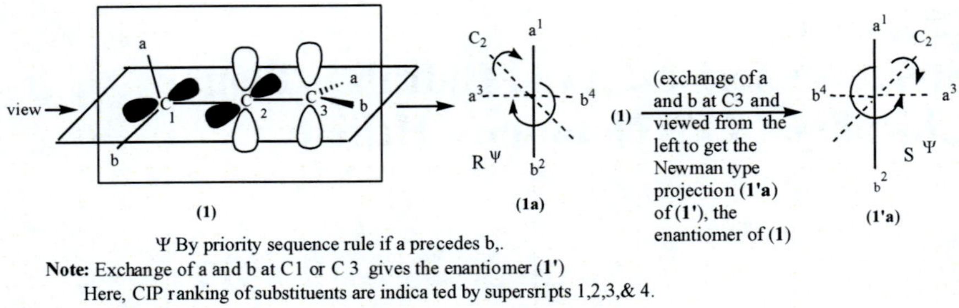

Fig. 10.1 Orbital picture of an allene, its Newman projection, its enantiomer, and their (*R*, *S*) nomenclatures

The structure (**1**) can be projected to a Newman-like projection formula (**1a**) to *R*-configuration, as shown, viewed from the left side with the front ligands in a vertical plane and the rear ligands in a horizontal plane (Fig. 10.1). The (*R, S*) nomenclature is independent of the direction of viewing, and the same specification will follow when viewed from the right side. The allenes of the general formula $C_{ab} = C = C_{ab}$ possess a C_2 axis, but no σ-plane, and belong to the point group $\mathbf{C_2}$. If the three or four of the ligands are different as in $C_{ab} = C = C_{ac}$ or $C_{ab} = C = C_{cd}$, the C_2 axis disappears, and the molecules are totally asymmetric and possess $\mathbf{C_1}$ point group. This is true for all axially chiral molecules.

For configurational nomenclature of allenes and other axially chiral molecules, the standard subrule 0, which states that "the near groups precede the far groups," is considered first, ahead of other subrules. *R, S* nomenclature in cases of some specific examples of a few optically active allenes and their enantiomers are illustrated in Fig. 10.2.

The molecule is viewed from any end of the chiral axis, and Newman-like projections are drawn; the groups near the viewer are numbered 1 and 2, whereas the groups at the far end are numbered 3 and 4, following the priority sequence rule. The order $1 \rightarrow 2 \rightarrow 3$, clockwise or anticlockwise, gives the configuration as *R* or *S*, respectively. This type of nomenclature is applicable for other types of compounds with axial chirality, to be discussed in the sequel. Interchange of the two geminal groups at any end in these molecules leads to the enantiomer.

Maitland and Mills [1, 2] prepared the first optically active allene (**3**), 60 years after Van't Hoff's prediction. 1,3-Diphenyl-1,3-di-α-naphthyl-2-propen-1-ol (Fig. 10.2) was separately dehydrated with (±)-, (+)-, and (−)-camphorsulfonic acid to give (±)-, (+)-(~5% *ee*), (−)-(~5% *ee*), respectively. The absolute configuration of the (+)- or (−)-enantiomer was, however, not determined.

Naturally occurring (−)-glutinic acid and the antibiotic mycomycin [3] are also examples of optically active allenes (Fig. 10.2) [4].

For the (R, S) specification of other types of axially chiral molecules the same procedure# is followed (see Figs. 10.3, 10.5, 10.6, 10.7, 10.11, and 10.12).

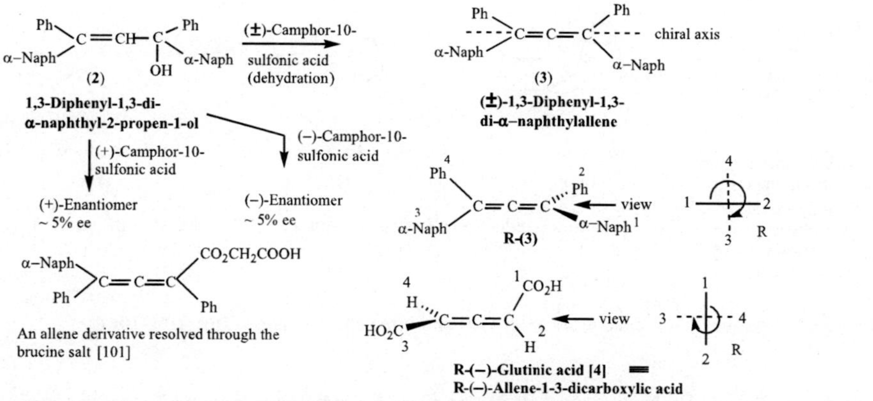

Mycomycin (**(5)** is an α-substituted acetic acid. Its molecule contains two alkynes, one allene, and a butadiene moieties - all in conjugation. The allene part is res ponsible for enantiomerism. It is a natural antibiotic.)

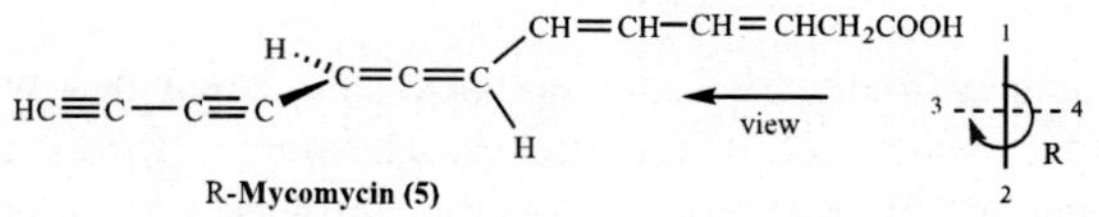

Nomenclature: One has to look along the chiral axis of the bi aryl (or any axially chiral) molecule and project on a plane orthogonal to the chiral axis; now, after assigning the pr iority sequence 1 (or a) and 2 (or b) to the near ligands shown by a solid line, and 3 (or c) and 4 (or d) to the fa r ligands shown by a dotted line the (R,S) specification can be made as shown in the examples given above.

Fig. 10.2 A few optically active allenes and their (*R*, *S*) specification

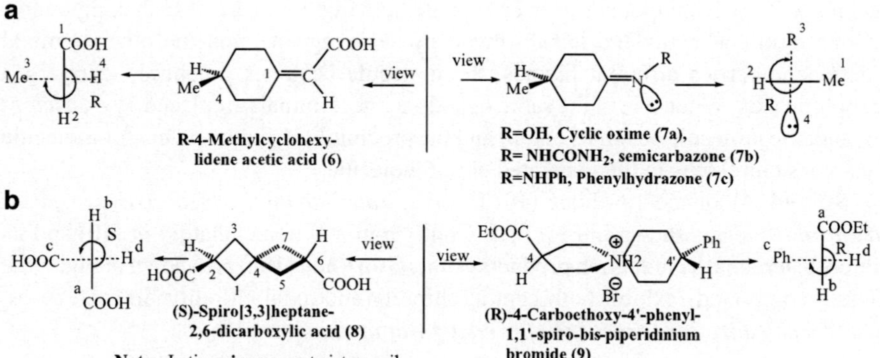

Note: Latin **spira** means twist or coil

Fig. 10.3 Configurational nomenclatures of some known resolved axially chiral compounds **a** derivatives of an alkalydenecyclohexane and 4-carboxycyclohexylidene imine, **b** two spiranes

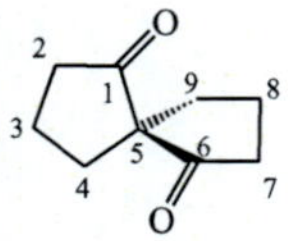

R-Spiro[4,4]nonane-1,6-dione (10)
C_5 is the chiral center
Priority sequence;
C1> C6> C4> C9
So, R

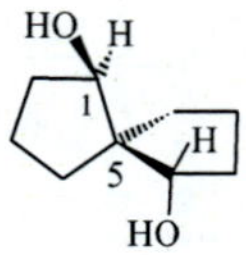

1R,5R, 6S
(10a)

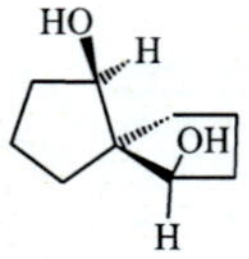

1R, 5R, 6R
(10b)

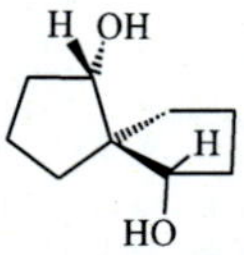

1S, 5R, 6S
(10c)

In all the reduction products the spiro-carbon C5 has R configuration

Fig. 10.4 Spiranes with one chiral center **(10)** and three chiral centers **(10a, 10b, 10c)**

10.3 Chiral Spiranes and Analogs. Configurational Nomenclature

Spiranes, alkylidenecycloalkanes, adamantanes, and catenanes with appropriate substituents can be dissymmetric like allenes. The name "*spirane*," derived from the Latin *spira* meaning twist or coil, implies that spiranes are nonplanar. One of the double bonds of an allene upon replacement by a ring, such as (**6**) (Fig. 10.3), gives rise to alkyledenecycloalkanes, which are also called hemispiranes. When both the double bonds of an allene are replaced by two rings, a *spirane*, e.g., (8), is generated. Here, also the two terminal disymmetrically substituted methylene planes are orthogonal to each other. The compounds having pairs of nonequivalent geminal substituents at both ends will exhibit enantiomerism, and (*R, S*) specification is achieved in such compounds as in case of allenes (Fig. 10.3). In all such compounds, if one terminal methylene is substituted by same ligands, and the other terminal methylene carries different ligands, the molecule becomes prochiral—carrying a prochiral axis. When one of the same ligands at one terminal is replaced by a different ligand, the molecule becomes chiral, and the prochiral axis in the precursor molecule becomes chiral axis in the generated chiral molecule.

Spiro[4, 4]nonane-1,6-dione (**10**) [5], *a spirane having a chiral axis as well as a chiral center*, is shown in Fig. 10.4. Configurational nomenclature of (**10**) and its three distereomeric reduction products (**10a**), (**10b**), and (**10c**) are shown in Fig. 10.4. These compounds exhibit both central chirality and axial chirality; in such cases, *central chirality has precedence for configurational nomenclature*.

10.4 Chiral Adamantoids. Configurational Nomenclature

Adamantane-2,6-dicarboxylic acid (**11**) (Fig. 10.5) is an axially chiral compound. In it, C2 and C6 methylenes are dissymmetrically substituted and exist in orthogonal planes. The adamantoid (**11**) is shown to have *S* configuration. The imaginary chiral

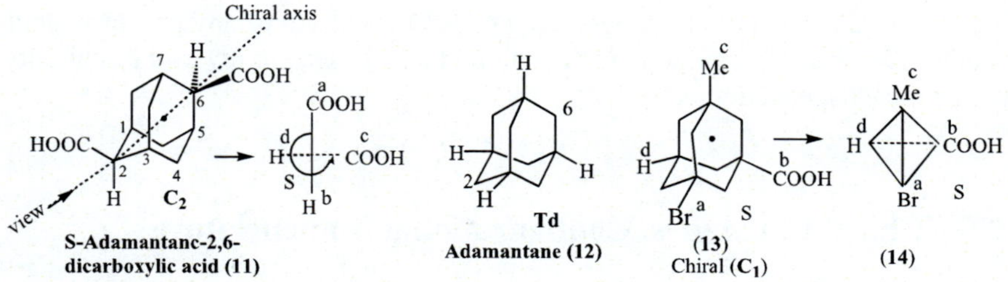

Fig. 10.5 Adamantane, chiral adamantanoids, and specification of chirality

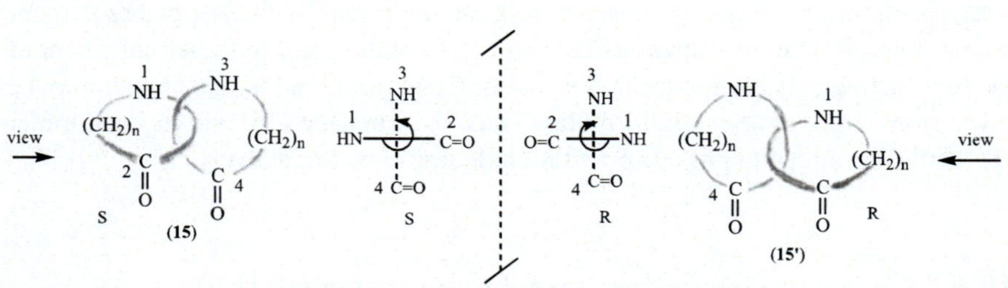

Fig. 10.6 Configurational nomenclatures of a chiral catenane and its enantiomer

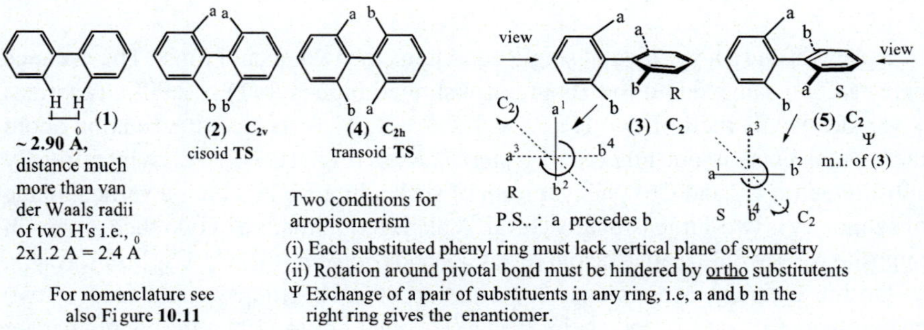

Fig. 10.7 Principle of atropisomerism. *R, S* nomenclature

axis in adamantoids passes through the substituted terminal carbon atoms C2 and C6 and the geometrical center of the ring system.

It may be mentioned here that if the four bridgehead carbon atoms of adamantane (**12**) bear four different ligands as in (**13**), the molecule becomes chiral. Adamantane [6] is a highly symmetrical molecule having $\mathbf{T_d}$ symmetry like CH_4. On the other hand, the adamantane derivative (**13**), carrying four different chiral centers at the four bridgeheads, becomes completely asymmetric with the point group $\mathbf{C_1}$ and exists only as a (±)-racemate pair. The four different ligands in (**13**) form a tetrahedral arrangement like (**14**), and the chirality of the molecule may be associated with a center, represented by a dot, in the unoccupied space of the adamantane skeleton

(Fig. 10.5). Like a centrodissymmetric compound exchange of any pair of ligands abcd in (**13**), or of the pair ab at C2 or C6 in (**11**), will lead to reversal of chirality and will give the enantiomer.

10.5 Chiral Catenanes. Configurational Nomenclature

The name catenane (from Latin *catena* meaning chain) was coined by Wasserman [7] to molecules of a type of unusual topology containing two or more *dissimilar* intertwined (or knotted) rings. The two rings are *to be held* with their planes perpendicular to each other as shown in (**15**) (Fig. 10.6) with regard to the arrangement of the four distinguishable groups in the chains. Configurational nomenclature may be given from similar projection formulas. Thus, the catenane (**15**) and its enantiomer (**15′**) are determined to possess *S* and *R* configurations, respectively.

10.6 Biphenyl Derivatives and Atropisomerism [8]

10.6.1 Introduction

In biphenyls an sp^2–sp^2 pivotal single bond joins the two phenyl rings. The distance between two same side ortho Hs in unsubstituted biphenyl (**1**) is 2.90 Å (approx.) > van der Waals radii of two H's, 2×1.2 Å $= 2.24$ Å. Hence, the rotation about the pivotal bond is not hindered by steric factor (Fig. 10.7). A dissymmetrically substituted (at C2 and C6) phenyl group of such a biphenyl lacking a vertical plane of symmetry is two-dimensionally chiral. A planar combination of two such groups in opposite ways would lead to cisoid [(**2**), $\mathbf{C_{2v}}$] and *transoid* [(4), $\mathbf{C_4}$, $\mathbf{C_{2h}}$] conformers.

On the contrary, a nonplanar combination of such groups would lead to two enantiomers (**3**) and (**5**), each belonging to $\mathbf{C_2}$ point group. When the energy barrier exceeds 19–20 kcal/mol (80–85 kJ mol^{-1}), the enantiomers are separable at room temperature. This type of enantiomerism was first discovered by Christie and Kenner in 1922 in the case of 6,6′-dinitro-2,2′-diphenic acid [9] (Fig. 10.9). Richard Kuhn later in 1933 called it *atropisomerism* and such enantiomers as *atropisomers* (from Greek *a* meaning "not" and *tropos* meaning "turn"), since such molecules do not turn around the molecular axis (due to steric hindrance). Thus, atropisomers are simply rotational energy barrier isomers and are classified as enantiomeric, from the viewpoint of internal energy, since the internal energies are the same.

10.6.2 Energy Profile Diagram

Energy Profile Diagram (approx.) is shown in Fig. 10.8 for a 360° displacement of torsion angle (ω) about the pivotal bond or showing the interplanar angle in *ortho*-substituted biphenyls. The following points are to be noted:

(i) Nonplanarity of the two aromatic planes is caused by the steric demands of the *ortho* substituents. This is opposed by π electron overlap of the two rings; maximum stabilization occurs when the rings are coplanar. Even the biphenyl itself is nonplanar in the ground state (the inter-ring torsion angle being 44° in the vapor phase, but in the crystal phase the rings are coplanar because of packing forces [10].

(ii) The planar diastereomeric conformations (**2**) and (**4**) represent the energy maxima due to severe steric interference of the ortho substituents on either side; the former [*cisoid* **TS**, (**2**)] having similar groups on the same side has higher energy than the *transoid* **TS** (**4**) having dissimilar groups on the same side.

(iii) Recemization of the two enantiomers (**3**) and (**5**) takes place with greater ease through the transoid conformation (**4**) involving less activation energy than that involving the cisoid conformation (**2**).

(iv) The bulkier the *ortho* substituents are, the higher is the energy barrier between the enantiomers.

Fig. 10.8 Energy profile diagram of *ortho*-substituted biphenyl

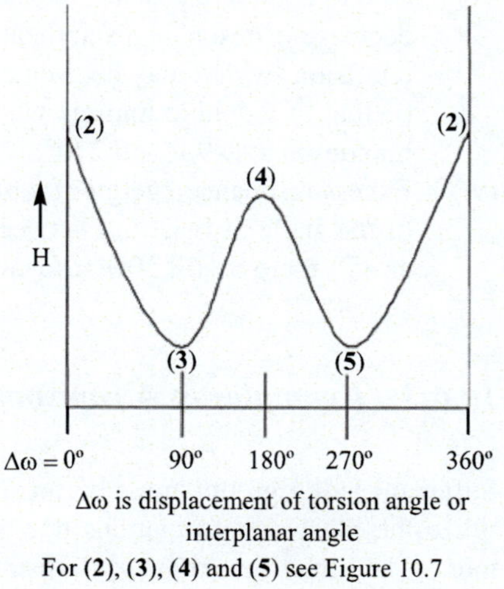

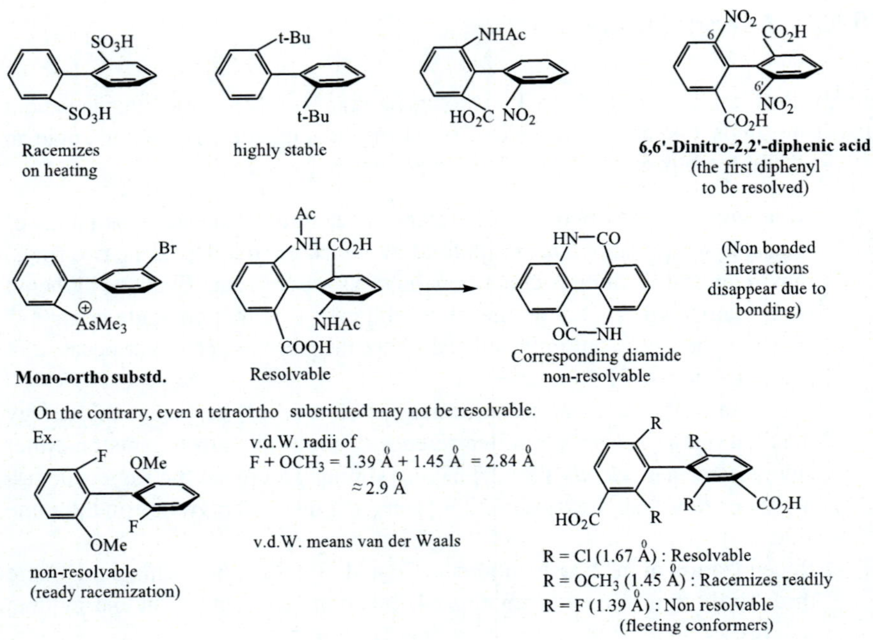

Fig. 10.9 Resolvable (stable) atropisomers

(v) Due to complete absence of resonance stabilization at $\omega = 90°$ or $270°$, the decrease of resonance stabilization may be more than the decrease in steric repulsion (which may be even absent) at these positions and a situation may be that of a double minima when ω is around $90°$ and $270°$, and of a small maximum at ω $90°$ and $270°$.

(vi) Excess resonance energy of a biphenyl over two benzene rings is a function of the angle of twist (ω) between the phenyl groups. At even as large twists as $45°$, there is still 50% resonance energy.

10.6.3 Examples of Atropisomerism

Sufficient bulky groups must be present in *ortho* positions of biphenyls to destabilize the planar conformations due to steric repulsion and to generate atropisomers. They may be *ortho ortho* disubstituted, trisubstituted, tetrasubstituted, and even monosubstituted. Examples are given in Fig. 10.9.

Order of Steric hindrance : Br > Me > Cl > NO_2 > COOH > OMe > F > H > D
roughly corresponds to v. d. W. radii of atoms or groups

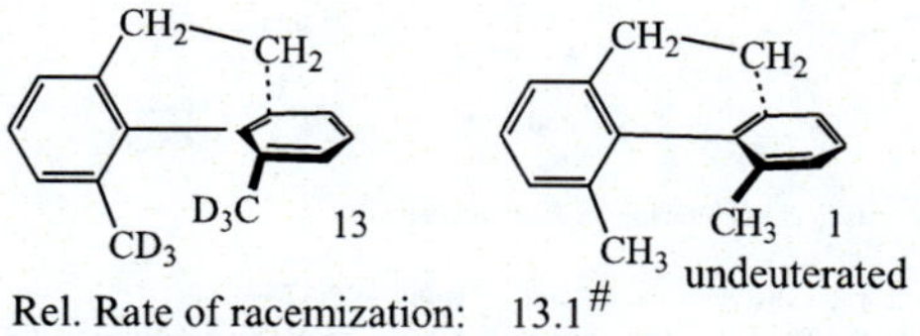

Rel. Rate of racemization: 13.1#

since D has smaller van der Waals radius than that of H, because of the lower zero point vibration frequency of D

*Buttressing Effect:

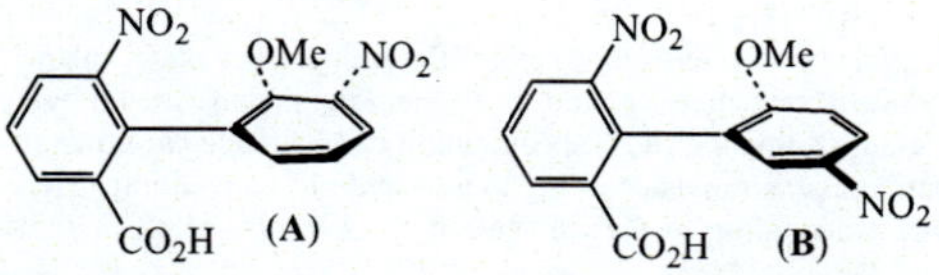

Rate of racemization of (**A**) is much lower than that of (**B**) due to the buttressing effect of the bulky NO_2 group in (**A**) adjacent to the OMe group.

*Order: NO_2 > Br > Cl > Me (different from bulk order)

Fig. 10.10 Order of steric hindrance and buttressing effect on configurational stability of biphenyls

10.6.4 Orders of Steric Hindrance and of Buttressing Effect [11]

These are depicted in Fig. 10.10.

*10.6.5 Configurational Nomenclature of Chiral Biphenyls (**R, S** or **aR, aS**)*

Selection rules (new convention*, 1966 [12]) for axial chirality have been introduced in case of atropisomers.

Subrule 0: Proximity rule: Groups about the near end of the chiral (molecular) axis precede over groups about the far end (same as in other axially chiral molecules). Either end can be taken as the far end.

a*Ordering of groups: Only the four atoms C2, C6, C2′, and C6′ which correspond to the four vertices of the *elongated tetrahedron* and which contribute more to the chirality properties of the molecule are considered for sequencing, as shown in Fig. 10.11. Thus, ordering of groups starts from the first off-chiral axis atom (near the observer) in chiral biphenyls or chiral allenes.

Thus, the fiducial or identifying groups are pairs directly bonded to the chiral axis (with minimum effective range), one pair of which lie in each plane, orthogonal to each other. They are most concerned with producing chirality,

c. of Fig. 10.11 ***P, M-Nomenclature.*** Alternatively, molecules with chiral axes may be viewed as helicenes (since they resemble the helicenes to be discussed in the sequel). Their configuration may be designated as *P* or *M*. If the turn from the higher priority front ligand "a" to

a Ordering of groups

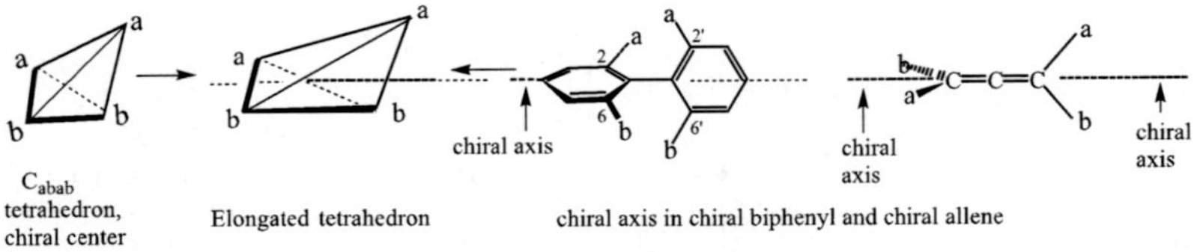

Ordering starts from the first off axis chiral biphenyl atoms (near the observer) i.e., 2nd and 6th atoms of the front ring.

- The priority sequence is fixed by the four ortho carbons after due complementation for quadriligancy proceeding along the branch of higher priority.
- molecule viewed from either end leads to the same configurational descriptor. R or S.

b. Nomenclature : One has to look along the chiral axis of the biaryl molecule and project on a plane orthogonal to the chiral axis; now, after assigning the priority sequence 'a' and 'b' to the near ligands shown by a solid line, and 'c' and 'd' to the far ligands shown by a dotted line the (R,S) specification can be made as shown in the following examples. In many cases the nomenclature remains same according to new and old conventions. Here we are citing two examples where nomenclature changes according to the convention. Viewing from left or right gives the same specification.

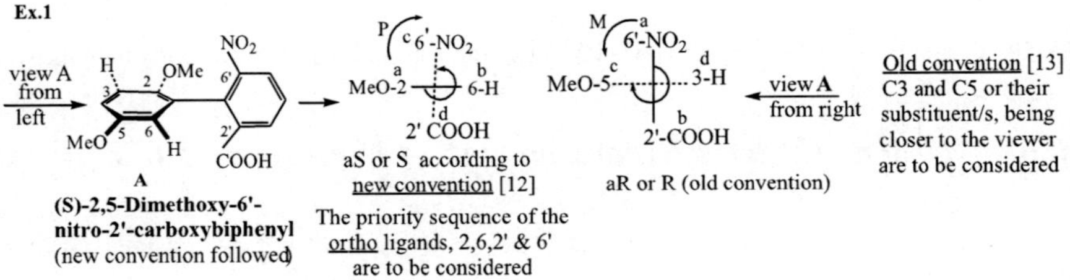

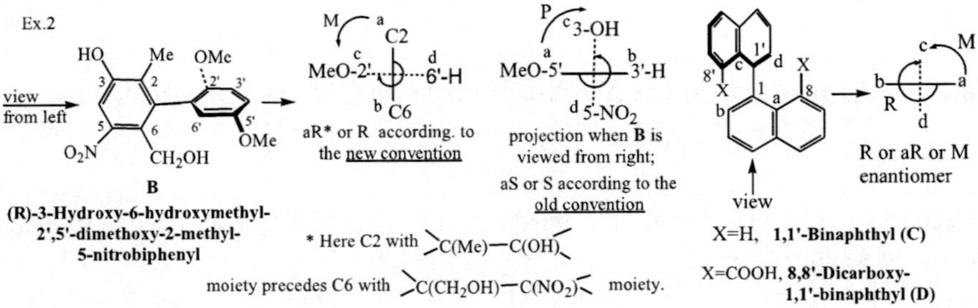

Fig. 10.11 Configurational nomenclatures (new and old conventions) of some chiral biphenyls and binaphthyls

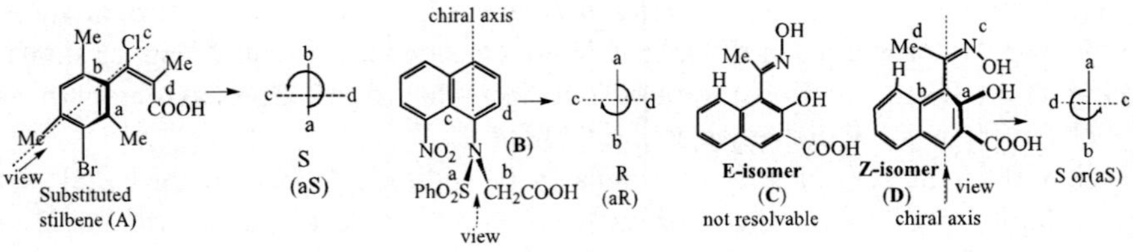

Fig. 10.12 Atropisomerism in acyclic analogs of biphenyl and in oximes

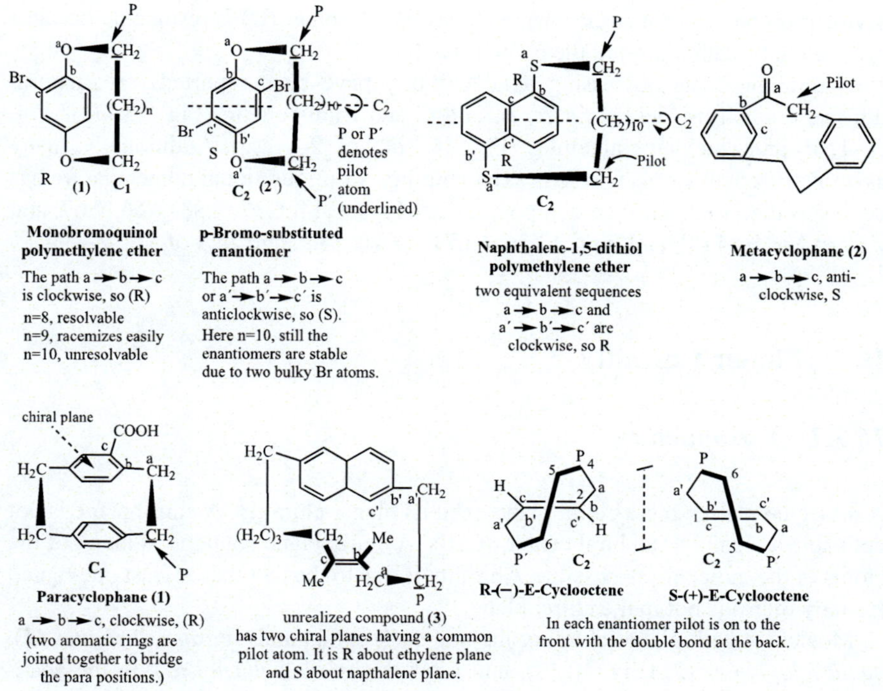

Note: (1), (2) and (3) are ansa compounds; in Latin ansa means handle.

Fig. 10.13 Some compounds of planar chirality and their *(R, S)* specification

the higher priority rear ligand "c" is clockwise, the configuration is *P*, and if anticlockwise it is *M* (see Fig. 10.11). In general, *aR* corresponds with *M* and *aS* with *P*. The compound **A** possesses *aS* or *P* configuration and **B** possesses *aR* or *M* configuration according to the new convention introduced in 1966 [12]. According to the old convention published in 1956 [13], the nomenclature is the opposite in these cases.

10.6.6 Some Interesting Examples of Axially Chiral Molecules Exhibiting Atropisomerism

A few examples are illustrated in Fig. 10.12. Molecules of the type in which one planar ring is replaced by a dissymmetrically substituted trigonal atom (two-dimensionally chiral) may exhibit atropisomerism if sufficient steric hindrance exists around the pivotal bond. A substituted stilbene (**A**) has been resolved. In the substituted naphthylamine (**B**), the peri nitro group gives rise to restricted rotation around the pivotal bond (chiral axis). Their *(R, S)* nomenclatures are also illustrated (Fig. 10.12).

Again, of the two oximes of 1-acetyl-2-hydroxynaphthalene-3-carboxylic acid, the *E*-isomer (**C**) (Fig. 10.12) is not resolvable (since the restriction around the pivotal aryl-carbon bond is not sufficient), whereas the Z-isomer (**D**) is resolvable because of restricted rotation around the pivotal bond.

Natural products possessing axial chirality have been reported, e.g., natural dimeric coumarins, desertorin A, B, and C and triumbelletin (Fig. 13.71 of Vol. 2—[10]) and colchicine alkaloids (Fig. 24.1 of Vol. 2—[10]). Additionally, many asymmetric catalysts also possess axial chirality. Configurational nomenclature has been specified for each such compound (*vide* Figs. 6.14, 6.37, 6.38, 7.66, 7.67, and 7.69 of Vol.1—[10], and Figs. 13.56, 13.71, 17.10, 18.9, and 24.1 of Vol. 2—[10].

10.7 Planar Chirality

10.7.1 Introduction

A molecule possessing a chiral plane exhibits planar chirality. A chiral plane is not definable as easily as a chiral center or axis. A chiral plane contains as many of the atoms of the molecule as possible. Chirality is due to the fact that at least one ligand (usually more) is not in the chiral plane.

Moelcules with planar chirality include *ansa* compounds, *paracyclophanes (1), metacyclophanes (2)* (Fig. 10.13), and a few *trans-cycloalkenes (3).* In the enantiomers, the methylene chain is on either side of the aryl ring or C = C bond. The interconversion between the enantiomers is prevented by the inability of the alicyclic ring, being too small, to swing from one side to the other of the aryl or olefinic plane.

10.7.2 The (R, S) Specification of Planar Chirality

The specification of these compounds (Fig. 10.13) is done by application of the following *selection rule.*

Selection rule [13]: The most preferred atom directly bound to atoms in the chiral plane is selected as the *pilot atom* (spectator point). It is the first out-of-plane atom linked to the sequence-preferred end of the chiral plane. The *sequencing starts with the first in-plane atom directly bound to the pilot atom* (the underlined *C*) and going along the in-plane sequence (marked as a, b, and c) involving the more preferred atom at each branch. The order in which a, b, and c appear when seen from the pilot atom specifies the absolute configuration, i.e., *R* for clockwise and *S* for anticlockwise order (see Fig. 10.13).

The compounds of planar chirality: (**1**), (**2**), and (**3**) are *ansa* compounds, (**4**) and (**5**) are cyclophanes, (**6**) is an imaginary compound containing two chiral planes having a common pilot atom. Cyclooctene is the smallest ring which can accommodate a *trans*-double bond. Two enantiomeric configurations (**7**) and (**8**) are possible.

The interconversion of the two enantiomers (**7**) and (**8**) of *trans*- or *E*-cyclooctene requires the swinging of the tetramethylene chain over and below the plane of the double bond (chiral plane) which is opposed by angle strain (ring strain). The two enantiomers have been resolved by Cope and coworkers [14, 15] who determined the absolute configuration of the enantiomers [16]; the (−)-enantiomer has been shown to be *R* and the (+)-enantiomer *S*. The molecule possesses $\mathbf{C_2}$ point group, having only the C_2 axis passing through the center of the double bond and bisecting the 5–6 bond. In the higher homologues increase in the mobility of the polymethylene chain decreases the rotational barrier. Thus, *trans*-cyclononene exists in enantiomeric forms only at −80 °C, and *trans*-cyclodecene is an extremely mobile system.

10.8 Helicity and *P*,*M*-Designation

A helix is inherently chiral and may be considered as displaying axial chirality—its axis serving as the chiral axis. However, it is more convenient to discuss such type of chirality as helicity. A helix possessing a C_2 axis orthogonal to the chiral axis is called *palindromic*, and it looks same from either end. If a helix moves from one end to the other in a clockwise direction, it is designated *P* (plus) and if it moves in an anticlockwise direction, it is designated *M* (minus) (Fig. 10.14). It is known that a polypeptide chain derived from natural L-amino acids often coil to form an α-helix with *P* helicity, but it is not palindromic since its two ends are nonequivalent, and hence, it lacks a C_2 axis. Its point group is $\mathbf{C_1}$. The nucleic acids present in DNA or RNA possess the skeleton of a double *P*-helix, preserved by strong intramolecular H-bondings [17].

Molecular overcrowding [18] is exemplified by hexahelicine (*A*) [19] and 4,5-disubstituted phenanthrenes as in (***B***) [20], (***C***) [21], and (***D***) and (***E***) [22] (Fig. 10.14).

The ring structures, due to overcrowding, assume helicity since the two terminal rings, as in (**A**), or the substituents at 4 and 5 positions of phenanthrene, as in (**B**) to (**E**), are large enough to prevent their existence in the same plane as that of the aromatic rings, thus giving rise to this type of optical isomerism, due to molecular asymmetry [17] arising out of helicity. With respect to some parts of the central rings,

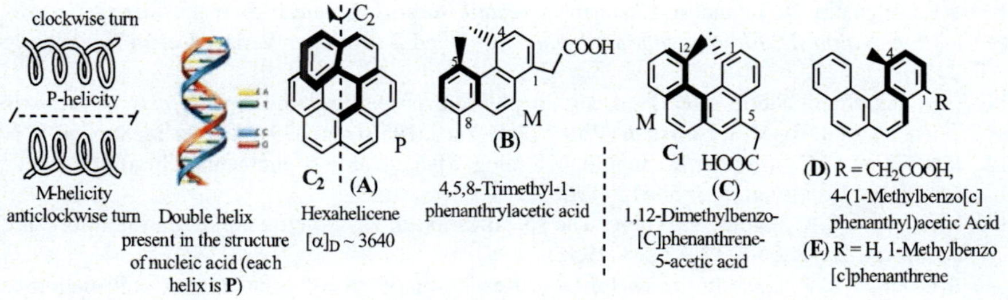

Fig. 10.14 Helical structures. Helicity due to molecular overcrowding

the ring planes on the upper right side, as written in the Figure, are gradually bent up, and the ring planes on the upper left side are gradually bent down by small angles, (10°–14°) in case of *P* helicity, as have been evident from X-ray diffraction analysis [17]. In case of *M*-helicity the ring planes will bend just in opposite direction. Thus, in each case two helical enantiomers exist. The synthetic racemates (**A**) to (**E**) have been resolved. The *P*, *M* nomenclature and the point groups of these molecules are depicted in Fig. 10.14.

Hexahelicene having very high rotation, $[\alpha]_D$ ~ 3640, racemizes slowly at its mp 266 °C. Higher helicenes readily racemize at their mps (about 200°) [23]. The absolute configuration, however, has not been assigned. As shown earlier in **c** of Fig. 10.11, chiral biphenyls may also be viewed as helicenes and designated as *P* or *M*. For the relationship between the sign of the CE and absolute configuration of helicenes; see Sect. 12.2.

References

1. P. Maitland, W.H. Mills, Experimental demonstration of the allene asymmetry. Nature **135**, 994–994 (1935)
2. P. Maitland, W.H. Mills, Resolution of allene hydrocarbon into optical antipodes. J. Chem. Soc. 987–988 (1936)
3. W.D. Celmer, I.A. Solomon, Mycomycin. III. The structure of Mycomycin, an Antibiotic Containing Allene, Diacetylene and *cis*, *trans*- Diene Groupings. J. Am. Chem. Soc. **75**, 1372–1376 (1953)
4. E.P. Kohler, J.T. Walker, M. Tishler, The resolution of an Allenic Compound. J. Am. Chem. Soc. **57**, 1743–1745 (1935)
5. H. Gerlach, Űber die Chiralität der enantiomeren Spiro[4,4]nonan-1,6-dione. Helv. Chim. Acta **51**, 1587–1593 (1968)
6. H.W. Whitlock, M.W. Siefken, Tricyclo[4.4.0.0^{3,8}] decane to adamantane rearrangement. J. Am. Chem. Soc. **90**, 4929–4939 (1968)
7. E. Wasserman, The preparation of interlocking rings: a catenane. J. Am. Chem. Soc. **82**, 4433–4434 (1960)
8. M. Oki, Recent advances in atropisomerism, in *Topics in Atereochemistry,* ed. By E. L. Eliel, N. L. Allinger, S. W. Wilen, vol. 14, (John Wiley and Sons, 1983), pp. 1–78
9. G.H. Christie, J. Kenner, The molecular configurations of polynuclear aromatic compounds. Part I. The resolution of r-6:6′-dinitro and 4:6:4′:6′-tetranitro-diphenic acids into optically active components. J. Chem. Soc. **121**, 614–620 (1922)
10. S.K. Talapatra, B. Talapatra, *Chemistry of Plant Natural Products: Stereochemistry, Conformation, Synthesis, Biology, and Medicine,* vol. 1 and 2 (Springer-Verlag, Berlin Heidelberg, 2015)
11. L.L. Ingraham, Steric effects on certain physical properties, in *Steric Effects in Organic Chemistry*, ed. By M.S. Newman (Wiley, New York, 1956), pp. 479–522, pertinent page 483
12. R.S. Cahn, S. Christopher, K. Ingold, V. Prelog, Specification of molecular chirality. Angew. Chem. Int. Edn. Engl. **5**, 385–415 (1966)
13. R.S. Cahn, C.K. Ingold, V. Prelog, The specification of asymmetric configuration in organic chemistry. Experientia **12**, 81–94 (1956)
14. A.C. Cope, C.R. Ganellin, H.W. Johnson, Resolution of *trans*-Cycloőctene; Confirmation of the Asymmetry of *cis-trans*-1,5-Cycloőctadiene. J. Am. Chem. Soc. **84**, 3191–3192 (1962)

15. A.C. Cope, C.R. Ganellin, H.W. Johnson, T.V. Van Auken, H.J.S. Winkler, Molecular Asymmetry of Olefins. I. Resolution of *trans*-Cyclooctene. J. Am. Chem. Soc. **85**, 3276–3279 (1963)
16. A.C. Cope, A.S. Mehta, Molecular asymmetry of olefins II. The absolute configuration of *trans*-cyclooctene. J. Am. Chem. Soc. **86**, 5626–5630 (1964)
17. M. Goodman, A.S. Vardini, N.S. Choi, Y. Matsuda, Polypeptide Stereochemistry. Top. Stereochem. **5**, 69–166 (1970)
18. M.S. Newman, Molecular complexes and molecular asymmetry, in *Steric Effects in Organic Chemistry*, ed. by M.S. Newman (John Wiley, New York, 1956), Chapter 10, pp. 471–478
19. M.S. Newman, D. Lednicer, The synthesis and resolution of hexahelicene. J. Am. Chem. Soc. **78**, 4765–4770 (1956)
20. M.S. Newman, A.S. Hussey, The synthesis and optical resolution of 4,5,8-Trimethyl-1-phenanthrylic Acid. J. Am. Chem. Soc. **69**, 3023–3027 (1947)
21. M.S. Newman, R.M. Wise, The synthesis and resolution of 1,12-Dimethylbenzo[c]phenanthrene-5-acetic acid. J. Am. Chem. Soc. **78**, 450–454 (1956)
22. M.S. Newman, W.B. Wheatley, Optical Activity of the 4,5-Phenanthrene Type: 4-(1-Methylbenzo[c]phenanthryl)-Acetic Acid and 1-Methylbenzo[c]phenanthrene. J. Am. Chem. Soc. **70**, 1913–1916 (1948)
23. R.H. Martin, M.-J. Marchant, Thermal Racemization of **6**, **7**, **8**, **9** Helicenes. Tetrahedron Lett. 3707–3708 (1972)

Chapter 11
Chiroptical Properties I: Optical Rotation. ORD, CD

Abstract In this chapter, the following topics have been discussed and explained: origin of optical rotation, circular birefringence, its effects, LP wave—resultant of RCP and LCP, optical rotatory dispersion (ORD), plain curve; circular birefringence and circular dichroism (CD), cotton effect (CE); axial haloketone rule and its applications, namely—position of halogen substituent, absolute configuration and conformational mobility; the octant rule and its applications with examples; absolute configuration of *trans*-decalones through octant rule; CE signs (observed—same as predicted) of octant projection diagrams (opds) of a few perhydrophenanthrenones and perhydroanthracenones, some cholestanones and coprostanones.

11.1 Origin of Optical Rotation. Circular Birefringence, Its Effect

Optical rotation and optical isomerism have been briefly discussed in Sect. 2.2.8 [1–4]. Now we would dwell on the origin of optical rotation. The electric field associated with the light wave in ordinary radiation oscillates in all directions perpendicular to the direction of propagation along the *z*-axis. Such radiation is isotropic or unpolarized. If the radiation is filtered to remove all oscillations other than in one direction, say in the *xz* plane, then the light becomes linearly polarized, since the projection in *xy* plane is linear, and the light becomes anisotropic.

When a plane-polarized (better called linearly polarized or LP) monochromatic light wave passes through a dissymmetric medium (nonracemic sample of a chiral substance), the plane of polarization rotates giving an optical rotation—a chiroptical property exhibited by nearly all chiral molecules. Two enantiomers exhibit optical rotations, equal in magnitude, but opposite in sign.

The electrical field vector (E) of an LP wave oscillates in the *xy* plane along the direction of propagation (*z*-axis). The LP wave is the resultant of two chiral components—a right circularly polarized (RCP) ray and a left circularly polarized (LCP) ray, whose projections on the *xy* plane are circles (Fig. 11.1).

S. K. Talapatra and B. Talapatra, *Basic Concepts in Organic Stereochemistry*,
https://doi.org/10.1007/978-3-030-95990-6_11

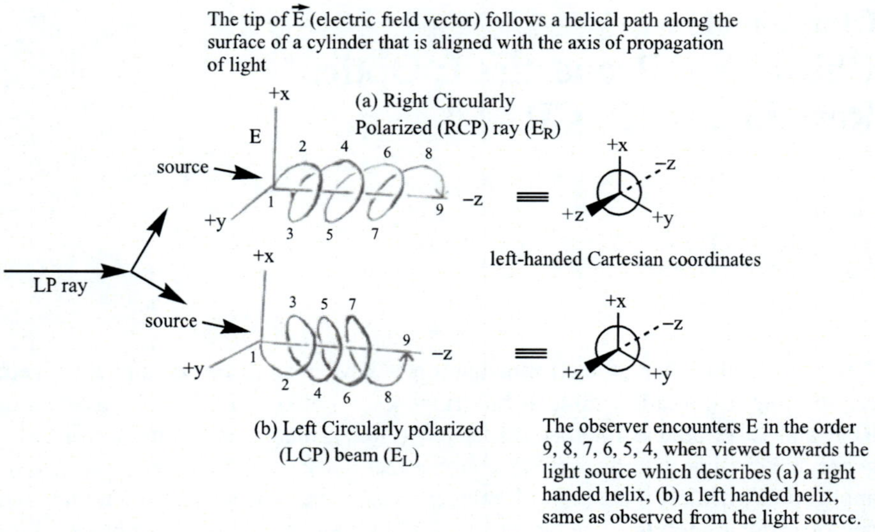

Fig. 11.1 LP wave is the resultant of RCP and LCP rays (*E* electric field vector)

In a symmetric or isotropic medium, the two components RCP and LCP travel at the same velocity. The resultant of the two rays in phase constitutes the linearly polarized light.

RCP ray + LCP ray = LP ray

These rays make diastereomeric relationship with the two enantiomers and so interact differently. When the linearly polarized light is passed through a nonracemic sample of a chiral or dissymmetric compound (or medium), the velocities of LCP and RCP rays become different; the two circularly polarized rays will have different refractive indices, viz. n_L and n_R, being the refractive indices of left and right circularly polarized rays; this is known as *circular birefringence.* If $n_L > n_R$, $V_R > V_L$ since $n = V_O/V$, V_O and V are velocities in vacuum and the medium, respectively; thus, the plane of polarization or the LP wave will rotate toward right, i.e., the observed rotation α will be positive. The compound is said to be dextrorotatory. Thus, if $n_R > n_L$, $V_L > V_R$ and hence the observed rotation α will be levorotatory.

The observed rotation changes continuously as the LP proceeds (α varies directly with length l). Optical rotation due to circular birefringence for 1 cm path length is given by the expression $\alpha = (n_L - n_R)$. π/λ radian = Δn 180/λ degree. The specific rotation at a wavelength λ may be given by the expression $[\alpha]_\lambda$ in degrees = $\alpha \times 10/c$ for 1 *dm* path length, c = concentration in g/ml. The optical rotation of a chiral compound is usually reported as the specific rotation at sodium D-line (589 Å), i.e., as $[\alpha]_D$. The molar rotation at a wavelength λ, $[M]_\lambda$, or $[\Phi]_\lambda = ([\alpha]_\lambda.\ M)/100$ (cf. Sect. 1.2.9).

11.2 Optical Rotatory Dispersion. Plain Curve

The specific rotation [α] of a chiral compound depends upon the wavelength of the monochromatic light wave. The measurement of specific rotation as a function of wavelength is called optical rotatory dispersion (ORD). Specific rotations of chiral compounds are generally reported for sodium D-line (589 *nm* wavelength), which is quite far from UV region. Consequently, the values of specific rotations are much lower than those recorded near the UV absorption maximum; if $[\alpha]_D$ is too low to be detected, and it is not possible to ascertain whether such a compound is dextrorotatory or levorotatory, the enantiomer may be mistaken as a racemic variety. In old literature, the $[\alpha]_D = \pm 0$ has been assigned to such chiral compounds whose rotation at 589 nm could not be measured.

Specific rotations of such compounds having very low $[\alpha]_D$ values or any chiral compound having no UV absorption if measured at shorter wavelengths undergo many fold increment as one approaches the absorption maximum below 210 nm (due to $\sigma \rightarrow \sigma^*$ transition). Such an ORD curve is called a plain curve (Fig. 11.2). A plain curve results when measurement is done at wavelengths away from the absorption maximum (λ_{max}), and only circular birefringence is operative. By measuring a plain curve, one can assay in a precise way the enantiomeric purity of a natural product or a bioorganic compound having no UV absorption. A chiral compound, if racemic, remains inactive throughout the measureable wavelength.

A plain curve for certain molecules may or may not cross the zero rotation axis. Thus, the *ortho* isomer of α-(iodophenoxy)-propionic acid shows levorotation at the D-line, whereas the *meta* and *para* isomers show $[\alpha]_D$ positive, although all three isomers have the same absolute configuration (Fig. 11.2) [1]. Each of the three position isomers shows plain positive curve; the observed positive rotation increases

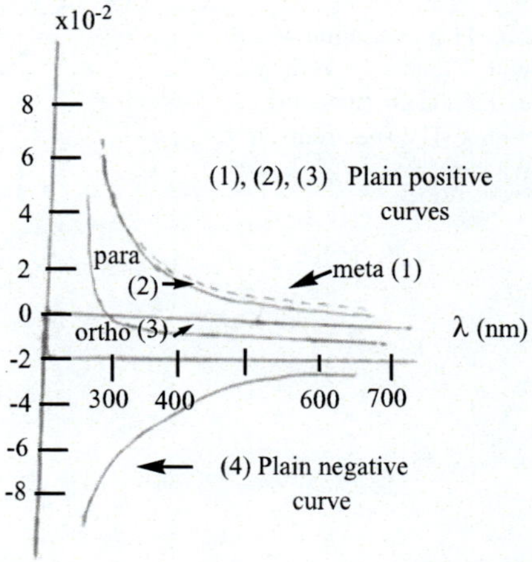

Fig. 11.2 ORD curves α-(*m*-, *p*-, and *o*-iodophenoxy)-propionic acids (*from C. Djerassi "Optical Rotatory Dispersion"*, **1960**, *McGraw-Hill Book Company*). Plain positive and negative curves

with decrease of wavelength. In case of negative plain curve, the observed negative rotation increases with decrease of wavelength.

11.3 Circular Birefringence and Circular Dichroism. Cotton Effect

It has already been mentioned that when the LP rays of different wavelengths pass through a medium having a λ_{max} in the region studied, in addition to circular birefringence, the intensity of emergent light diminishes due to absorption. The two circularly polarized light rays, RCP and LCP, are absorbed to different extents; this phenomenon is termed as circular dichroism. Both these phenomena are wavelength dependent. Absorptions of RCP and LCP rays are maximum at the absorption maximum (λ_{max}). The refractive index increases with the decrease of the wavelength but rapidly falls to a minimum in the region of absorption showing abnormal dispersion. At the absorption maximum, there is no effect on refractive index ($n = 1$ for both RCP and LCP rays). Figure 11.3 shows the variation of Δn ($=n_L - n_R$), refractive indices dispersion (I) of LCP and RCP, and of absorbance (A) (II) over a range of wavelength (including the absorption maximum, λ_{max}); Fig. 11.4 shows ORD curve with positive Cotton effect (III) and a positive CD spectrum (IV).

If an optically active compound absorbs in the UV or visible region, the two components RCP and LCP of the LP will be absorbed to different extents (anisotropic absorption), and this phenomenon of differential absorption is called *circular dichroism*, usually abbreviated as CD. The differential absorption is caused by electronic (along with vibrational) transition of energy states associated with a chiroptic chromophore in a chiral molecule. Let A_L and A_R represent the absorbances of LCP

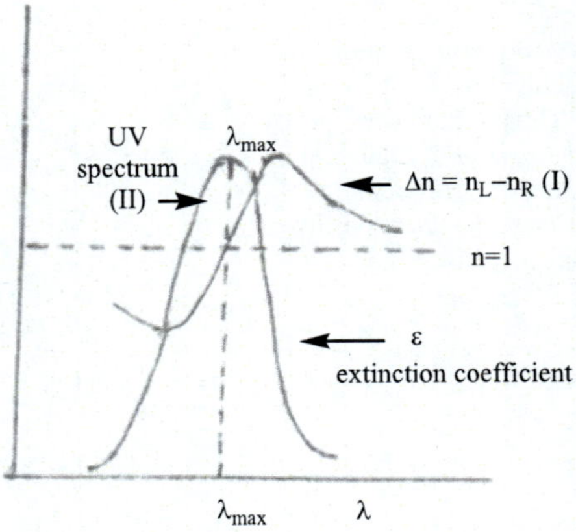

Fig. 11.3 Variation of Δn with λ (curve I). Variation of extinction coefficient (ε) with λ (UV spectrum, curve II)